Graduate Texts in Mathematics 112

Graduate Texts in Mathematics

1 TAKEUTI/ZARING. Introduction to Axiomatic Set Theory. 2nd ed.
2 OXTOBY. Measure and Category. 2nd ed.
3 SCHAEFFER. Topological Vector Spaces.
4 HILTON/STAMMBACH. A Course in Homological Algebra.
5 MACLANE. Categories for the Working Mathematician.
6 HUGHES/PIPER. Projective Planes.
7 SERRE. A Course in Arithmetic.
8 TAKEUTI/ZARING. Axiomatic Set Theory.
9 HUMPHREYS. Introduction to Lie Algebras and Representation Theory.
10 COHEN. A Course in Simple Homotopy Theory.
11 CONWAY. Functions of One Complex Variable. 2nd ed.
12 BEALS. Advanced Mathematical Analysis.
13 ANDERSON/FULLER. Rings and Categories of Modules.
14 GOLUBITSKY/GUILLEMIN. Stable Mappings and Their Singularities.
15 BERBERIAN. Lectures in Functional Analysis and Operator Theory.
16 WINTER. The Structure of Fields.
17 ROSENBLATT. Random Processes. 2nd ed.
18 HALMOS. Measure Theory.
19 HALMOS. A Hilbert Space Problem Book. 2nd ed., revised.
20 HUSEMOLLER. Fibre Bundles. 2nd ed.
21 HUMPHREYS. Linear Algebraic Groups.
22 BARNES/MACK. An Algebraic Introduction to Mathematical Logic.
23 GREUB. Linear Algebra. 4th ed.
24 HOLMES. Geometric Functional Analysis and its Applications.
25 HEWITT/STROMBERG. Real and Abstract Analysis.
26 MANES. Algebraic Theories.
27 KELLEY. General Topology.
28 ZARISKI/SAMUEL. Commutative Algebra. Vol. I.
29 ZARISKI/SAMUEL. Commutative Algebra. Vol. II.
30 JACOBSON. Lectures in Abstract Algebra I: Basic Concepts.
31 JACOBSON. Lectures in Abstract Algebra II: Linear Algebra.
32 JACOBSON. Lectures in Abstract Algebra III: Theory of Fields and Galois Theory.
33 HIRSCH. Differential Topology.
34 SPITZER. Principles of Random Walk. 2nd ed.
35 WERMER. Banach Algebras and Several Complex Variables. 2nd ed.
36 KELLEY/NAMIOKA et al. Linear Topological Spaces.
37 MONK. Mathematical Logic.
38 GRAUERT/FRITZSCHE. Several Complex Variables.
39 ARVESON. An Invitation to C^*-Algebras.
40 KEMENY/SNELL/KNAPP. Denumerable Markov Chains. 2nd ed.
41 APOSTOL. Modular Functions and Dirichlet Series in Number Theory.
42 SERRE. Linear Representations of Finite Groups.
43 GILLMAN/JERISON. Rings of Continuous Functions.
44 KENDIG. Elementary Algebraic Geometry.
45 LOÈVE. Probability Theory I. 4th ed.
46 LOÈVE. Probability Theory II. 4th ed.
47 MOISE. Geometric Topology in Dimensions 2 and 3.

continued after Index

Serge Lang

Elliptic Functions

Second Edition

Springer-Verlag
New York Berlin Heidelberg
London Paris Tokyo

Serge Lang
Department of Mathematics
Yale University
New Haven, CT 06520
U.S.A.

AMS Classifications: 10D05, 12B25

Library of Congress Cataloging in Publication Data
Lang, Serge
Elliptic functions.
(Graduate texts in mathematics; 112)
Bibliography: p.
1. Functions, Elliptic. I. Title.
QA343.L35 1987 515.9′83 87-4514

The first edition of this book was published by Addison-Wesley Publishing Company, Inc., Reading, MA, in 1973.

Printed and bound by R. R. Donnelley & Sons, Harrisonburg, Virginia.
Printed in the United States of America.

9 8 7 6 5 4 3 2 1

ISBN 0-387-96508-4 Springer-Verlag New York Berlin Heidelberg
ISBN 3-540-96508-4 Springer-Verlag Berlin Heidelberg New York

Preface

Elliptic functions parametrize elliptic curves, and the intermingling of the analytic and algebraic-arithmetic theory has been at the center of mathematics since the early part of the nineteenth century.

Some new techniques and outlooks have recently appeared on these old subjects, continuing in the tradition of Kronecker, Weber, Fricke, Hasse, Deuring. Shimura's book *Introduction to the arithmetic theory of automorphic functions* is a splendid modern reference, which I found very helpful myself to learn some aspects of elliptic curves. It emphasizes the direction of the Hasse-Weil zeta function, Hecke operators, and the generalizations due to him to the higher dimensional case (abelian varieties, curves of higher genus coming from an arithmetic group operating on the upper half plane, bounded symmetric domains with a discrete arithmetic group whose quotient is algebraic). I refer the interested reader to his book and the bibliography therein.

I have placed a somewhat different emphasis in the present exposition. First, I assume less of the reader, and start the theory of elliptic functions from scratch. I do not discuss Hecke operators, but include several topics not covered by Shimura, notably the Deuring theory of ℓ-adic and p-adic representations; the application to Ihara's work; a discussion of elliptic curves with non-integral invariant, and the Tate parametrization, with the applications to Serre's work on the Galois group of the division points over number fields, and to the isogeny theorem; and finally the Kronecker limit formula and the discussion of values of special modular functions constructed as quotients of theta functions, which are better than values of the Weierstrass function because they are units when properly normalized, and behave in a specially good way with respect to the action of the Galois group.

Thus the present book has a very different flavor from Shimura's. It was unavoidable that there should be some non-empty overlapping, and I have chosen to redo the complex multiplication theory, following Deuring's algebraic method, and reproducing some of Shimura's contributions in this line (with some simplifications, e.g. to his reciprocity law at fixed points, and with another proof for the theorem concerning the automorphisms of the modular function field).

I do not emphasize elliptic curves in characteristic p, except as they arise by reduction from characteristic 0. Thus I have omitted most of the theory proper to characteristic p, especially the finer theory of supersingular invariants. The reader should be warned, however, that this theory is important for the deeper analysis of the arithmetic theory of elliptic curves. The two appendices should help the reader get into the literature.

I thank Shimura for his patience in explaining to me some facts about his research; Eli Donkar for his notes of a course which provided the basis for the present book; Swinnerton-Dyer and Walter Hill for their careful reading of the manuscript.

New Haven, Connecticut SERGE LANG

Note for the Second Edition

I thank Springer-Verlag for keeping the book in print. It is unchanged except for the corrections of some misprints, and two items:

1. John Coates pointed out to me a mistake in Chapter 21, dealing with the L-functions for an order. Hence I have eliminated the reference to orders at that point, and deal only with the absolute class group.
2. I have renormalized the functions in Chapter 19, following Kubert–Lang. Thus I use the Klein forms and Siegel functions as in that reference. Actually, the final formulation of Kronecker's Second Limit Formula comes out neater under this renormalization.

S. L.
November 1986

Contents

PART FOUR THETA FUNCTIONS AND KRONECKER LIMIT FORMULA

APPENDICES ELLIPTIC CURVES IN CHARACTERISTIC p

Part One
General Theory

In this part we study elliptic curves, which can be defined by the Weierstrass equation $y^2 = 4x^3 - g_2x - g_3$. We shall see that their complex points form a commutative group, which is complex analytically isomorphic to a complex torus $\mathbf{C}/L$, where L is a lattice in $\mathbf{C}$. We study these curves in general, especially those which are "generic". We consider their homomorphisms, isomorphisms, and their points of finite order in general. We also relate such curves with modular functions, and show how to parametrize isomorphism classes of curves by points in the upper half plane modulo $SL_2(\mathbf{Z})$. We constantly interrelate the transcendental parametrizations with the algebraic properties involved. Our policy is to tell the reader what is true in arbitrary characteristic (due to Hasse), and give the short proofs mostly only in characteristic 0, using the transcendental parametrization.

1 *Elliptic Functions*

§1. THE LIOUVILLE THEOREMS

By a **lattice** in the complex plane **C** we shall mean a subgroup which is free of dimension 2 over **Z**, and which generates **C** over the reals. If ω_1, ω_2 is a basis of a lattice L over **Z**, then we also write $L = [\omega_1, \omega_2]$. Such a lattice looks like this:

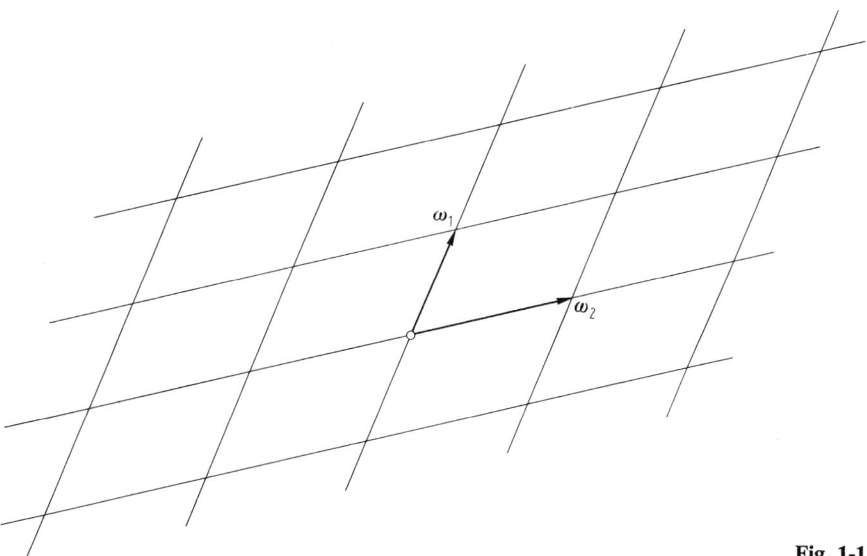

Fig. 1-1

Unless otherwise specified, we also assume that $\mathrm{Im}(\omega_1/\omega_2) > 0$, i.e. that ω_1/ω_2 lies in the upper half plane $\mathfrak{H} = \{x + iy, y > 0\}$. An **elliptic function** f (with respect to L) is a meromorphic function on **C** which is L-periodic, i.e.

$$f(z + \omega) = f(z)$$

for all $z \in \mathbf{C}$ and $\omega \in L$. Note that f is periodic if and only if

$$f(z + \omega_1) = f(z) = f(z + \omega_2).$$

An elliptic function which is entire (i.e. without poles) must be constant, because it can be viewed as a continuous function on $\mathbf{C}/L$, which is compact (homeomorphic to a torus), whence the function is bounded, and therefore constant.

If $L = [\omega_1, \omega_2]$ as above, and $\alpha \in \mathbf{C}$, we call the set consisting of all points

$$\alpha + t_1\omega_1 + t_2\omega_2, \qquad 0 \leqq t_i \leqq 1$$

a **fundamental parallelogram** for the lattice (with respect to the given basis). We could also take the values $0 \leqq t_i < 1$ to define a fundamental parallelogram, the advantage then being that in this case we get unique representatives for elements of $\mathbf{C}/L$ in $\mathbf{C}$.

Theorem 1. *Let P be a fundamental parallelogram for L, and assume that the elliptic function f has no poles on its boundary ∂P. Then the sum of the residues of f in P is* 0.

Proof. We have

$$2\pi i \sum \operatorname{Res} f = \int_{\partial P} f(z)\, dz = 0,$$

this last equality being valid because of the periodicity, so the integrals on opposite sides cancel each other.

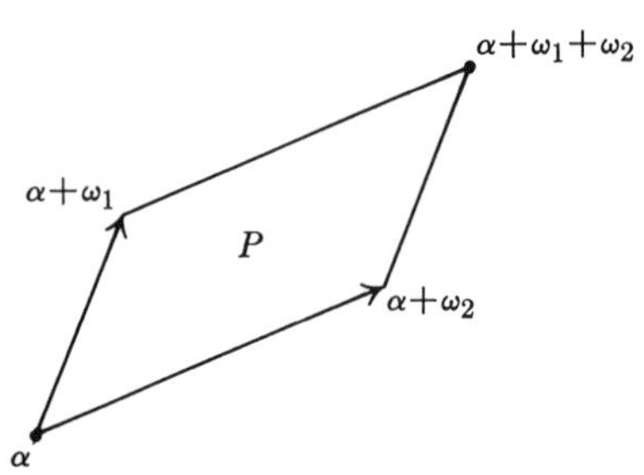

Fig. 1-2

An elliptic function can be viewed as a meromorphic function on the torus $\mathbf{C}/L$, and the above theorem can be interpreted as saying that the sum of the residues on the torus is equal to 0. Hence:

Corollary. *An elliptic function has at least two poles (counting multiplicities) on the torus.*

Theorem 2. *Let P be a fundamental parallelogram, and assume that the elliptic function f has no zero or pole on its boundary. Let $\{a_i\}$ be the singular points (zeros and poles) of f inside P, and let f have order m_i at a_i. Then*

$$\sum m_i = 0.$$

Proof. Observe that f elliptic implies that f' and f'/f are elliptic. We then obtain

$$0 = \int_{\partial P} f'/f(z)\,dz = 2\pi\sqrt{-1}\sum \text{Residues} = 2\pi\sqrt{-1}\sum m_i,$$

thus proving our assertion.

Again, we can formulate Theorem 2 by saying that the sum of the orders of the singular points of f on the torus is equal to 0.

Theorem 3. *Hypotheses being as in Theorem 2, we have*

$$\sum m_i a_i \equiv 0 \ (\text{mod } L).$$

Proof. This time, we take the integral

$$\int_{\partial P} z\frac{f'(z)}{f(z)}\,dz = 2\pi\sqrt{-1}\sum m_i a_i,$$

because

$$\text{res}_{a_i}\, z\frac{f'(z)}{f(z)} = m_i a_i.$$

On the other hand we compute the integral over the boundary of the parellelogram by taking it for two opposite sides at a time. One pair of such integrals is equal to

$$\int_{\alpha}^{\alpha+\omega_1} z\frac{f'(z)}{f(z)}\,dz - \int_{\alpha+\omega_2}^{\alpha+\omega_1+\omega_2} z\frac{f'(z)}{f(z)}\,dz.$$

We change variables in the second integral, letting $u = z - \omega_2$. Both integrals are then taken from α to $\alpha + \omega_1$, and after a cancellation, we get the value

$$-\omega_2\int_{\alpha}^{\alpha+\omega_1} \frac{f'(u)}{f(u)}\,du = 2\pi\sqrt{-1}\,k\omega_2,$$

for some integer k. The integral over the opposite pair of sides is done in the same way, and our theorem is proved.

§2. THE WEIERSTRASS FUNCTION

We now prove the existence of elliptic functions by writing some analytic expression, namely the Weierstrass function

$$\wp(z) = \frac{1}{z^2} + \sum_{\omega\in L'}\left[\frac{1}{(z-\omega)^2} - \frac{1}{\omega^2}\right],$$

where the sum is taken over the set of all non-zero periods, denoted by L'. We have to show that this series converges uniformly on compact sets not including the lattice points. For bounded z, staying away from the lattice points, the expression in the brackets has the order of magnitude of $1/|\omega|^3$. Hence it suffices to prove:

Lemma. If $\lambda > 2$, then $\sum_{\omega \in L'} \frac{1}{|\omega|^\lambda}$ converges.

Proof. The partial sum for $|\omega| \leqq N$ can be decomposed into a sum for ω in the annulus at n, i.e. $n - 1 \leqq |\omega| \leqq n$, and then a sum for $1 \leqq n \leqq N$. In each annulus the number of lattice points has the order of magnitude n. Hence

$$\sum_{|\omega| \leqq N} \frac{1}{|\omega|^\lambda} \ll \sum_1^\infty \frac{n}{n^\lambda} \ll \sum_1^\infty \frac{1}{n^{\lambda - 1}}$$

which converges for $\lambda > 2$.

The series expression for $\wp$ shows that it is meromorphic, with a double pole at each lattice point, and no other pole. It is also clear that $\wp$ is even, i.e.

$$\wp(z) = \wp(-z)$$

(summing over the lattice points is the same as summing over their negatives). We get $\wp'$ by differentiating term by term,

$$\wp'(z) = -2 \sum_{\omega \in L} \frac{1}{(z - \omega)^3},$$

the sum being taken for all $\omega \in L$. Note that $\wp'$ is clearly periodic, and is odd, i.e.

$$\wp'(-z) = -\wp'(z).$$

From its periodicity, we conclude that there is a constant C such that

$$\wp(z + \omega_1) = \wp(z) + C.$$

Let $z = -\omega_1/2$ (not a pole of $\wp$). We get

$$\wp\left(\frac{\omega_1}{2}\right) = \wp\left(-\frac{\omega_1}{2}\right) + C,$$

and since $\wp$ is even, it follows that $C = 0$. Hence $\wp$ is itself periodic, something which we could not see immediately from its series expansion.

It is clear that the set of all elliptic functions (with respect to a given lattice L) forms a field, whose constant field is the complex numbers.

Theorem 4. *The field of elliptic functions (with respect to* L*) is generated by* $\wp$ *and* $\wp$ *.*

Proof. If f is elliptic, we can write f as a sum of an even and an odd elliptic function as usual, namely

$$f(z) = \frac{f(z) + f(-z)}{2} + \frac{f(z) - f(-z)}{2}.$$

If f is odd, then the product $f\wp'$ is even, so it will suffice to prove that $\mathbf{C}(\wp)$ is the field of even elliptic functions, i.e. if f is even, then f is a rational function of $\wp$.

Suppose that f is even and has a zero of order m at some point u. Then clearly f also has a zero of the same order at $-u$ because

$$f^{(k)}(u) = (-1)^k f^{(k)}(-u).$$

Similarly for poles.

If $u \equiv -u \pmod{L}$, then the above assertion holds in the strong sense, namely f has a zero (or pole) of even order at u.

Proof. First note that $u \equiv -u \pmod{L}$ is equivalent to

$$2u \equiv 0 \pmod{L}.$$

On the torus, there are exactly four points with this property, represented by

$$0, \frac{\omega_1}{2}, \frac{\omega_2}{2}, \frac{\omega_1 + \omega_2}{2}$$

in a period parallelogram. If f is even, then f' is odd, i.e.

$$f'(u) = -f'(-u).$$

Since $u \equiv -u \pmod{L}$ and f' is periodic, it follows that $f'(u) = 0$, so that f has a zero of order at least 2 at u. If $u \not\equiv 0 \pmod{L}$, then the above argument shows that the function

$$g(z) = \wp(z) - \wp(u)$$

has a zero of order at least 2 (hence exactly 2 by Theorem 2 and the fact that $\wp$ has only one pole of order 2 on the torus). Then f/g is even, elliptic, holomorphic at u. If $f(u)/g(u) \neq 0$ then $\mathrm{ord}_u f = 2$. If $f(u)/g(u) = 0$ then f/g again has a zero of order at least 2 at u and we can repeat the argument. If $u \equiv 0 \pmod{L}$ we use $g = 1/\wp$ and argue similarly, thus proving that f has a zero of even order at u.

Now let u_i $(i = 1, \ldots, r)$ be a family of points containing one representative from each class $(u, -u) \pmod{L}$ where f has a zero or pole, other than the class of L itself. Let

$$m_i = \mathrm{ord}_{u_i} f \qquad \text{if} \quad 2u_i \not\equiv 0 \pmod{L},$$
$$m_i = \tfrac{1}{2}\,\mathrm{ord}_{u_i} f \qquad \text{if} \quad 2u_i \equiv 0 \pmod{L}.$$

Our previous remarks show that for $a \in \mathbf{C}$, $a \not\equiv 0 \pmod{L}$, the function

$\wp(z) - \wp(a)$ has a zero of order 2 at a if and only if $2a \equiv 0 \pmod{L}$, and has distinct zeros of order 1 at a and $-a$ otherwise. Hence for all $z \not\equiv 0 \pmod{L}$ the function

$$\prod_{i=1}^{r} [\wp(z) - \wp(u_i)]^{m_i}$$

has the same order at z as f. This is also true at the origin because of Theorem 2 applied to f and the above product. The quotient of the above product by f is then an elliptic function without zero or pole, hence a constant, thereby proving Theorem 4.

Next, we obtain the power series development of $\wp$ and $\wp'$ at the origin, from which we shall get the algebraic relation holding between these two functions. We do this by brute force.

$$\begin{aligned}\wp(z) &= \frac{1}{z^2} + \sum_{\omega \in L'} \left[\frac{1}{\omega^2}\left(1 + \frac{z}{\omega} + \left(\frac{z}{\omega}\right)^2 + \cdots\right)^2 - \frac{1}{\omega^2}\right] \\ &= \frac{1}{z^2} + \sum_{\omega \in L'} \sum_{m=1}^{\infty} (m+1)\left(\frac{z}{\omega}\right)^m \frac{1}{\omega^2} \\ &= \frac{1}{z^2} + \sum_{m=1}^{\infty} c_m z^m\end{aligned}$$

where

$$c_m = \sum_{\omega \neq 0} \frac{m+1}{\omega^{m+2}}.$$

Note that $c_m = 0$ if m is odd.

Using the notation

$$s_m(L) = s_m = \sum_{\omega \neq 0} \frac{1}{\omega^m}$$

we get the expansion

$$\boxed{\wp(z) = \frac{1}{z^2} + \sum_{n=1}^{\infty} (2n+1) s_{2n+2}(L) z^{2n},}$$

from which we write down the first few terms explicitly:

$$\boxed{\wp(z) = \frac{1}{z^2} + 3s_4 z^2 + 5s_6 z^4 + \cdots}$$

and differentiating term by term,

$$\wp'(z) = \frac{-2}{z^3} + 6s_4 z + 20s_6 z^3 + \cdots.$$

Theorem 5. *Let* $g_2 = g_2(L) = 60s_4$ *and* $g_3 = g_3(L) = 140s_6$. *Then*

$$\wp'^2 = 4\wp^3 - g_2\wp - g_3.$$

Proof. We expand out the function

$$\varphi(z) = \wp'(z)^2 - 4\wp(z)^3 + g_2\wp(z) + g_3$$

at the origin, paying attention only to the polar term and the constant term. This is easily done, and one sees that there is enough cancellation so that these terms are 0, in other words, $\varphi(z)$ is an elliptic function without poles, and with a zero at the origin. Hence φ is identically zero, thereby proving our theorem.

The preceding theorem shows that the points $(\wp(z), \wp'(z))$ lie on the curve defined by the equation

$$y^2 = 4x^3 - g_2 x - g_3.$$

The cubic polynomial on the right-hand side has a discriminant given by

$$\Delta = g_2^3 - 27g_3^2.$$

We shall see in a moment that this discriminant does not vanish.

Let

$$e_i = \wp\left(\frac{\omega_i}{2}\right), \qquad i = 1, 2, 3,$$

where $L = [\omega_1, \omega_2]$ and $\omega_3 = \omega_1 + \omega_2$. Then the function

$$h(z) = \wp(z) - e_i$$

has a zero at $\omega_i/2$, which is of even order so that $\wp'(\omega_i/2) = 0$ for $i = 1, 2, 3$, by previous remarks. Comparing zeros and poles, we conclude that

$$\wp'^2(z) = 4(\wp(z) - e_1)(\wp(z) - e_2)(\wp(z) - e_3).$$

Thus e_1, e_2, e_3 are the roots of $4x^3 - g_2 x - g_3$. Furthermore, $\wp$ takes on the value e_i with multiplicity 2 and has only one pole of order 2 mod L, so that $e_i \neq e_j$ for $i \neq j$. This means that the three roots of the cubic polynomial are distinct, and therefore

$$\Delta = g_2^3 - 27g_3^2 \neq 0.$$

§3. THE ADDITION THEOREM

Given complex numbers g_2, g_3 such that $g_2^3 - 27g_3^2 \neq 0$, one can ask whether there exists a lattice for which these are the invariants associated to the lattice as in the preceding section. The answer is yes, and we shall prove this in chapter 3. For the moment, we consider the case when g_2, g_3 are given as in the preceding section, i.e. $g_2 = 60s_4$ and $g_3 = 140s_6$.

We have seen that the map

$$z \mapsto (1, \wp(z), \wp'(z))$$

parametrizes points on the cubic curve A defined by the equation

$$y^2 = 4x^3 - g_2 x - g_3.$$

This is an affine equation, and we put in the coordinate 1 to indicate that we also view the points as embedded in projective space. Then the mapping is actually defined on the torus $\mathbf{C}/L$, and the lattice points, i.e. 0 on the torus, are precisely the points going to infinity on the curve. Let $A_{\mathbf{C}}$ denote the complex points on the curve. We in fact get a bijection

$$\mathbf{C}/L - \{0\} \to A_{\mathbf{C}} - \{\infty\}.$$

This is easily seen: For any complex number α, $\wp(z) - \alpha$ has at most two zeros, and at least one zero, so that already under $\wp$ we cover each complex number α. It is then verified at once that using $\wp'$ separates the points of $\mathbf{C}/L$ lying above α, thus giving us the bijection. If you know the terminology of algebraic geometry, then you know that the curve defined by the above equation is non-singular, and that our mapping is actually a complex analytic isomorphism between $\mathbf{C}/L$ and $A_{\mathbf{C}}$.

Furthermore, $\mathbf{C}/L$ has a natural group structure, and we now want to see what it looks like when transported to A. We shall see that it is algebraic. In other words, if

$$P_1 = (x_1, y_1), \qquad P_2 = (x_2, y_2), \qquad P_3 = (x_3, y_3)$$

and

$$P_3 = P_1 + P_2,$$

then we shall express x_3, y_3 as rational functions of (x_1, y_1) and (x_2, y_2). We shall see that P_3 is obtained by taking the line through P_1, P_2, intersecting it with the curve, and reflecting the point of intersection through the x-axis, as shown on Fig. 3.

Select $u_1, u_2 \in \mathbf{C}$ and $\notin L$, and assume $u_1 \not\equiv u_2 \pmod{L}$. Let a, b be complex numbers such that

$$\wp'(u_1) = a\wp(u_1) + b$$
$$\wp'(u_2) = a\wp(u_2) + b,$$

in other words $y = ax + b$ is the line through $(\wp(u_1), \wp'(u_1))$ and $(\wp(u_2), \wp'(u_2))$. Then

$$\wp'(z) - (a\wp(z) + b)$$

has a pole of order 3 at 0, whence it has three zeros, counting multiplicities, and two of these are at u_1 and u_2. If, say, u_1 had multiplicity 2, then by Theorem 3 we would have

$$2u_1 + u_2 \equiv 0 \pmod{L}.$$

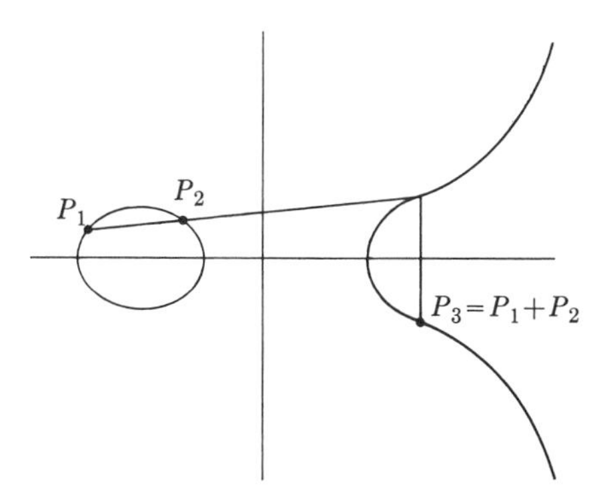

Fig. 1-3

If we fix u_1, this can hold for only one value of u_2. Let us assume that we do not deal with this value. Then both u_1, u_2 have multiplicity 1, and the third zero lies at

$$u_3 \equiv -(u_1 + u_2) \pmod{L}$$

again by Theorem 3. So we also get

$$\wp'(u_3) = a\wp(u_3) + b.$$

The equation

$$4x^3 - g_2x - g_3 - (ax + b)^2 = 0$$

has three roots, counting multiplicities. They are $\wp(u_1)$, $\wp(u_2)$, $\wp(u_3)$, and the left-hand side factors as

$$4(x - \wp(u_1))(x - \wp(u_2))(x - \wp(u_3)).$$

Comparing the coefficient of x^2 yields

$$\wp(u_1) + \wp(u_2) + \wp(u_3) = \frac{a^2}{4}.$$

But from our original equations for a and b, we have

$$a(\wp(u_1) - \wp(u_2)) = \wp'(u_1) - \wp'(u_2).$$

Therefore from

$$\wp(u_3) = \wp(-(u_1 + u_2)) = \wp(u_1 + u_2)$$

we get

$$\wp(u_1 + u_2) = -\wp(u_1) - \wp(u_2) + \frac{1}{4}\left(\frac{\wp'(u_1) - \wp'(u_2)}{\wp(u_1) - \wp(u_2)}\right)^2$$

or in algebraic terms,

$$x_3 = -x_1 - x_2 + \frac{1}{4}\left(\frac{y_1 - y_2}{x_1 - x_2}\right)^2.$$

Fixing u_1, the above formula is true for all but a finite number of $u_2 \not\equiv u_1 \pmod{L}$. whence for all $u_2 \not\equiv u_1 \pmod{L}$ by analytic continuation.

For $u_1 \equiv u_2 \pmod{L}$ we take the limit as $u_1 \to u_2$ and get

$$\wp(2u) = -2\wp(u) + \frac{1}{4}\left(\frac{\wp''(u)}{\wp'(u)}\right)^2.$$

These give us the desired algebraic addition formulas. Note that the formulas involve only g_2, g_3 as coefficients in the rational functions.

This is as far as we shall push the study of the $\wp$-function in general, except for a Fourier expansion formula in Chapter 4. For further information, the reader is referred to Fricke [B2]. For instance one can get formulas for $\wp(nz)$, one can get a continued fraction expansion (done by Frobenius), etc. Classics like Fricke still contain much information which has not yet reappeared in more modern books, nor been made much use of, although history shows that everything that has been discovered along those lines ultimately returns to the center of the stage at some point.

§4. ISOMORPHISM CLASSES OF ELLIPTIC CURVES

Theorem 6. *Let L, M be two lattices in* **C** *and let*

$$\lambda: \mathbf{C}/L \to \mathbf{C}/M$$

be a complex analytic homomorphism. Then there exists a complex number α such that the following diagram is commutative.

$$\begin{array}{ccc} \mathbf{C} & \xrightarrow{\alpha} & \mathbf{C} \\ \downarrow & & \downarrow \\ \mathbf{C}/L & \xrightarrow[\lambda]{} & \mathbf{C}/M \end{array}$$

The top map is multiplication by α, and the vertical maps are the canonical homomorphisms.

Proof. Locally near 0, λ can be expressed by a power series,

$$\lambda(z) = a_0 + a_1 z + a_2 z^2 + \cdots,$$

and since a complex number near 0 represents uniquely its class mod L, it follows from the formula

$$\lambda(z + z') \equiv \lambda(z) + \lambda(z') \pmod{M}$$

that the congruence can actually be replaced by an equality. Hence we must have

$$\lambda(z) = a_1 z,$$

for z near 0. But z/n for arbitrary z and large n is near 0, and from this one concludes that for any z we must have

$$\lambda(z) \equiv a_1 z \pmod{M}.$$

This proves our theorem.

We see that λ is represented by a multiplication α, and that

$$\alpha L \subset M.$$

Conversely, given a complex number α and lattices L, M such that $\alpha L \subset M$, multiplication by α induces a complex analytic homomorphism of $\mathbf{C}/L$ into $\mathbf{C}/M$.

Two complex toruses $\mathbf{C}/L$ and $\mathbf{C}/M$ are isomorphic if and only if there exists a complex number α such that $\alpha L = M$. We shall say that two lattices L, M are **linearly equivalent** if this condition is satisfied. In the next chapter, we shall find an analytic invariant for equivalence classes of lattices.

By an **elliptic curve,** or **abelian curve** A, one means a complete non-singular curve of genus 1, and a special point O taken as origin. The Riemann–Roch theorem defines a group law on the group of divisor classes of A. Actually, if P, P' are points on A, then there exists a unique point P'' such that

$$(P) + (P') \sim (P'') + (O),$$

where $\sim$ means linear equivalence, i.e. the left-hand side minus the right-hand side is the divisor of a rational function on the curve. The group law on A is then $P + P' = P''$. In characteristic $\neq 2$ or 3, using the Riemann–Roch theorem, one finds that the curve can be defined by a Weierstrass equation

$$y^2 = 4x^3 - g_2 x - g_3,$$

with g_2, g_3 in the ground field over which the curve is defined. Conversely, any homogeneous non-singular cubic equation has genus 1 and defines an abelian curve in the projective plane, once the origin has been selected. These facts depend on elementary considerations of curves. A curve defined by equations in projective space is said to be **defined over a field** k if the coefficients of these equations lie in k. For the Weierstrass equation, this means $g_2, g_3 \in k$.

For our purposes, if the reader is willing to exclude certain special cases, it will always suffice to visualize an elliptic curve as a curve defined by the above equation, with the addition law given by the rational formulas obtained from the addition theorem of the $\wp$ function. The origin is then the point at infinity. If A is defined over k, we denote by A_k the set of points (x, y) on the curve with $x, y \in k$, together with infinity, and call it the group of **k-rational points** on the curve. It is a group because the addition is rational, with coefficients in k.

If A, B are elliptic curves, one calls a **homomorphism** of A into B a group homomorphism whose graph is algebraic in the product space. If $\lambda: A \to B$ is such a homomorphism, and the curves are defined over the complex numbers, then λ induces a complex analytic homomorphism also denoted by λ,

$$\lambda: A_{\mathbf{C}} \to B_{\mathbf{C}},$$

viewing the groups of complex points on A and B as complex analytic groups. Suppose that the curves are obtained from lattices L and M in $\mathbf{C}$ respectively, i.e. we have maps

$$\varphi: \mathbf{C}/L \to A_{\mathbf{C}} \quad \text{and} \quad \psi: \mathbf{C}/M \to B_{\mathbf{C}}$$

which are analytic isomorphisms. As we saw above, our homomorphism λ is then induced by a multiplication by a complex number.

Conversely, it can be shown that any complex analytic homomorphism $\gamma: \mathbf{C}/L \to \mathbf{C}/M$ induces an algebraic one, i.e. there exists an algebraic homomorphism λ which makes the following diagram commutative.

$$\begin{array}{ccc} \mathbf{C}/L & \xrightarrow{\gamma} & \mathbf{C}/M \\ \varphi \downarrow & & \downarrow \psi \\ A_{\mathbf{C}} & \xrightarrow[\lambda]{} & B_{\mathbf{C}} \end{array}$$

We shall make a table of the effect of an isomorphism on the coefficients of the equations for elliptic curves, and their coordinates.

Let us agree that if A is an elliptic curve parametrized by the Weierstrass functions, for the rest of this section,

$$\varphi_A: \mathbf{C}/L \to A_{\mathbf{C}}$$

is the map such that

$$\varphi_A(z) = (1, \wp(z), \wp'(z)).$$

The $\wp$ function depends on L, and we shall denote it by

$$\wp(z, L).$$

Similarly for $\wp'(z, L)$. These satisfy the homogeneity property

$$\wp(cz, cL) = c^{-2}\wp(z, L) \quad \text{and} \quad \wp'(cz, cL) = c^{-3}\wp'(z, L)$$

for any $c \in \mathbf{C}$, $c \neq 0$.

Suppose that we are given two elliptic curves with parametrizations

$$\varphi_A: \mathbf{C}/L \to A_{\mathbf{C}} \quad \text{and} \quad \varphi_B: \mathbf{C}/M \to B_{\mathbf{C}},$$

and suppose that

$$M = cL,$$

so that the curves are isomorphic, with an isomorphism

$$\lambda: A \to B$$

induced by the multiplication by c. Then the coefficients g_2, g_3 of these curves satisfy the transformation

$$g_2(cL) = c^{-4}g_2(L)$$
$$g_3(cL) = c^{-6}g_3(L).$$

We let x_A and x_B denote the x-coordinate in the Weierstrass equation satisfied by the curves, respectively. Thus in general,

$$x(\varphi(z)) = \wp(z),$$

and similarly

$$y(\varphi(z)) = \wp'(z).$$

If P is a point on A, then the homogeneity properties of the Weierstrass functions can then be expressed purely algebraically by the formulas

$$x_B(\lambda(P)) = c^{-2}x_A(P) \quad \text{and} \quad y_B(\lambda(P)) = c^{-3}y_A(P).$$

These same formulas are valid in all characteristic $\neq 2$ or 3, and one can give purely algebraic proofs. In other words:

Suppose that A, B are elliptic curves in arbitrary characteristic $\neq 2, 3$ and in Weierstrass form, defined by the equations

$$y^2 = 4x^3 - g_2x - g_3$$

and

$$y^2 = 4x^3 - g_2'x - g_3'$$

respectively. Let $\lambda: A \to B$ be an isomorphism, defined over a field k. Then there exists $c \in k$ such that

$$g_2' = c^4g_2, \qquad g_3' = c^6g_3$$

and if the points (x, y) and (x', y') correspond under λ then

$$x' = c^2x \quad \text{and} \quad y' = c^3y.$$

One can then define purely algebraically the invariant

$$J_A = \frac{g_2^3}{g_2^3 - 27g_3^2},$$

and using the above quoted result (proved in characteristic 0 by transcendental means) we see at once that A is isomorphic to B if and only if $J_A = J_B$ (in characteristic $\neq 2$ or 3). We shall later study the analytic properties of this function J.

The above discussion also shows:

If A, B are elliptic curves over a field k of characteristic $\neq 2, 3$, and if they become isomorphic over an extension of k, then they become isomorphic over an extension of k, of degree $\leqq 6$.

Proof. We put the elliptic curves in Weierstrass form as above. Then for some element c in the extension of k, we see that $c^4 = g_2'/g_2$ (if $g_2 \neq 0$) and $c^6 = g_3'/g_3$ (if $g_3' \neq 0$). Thus the isomorphism is defined over an extension of degree 6, and even an extension of degree 2 if $g_2'g_3' \neq 0$.

Example. There are a couple of examples with the special values of c taken as i and $-\rho$, where $\rho = e^{2\pi i/3}$, which are important. Suppose that A is given in Weierstrass form. Then multiplication by i on $\mathbf{C}$ induces the following changes:

$$(x, y) \mapsto (-x, iy), \qquad g_2 \mapsto g_2, \qquad g_3 \mapsto -g_3.$$

Multiplication by $-\rho$ induces the following changes:

$$(x, y) \mapsto (\rho x, -y), \qquad g_2 \mapsto \rho g_2 \qquad g_3 \mapsto g_3.$$

In particular, if $g_3 = 0$, then we see that the curve admits i as an automorphism and if $g_2 = 0$, we seen that it admits $-\rho$ as an automorphism.

In *arbitrary characteristic*, Deuring gave a complete description for the cases which can arise [4], and he also gives normal forms replacing the Weierstrass form [8]. A short "*formulaire*" in this direction was made available recently by Tate. It has been useful to many people, and is reproduced as an appendix. I thank Tate for letting me print it here for the first time.

Given a value for j, we can always find an equation for an elliptic curve with invariant j defined by a Weierstrass equation

$$y^2 = 4x^3 - cx - c$$

with

$$J = \frac{c^3}{c^3 - 27c^2} = \frac{c}{c - 27},$$

which we can solve for c, namely

$$c = \frac{27J}{J - 1},$$

provided that $J \neq 0, 1$. The two cases corresponding to $J = 0, 1$ are then special, and are associated with the values i, ρ in the upper half plane. From the algebraic point of view, the above equation "parametrizes" universally all elliptic curves (in characteristic $\neq 2, 3$) with J-invariant $\neq 0, 1$, i.e. such curves can be obtained by specializing the generic equation.

For the two special values, one can select a number of models, e.g.

$$y^2 = 4x^3 - 3x, \qquad \text{for } J = 1,$$
$$y^2 = 4x^3 - 1, \qquad \text{for } J = 0.$$

By a suitable normalization, one can define a function on an elliptic curve closely related to the x-function, but which is invariant under isomorphisms. Namely, if $g_2 g_3 \neq 0$, we define the **first Weber function**

$$h_A^1 = \frac{g_2 g_3}{\Delta} x_A.$$

The above relations immediately show that h_A is invariant under isomorphisms of A. When g_2 or $g_3 \neq 0$ we take:

$$h_A^2 = \frac{g_2^2}{\Delta} x_A^2 \qquad \text{if} \quad g_3 = 0,$$

$$h_A^3 = \frac{g_3}{\Delta} x_A^3 \qquad \text{if} \quad g_2 = 0.$$

We shall see later that the Weber functions play an important role in analyzing the fields generated by points of finite order on the curve.

Occasionally it is useful to normalize the Weber functions so that certain power series expansions have integral coefficients. In this case, one takes for the first Weber function the expression

$$-2^7 3^5 \frac{g_2 g_3}{\Delta} x.$$

The reader should keep in mind that except for the elegance of language, in what follows, this normalization will not be used, and wherever he sees such a normalization, he can forget about the factor $-2^7 3^5$. The important thing will be that except for that factor, the power series involved have integral coefficients, and this will be enough.

§5. ENDOMORPHISMS AND AUTOMORPHISMS

If $L = M$, we get all endomorphisms (complex analytic) of $\mathbf{C}/L$ by those complex α such that $\alpha L \subset L$. Those endomorphisms induced by ordinary integers are called **trivial.** In general, suppose that $L = [\omega_1, \omega_2]$ and $\alpha L \subset L$. Then there exist integers a, b, c, d such that

$$\alpha\omega_1 = a\omega_1 + b\omega_2,$$
$$\alpha\omega_2 = c\omega_1 + d\omega_2.$$

Therefore α is a root of the polynomial equation

$$\begin{vmatrix} x - a & -b \\ -c & x - d \end{vmatrix} = 0,$$

whence we see that α is quadratic irrational over $\mathbf{Q}$, and is in fact integral over $\mathbf{Z}$. Dividing $\alpha\omega_2$ by ω_2, we see that

$$\alpha = c\tau + d,$$

where $\tau = \omega_1/\omega_2$. Since ω_1, ω_2 span a lattice, their ratio cannot be real. If α is not an integer, then $c \neq 0$, and consequently

$$\mathbf{Q}(\tau) = \mathbf{Q}(\alpha).$$

Furthermore, α is not real, i.e. α is imaginary quadratic.

The ring R of elements $\alpha \in \mathbf{Q}(\tau)$ such that $\alpha L \subset L$ is a subring of the quadratic field $k = \mathbf{Q}(\tau)$, and is in fact a subring of the ring of all algebraic integers $\mathfrak{o}_k$ in k. The units in R represent the automorphisms of $\mathbf{C}/L$. It is well known and very easy to prove that in imaginary quadratic field, the only units of R are roots of unity, and a quadratic field contains roots of unity other than ± 1 if and only if

$$k = \mathbf{Q}(\sqrt{-1}) \qquad \text{or} \qquad k = \mathbf{Q}(\sqrt{-3}).$$

If R contains $i = \sqrt{-1}$, then $R = \mathbf{Z}[i]$ is the ring of all algebraic integers in k, which must be $\mathbf{Q}(i)$. If R contains a cube root of unity ρ, then $R = \mathbf{Z}[\rho]$ is the ring of all algebraic integers in k, which must be $\mathbf{Q}(\sqrt{-3})$. The units in this ring are the 6-th roots of unity, generated by $-\rho$.

We may view the Weber function as giving a mapping of A onto the projective line, and we shall now see that it represents the quotient of the elliptic curve by its group of automorphisms.

Theorem 7. *If an elliptic curve A (over the complex numbers) has only ± 1 as its automorphisms, let the Weber function be given for a curve isomorphic to A, in Weierstrass form, by the formula*

$$h(x, y) = \frac{g_2 g_3}{\Delta} x.$$

If A admits i as an automorphism, let the Weber function be

$$h(x, y) = \frac{g_2^2}{\Delta} x^2$$

and if A admits ρ as an automorphism, let the Weber function be

$$h(x, y) = \frac{g_3}{\Delta} x^3.$$

Let P, Q be two points on A. We have $h(P) = h(Q)$ if and only if there exists an automorphism ε of A such that $\varepsilon(P) = Q$.

Proof. We may assume that A is in Weierstrass form. In the first case, the only non-trivial automorphism of A is such that

(1) $$(x, y) \mapsto (x, -y),$$

and it is then clear that h has the desired property. If on the other hand A admits i as an automorphism, then multiplication by i in $\mathbf{C}/L$ corresponds to the mapping on points given by

(2) $$(x, y) \mapsto (-x, iy),$$

and it is then clear that $x^2(P) = x^2(Q)$ if and only if P, Q differ by some automorphism of A. Finally, if A admits ρ as an automorphism, then multiplication by-ρ in $\mathbf{C}/L$ corresponds to the mapping on points given by

(3) $$(x, y) \mapsto (\rho x, -y),$$

and it is again clear that $x^3(P) = x^3(Q)$ if and only if P, Q differ by some automorphism of A, as was to be shown.

2 Homomorphisms

§1. POINTS OF FINITE ORDER

Let A be an elliptic curve defined over a field k. For each positive integer N we denote by A_N the kernel of the map

$$t \mapsto Nt, \qquad t \in A,$$

i.e. it is the subgroup of points of order N. If A is defined over the complex numbers, then it is immediately clear from the representation $A_{\mathbf{C}} \approx \mathbf{C}/L$ that

$$A_N \approx \mathbf{Z}/N\mathbf{Z} \times \mathbf{Z}/N\mathbf{Z}.$$

The inverse image of these points in $\mathbf{C}$ occur as the points of the lattice $\frac{1}{N}L$, and their inverse image in $\mathbf{C}/L$ is therefore the subgroup

$$\frac{1}{N}L/L \subset \mathbf{C}/L.$$

Let

$$\varphi : \mathbf{C} \to A_{\mathbf{C}}$$

be an analytic representation of $A_{\mathbf{C}}$ as $\mathbf{C}/L$, and let $L = [\omega_1, \omega_2]$. If we let

$$t_1 = \varphi\left(\frac{\omega_1}{N}\right) \quad \text{and} \quad t_2 = \varphi\left(\frac{\omega_2}{N}\right),$$

then $\{t_1, t_2\}$ form a basis for A_N over $\mathbf{Z}/N\mathbf{Z}$, i.e. A_N is the direct sum of the cyclic groups of order N generated by t_1 and t_2, respectively.

If the elliptic curve is defined over a field of characteristic zero, say k, then we can embed k in $\mathbf{C}$ and apply the preceding result.

In general, suppose that A is defined over an arbitrary field k. Let $\delta = \delta_A$ be the identity mapping of A. Then $N\delta$ is an endomorphism of A. Hasse has shown algebraically that if N is not divisible by the characteristic, then $N\delta$ is separable and its kernel has exactly N^2 points, in fact again we have

$$A_N \approx \mathbf{Z}/N\mathbf{Z} \otimes \mathbf{Z}/N\mathbf{Z}.$$

If p is the characteristic, and p/N, then the map may be inseparable, but is still of degree N^2, cf. [17]. This will be discussed later.

Let A be an elliptic curve defined over a field k and let K be an extension of k. Let σ be an isomorphism of K, not necessarily identity on k. One defines A^σ to be the curve obtained by applying σ to the coefficients of the equation defining A. For instance, if A is defined by

$$y^2 = 4x^3 - g_2 x - g_3,$$

then A^σ is defined by

$$y^2 = 4x^3 - g_2^\sigma x - g_3^\sigma.$$

If P, Q are points of A in K, then we have the formula

$$(P + Q)^\sigma = P^\sigma + Q^\sigma.$$

The sum on the left refers to addition on A, and the sum on the right refers to addition on A^σ. This is obvious because the algebraic addition formula is given by rational functions in the coordinates, with coefficients in k. Of course, if $P = (x, y)$, then $P^\sigma = (x^\sigma, y^\sigma)$ is obtained by applying σ to the coordinates.

In particular, suppose that P is a point of finite order, so that $NP = O$. Since O is rational over k, we see that for any isomorphism σ of K over k we have $NP^\sigma = O$ also, whence P^σ is also a point of order N. Since the number of points of order N is finite, it follows in particular that the points of A_N are algebraic over k (i.e. their coordinates are algebraic over k).

If $P = (x, y)$, we let $k(P) = k(x, y)$ be the extension of k obtained by adjoining the coordinates of P. Similarly, we let

$$k(A_N)$$

be the compositum of all fields $k(P)$ for $P \in A_N$. Of course, we view all points of finite order as having coordinates in a fixed algebraic closure of k, which we denote by ${}^a k$ or k_a.

The above remarks show that the Galois group $\mathrm{Gal}(k_a/k)$ operates as a permutation group of A_N. Consequently $k(A_N)$ is a normal extension of k, and is Galois if N is not divisible by the characteristic of k. We call $k(A_N)$ the **field of N-division points** of A over k.

Furthermore, if σ is an automorphism of $k(A_N)$ over k, and if we let $\{t_1, t_2\}$ be a basis of A_N over $\mathbf{Z}/N\mathbf{Z}$, then σ can be represented by a matrix

$$\begin{pmatrix} a & b \\ c & d \end{pmatrix}$$

such that

$$\begin{pmatrix} \sigma t_1 \\ \sigma t_2 \end{pmatrix} = \begin{pmatrix} a t_1 + b t_2 \\ c t_1 + d t_2 \end{pmatrix} = \begin{pmatrix} a & b \\ c & d \end{pmatrix} \begin{pmatrix} t_1 \\ t_2 \end{pmatrix}.$$

Thus we get an injective homomorphism

$$\mathrm{Gal}(k(A_N)/k) \to GL_2(\mathbf{Z}/N\mathbf{Z}).$$

It is a basic problem of elliptic curves to determine which subgroup of GL_2 is obtained, for fields k, which are interesting from an arithmetic point of view: Number fields, p-adic fields, and the generic case, which will be treated later.

§2. ISOGENIES

We shall now relate points of finite order and homomorphisms of elliptic curves. Let A, B be elliptic curves and let

$$\lambda: A \to B$$

be a homomorphism (algebraic). If $\lambda \neq 0$, then the kernel of λ is finite. The algebraic argument is that both A, B are algebraic curves, so of dimension 1, and hence λ must be generically surjective, so of finite degree. Over the complex numbers, we have a simple analytic argument. Indeed, if $A_{\mathbf{C}} \approx \mathbf{C}/L$ and $B_{\mathbf{C}} \approx \mathbf{C}/M$, then λ is represented analytically by multiplication with a complex number α such that $\alpha L \subset M$, so that $L \subset \alpha^{-1}M$. The kernel of the homomorphism

$$\mathbf{C}/L \to \mathbf{C}/M$$

induced by λ is precisely $\alpha^{-1}M/L$, which is finite, because both $\alpha^{-1}M$ and L are of rank 2 over $\mathbf{Z}$.

We let $\mathrm{Hom}(A, B)$ be the group of homomorphisms of A into B. Let $\lambda \in \mathrm{Hom}(A, B)$ and $\lambda \neq 0$. Then $n\lambda \neq 0$ for any integer $n \neq 0$. This is obvious in characteristic 0 from the analytic representation, and is provable algebraically in any characteristic. If Γ is the graph of λ, then for any point $Q \in B$ we have

$$\lambda^{-1}(Q) = \sum_{i=1}^{N} (P_i) = \mathrm{proj}_A\,(\Gamma \cdot (A \times Q)),$$

the sum being a formal sum, and the inverse image being taken counting multiplicities which can be defined algebraically. However, don't worry about these for the most part because in characteristic 0, or if N is not divisible by the characteristic, then the multiplicities are 1, and the P_i are simply all the points in the set theoretic inverse image of Q by λ. Over the complex numbers, they are represented by $\alpha^{-1}M/L$ in the notation of the above paragraph. We call N the **degree** of λ, denoted by $\nu(\lambda)$ or $\deg \lambda$.

If $\nu(\lambda) = N$, then there always exists a homomorphism

$$\mu: B \to A$$

such that $\mu \circ \lambda = \mu\lambda = N\delta$.

The analytic proof is obvious. Viewing λ as a homomorphism of $\mathbf{C}/L$ into

$\mathbf{C}/M$, let L'/L be its kernel. Then L'/L has order N and $L' \subset \frac{1}{N}L$. Therefore we have a canonical homomorphism

$$\mathbf{C}/M \to \mathbf{C}\Big/\frac{1}{N}L$$

such that the composite homomorphism

$$\mathbf{C}/L \xrightarrow{\lambda} \mathbf{C}/M \to \mathbf{C}\Big/\frac{1}{N}L$$

has kernel $\frac{1}{N}L/L$, which represents A_N in $\mathbf{C}/L$. Now we have an isomorphism

$$\mathbf{C}\Big/\frac{1}{N}L \xrightarrow{N} \mathbf{C}/L$$

given by multiplication with N, and the composite

$$\mathbf{C}/M \to \mathbf{C}\Big/\frac{1}{N}L \xrightarrow{N} \mathbf{C}/L$$

is the desired homomorphism μ.

Note that $\mu\lambda = N\delta_A$, but that we also have $\lambda\mu = N\delta_B$, because

$$(\lambda\mu - N\delta) \circ \lambda = 0,$$

and λ is surjective.

Since $\mathrm{Hom}(A, B)$ has characteristic 0, we can form the tensor product

$$\mathbf{Q} \otimes \mathrm{Hom}(A, B) = \mathrm{Hom}(A, B)_{\mathbf{Q}},$$

i.e. introduce integral denominators formally. Then any non-zero element of $\mathrm{Hom}(A, B)_{\mathbf{Q}}$ has an inverse in $\mathrm{Hom}(B, A)_{\mathbf{Q}}$. In fact, if $\lambda \in \mathrm{Hom}(A, B)$ is of degree N, then

$$\lambda^{-1} = \frac{1}{N}\mu,$$

where μ is the element of $\mathrm{Hom}(B, A)$ such that $\mu\lambda = N\delta$.

We let $\mathrm{End}(A) = \mathrm{Hom}(A, A)$.

Proposition 1. *If* $\mathrm{End}(A)$ *or* $\mathrm{End}(B) \approx \mathbf{Z}$, *then either* $\mathrm{Hom}(A, B) = 0$ *or* $\mathrm{Hom}(A, B) \approx \mathbf{Z}$.

Proof. Say $\mathrm{End}(A) \approx \mathbf{Z}$ and suppose that there exists some homomorphism $\lambda\colon A \to B$, $\lambda \neq 0$. Let $\lambda\mu = N\delta$. The map

$$\alpha \mapsto \mu \circ \alpha$$

gives a homomorphism of $\mathrm{Hom}(A, B)$ into $\mathrm{End}(A)$, and this homomorphism must be injective, for if $\mu\alpha = 0$, then $N\alpha = \lambda\mu\alpha = 0$, whence $\alpha = 0$. This proves our proposition.

Two elliptic curves A, B are called **isogenous** if there exists a homomorphism from A onto B, and such a homomorphism is called an **isogeny**.

Proposition 2. *If A, B are isogenous and* $\mathrm{End}(A) \approx \mathbf{Z}$, *then* $\mathrm{End}(B) \approx \mathbf{Z}$. *Assuming that this is the case, if there exists an isomorphism $\lambda: A \to B$, then there is only one other isomorphism from A onto B, that is $-\lambda$.*

Proof. The argument is similar to that of Proposition 1, and is clear.

Let $\mathfrak{g}$ be a finite subgroup of A. Then there exists a homomorphism

$$\lambda: A \to B$$

whose kernel is precisely $\mathfrak{g}$, and in characteristic > 0 we can take λ to be separable, so that λ satisfies the universal mapping property for homomorphisms of A whose kernel contains $\mathfrak{g}$.

Again, over the complex numbers, this is obvious using the analytic representation. We sometimes write $B = A/\mathfrak{g}$.

Proposition 3. *Assume that* $\mathrm{End}(A) \approx \mathbf{Z}$ *and let $\mathfrak{g}$, $\mathfrak{g}'$ be finite subgroups of A, of the same order. Then $A/\mathfrak{g} \approx A/\mathfrak{g}'$ if and only if $\mathfrak{g} = \mathfrak{g}'$.*

Proof. Let $\lambda: A/\mathfrak{g} \to A/\mathfrak{g}'$ be an isomorphism, and let

$$\alpha: A \to A/\mathfrak{g} \quad \text{and} \quad \alpha': A \to A/\mathfrak{g}'$$

be the canonical maps. Then

$$\deg(\lambda \circ \alpha) = \deg \alpha = \operatorname{ord} \mathfrak{g} = \operatorname{ord} \mathfrak{g}' = \deg \alpha'.$$

Thus $\lambda\alpha$ and α' have the same degree. Since $\mathrm{Hom}(A, A/\mathfrak{g}') \approx \mathbf{Z}$, it follows that

$$\lambda\alpha = \pm \alpha',$$

whence α, α' have the same kernel, i.e. $\mathfrak{g} = \mathfrak{g}'$. The converse is of course obvious.

Let $\lambda: A \to B$ be an isogeny defined over a field K. Let σ be an isomorphism of K. The graph of λ is an algebraic variety, actually an elliptic curve isomorphic to A, and we can apply σ to it. If $P \in A_K$ is a K-rational point of A, then we have the formula

$$\lambda(P)^\sigma = \lambda^\sigma(P^\sigma).$$

Furthermore, the association $\lambda \mapsto \lambda^\sigma$ is an isomorphism

$$\mathrm{Hom}(A, B) \to \mathrm{Hom}(A^\sigma, B^\sigma).$$

These are elementary algebraic facts which we take for granted. Furthermore, suppose that A is defined over a field k and that $\mathfrak{g}$ is a finite subgroup of A such that the cycle

$$\sum_{P \in \mathfrak{g}} (P)$$

is rational over k. Then we also take for granted that $A/\mathfrak{g}$ is defined over k and that the canonical homomorphism

$$\lambda: A \to A/\mathfrak{g}$$

is defined over k.

§3. THE INVOLUTION

Let $\alpha: A \to A$ be an endomorphism of A. We denote by α' the endomorphism such that

$$\alpha\alpha' = \alpha'\alpha = \nu(\alpha)\delta,$$

where $\nu(\alpha)$ is the degree of α. It is clear that if $\alpha, \beta \in \mathrm{End}(A)$ then

$$(\alpha\beta)' = \beta'\alpha'.$$

Hasse proved algebraically in general that $(\alpha + \beta)' = \alpha' + \beta'$, so that

$$\alpha \mapsto \alpha'$$

is an anti-automorphism of $\mathrm{End}(A)$. The proof in the complex case is easy as usual. Indeed, suppose that $A_{\mathbf{C}} \approx \mathbf{C}/L$ as before. Then we may view α as a complex multiplication, such that $\alpha L \subset L$, and the degree of α satisfies

$$\nu(\alpha) = (L : \alpha L),$$

i.e. it is the index of αL in L. Furthermore, this index is the determinant $\det(\alpha)$, viewing α as an endomorphism of L, as free module of rank 2 over $\mathbf{Z}$. If α is non-trivial, we have already seen in Chapter 1, §5, that $\mathbf{Q}(\alpha)$ is imaginary quadratic, and the multiplication by α in L is the regular representation of the quadratic field. Hence

$$\alpha' = \nu(\alpha)\alpha^{-1}$$

is the complex conjugate of α, and $\nu(\alpha)$ is the norm of α.

3 The Modular Function

§1. THE MODULAR GROUP

By SL_2 we mean the group of 2×2 matrices with determinant 1. We write $SL_2(R)$ for those elements of SL_2 having coefficients in a ring R. In practice, the ring R will be $\mathbf{Z}$, $\mathbf{Q}$, $\mathbf{R}$. We call $SL_2(\mathbf{Z})$ the **modular group**.

If L is a lattice in $\mathbf{C}$, then we can always select a basis, $L = [\omega_1, \omega_2]$ such that $\omega_1/\omega_2 = \tau$ is an element of the upper half plane, i.e. has imaginary part > 0. Two bases of L can be carried into each other by an integral matrix with determinant ± 1, but if we normalize the bases further to satisfy the above condition, then the matrix will have determinant 1, in other words, it will be in $SL_2(\mathbf{Z})$. Conversely, transforming a basis as above by an element of $SL_2(\mathbf{Z})$ will again yield such a basis. This is based on a simple computation, as follows. If

$$\alpha = \begin{pmatrix} a & b \\ c & d \end{pmatrix}$$

is in $GL_2(\mathbf{R})$, i.e. is a real non-singular matrix, and $\operatorname{Im}(z) > 0$, then

$$\operatorname{Im} \frac{az + b}{cz + d} = \frac{(ad - bc) \operatorname{Im}(z)}{|cz + d|^2}.$$

We denote by $\mathfrak{H}$ the upper half plane, i.e. the set of complex numbers z with $\operatorname{Im} z > 0$. If α is a matrix as above, in $GL_2^+(\mathbf{R})$, (i.e. α has positive determinant), then we see that the element

$$\alpha(z) = \frac{az + b}{cz + d}$$

also lies in $\mathfrak{H}$, and one verifies by brute force that the association

$$(\alpha, z) \mapsto \alpha(z) = \alpha z$$

defines an operation of $GL_2^+(\mathbf{R})$ on $\mathfrak{H}$, i.e. is associative, and the unit matrix operates as the identity. In fact, all diagonal matrices aI ($a \in \mathbf{R}$) operate trivially, especially ± 1. Hence we have an operation of $SL_2(\mathbf{R})/\pm 1$ on $\mathfrak{H}$. For $\alpha \in SL_2(\mathbf{R})$, we have the often used relation

$$\operatorname{Im} \alpha(z) = \frac{\operatorname{Im} z}{|cz + d|^2}.$$

If f is a meromorphic function on $\mathfrak{H}$, then the function $f \circ \alpha$ such that

$$(f \circ \alpha)(z) = f(\alpha z)$$

is also meromorphic.

We let $\Gamma = SL_2(\mathbf{Z})$, so that Γ is a discrete subgroup of $SL_2(\mathbf{R})$. By a **fundamental domain** D for Γ in $\mathfrak{H}$ we shall mean a subset of $\mathfrak{H}$ such that every orbit of Γ has one element in D, and two elements of D are in the same orbit if and only if they lie on the boundary of D.

Theorem 1. *Let D consist of all $z \in \mathfrak{H}$ such that*

$$-\tfrac{1}{2} \leqq \operatorname{Re} z \leqq \tfrac{1}{2} \quad \text{and} \quad |z| \geqq 1.$$

Then D is a fundamental domain for Γ in $\mathfrak{H}$. Let

$$T = \begin{pmatrix} 1 & 1 \\ 0 & 1 \end{pmatrix} \quad \text{and} \quad S = \begin{pmatrix} 0 & -1 \\ 1 & 0 \end{pmatrix}.$$

Then S, T generate Γ.

Proof. We illustrate D on Fig. 1.

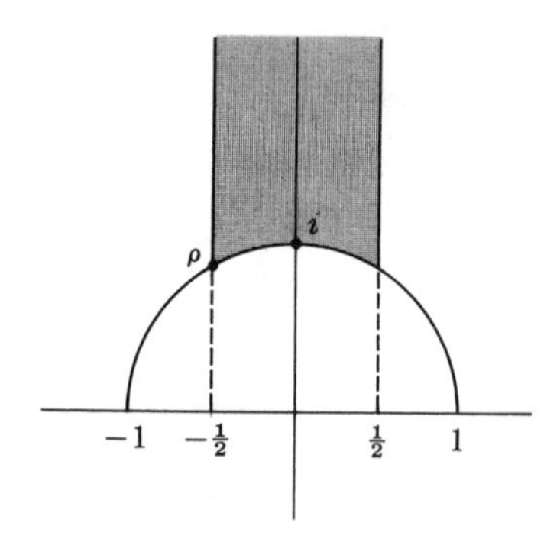

Fig. 3-1

On Fig. 1 we have indicated i and also the points where the vertical lines meet the circle of radius 1. The left-hand point is

$$\rho = e^{2\pi i/3} = \frac{-1 + \sqrt{-3}}{2},$$

i.e. the cube root of unity.

Let Γ' be the subgroup of Γ generated by S and T. Note that $-1 = S^2$ lies in Γ'. Given $z \in \mathfrak{H}$, iterating T on z shows that the orbit of z under powers of T contains an element whose real part lies in the interval $[-\frac{1}{2}, \frac{1}{2}]$. The formula giving the transformation of the imaginary part under Γ shows that the imaginary parts in an orbit of Γ are bounded from above, and tend to 0 as $\max(|c|, |d|)$ goes to infinity. In the orbit $\Gamma' z$ we can therefore select an element w whose imaginary part is maximal. If $|w| < 1$ then $Sw \in \Gamma' z$ and has greater imaginary part, so that $|w| \geqq 1$.

Next we prove that if $z, z' \in D$ are in the same orbit of Γ, then they arise from the obvious situation: Either they lie on the vertical sides and are translates by 1 or -1 of each other, or they lie on the base arc and are transforms of each other by S. We shall also prove that they are in the same orbit of Γ'.

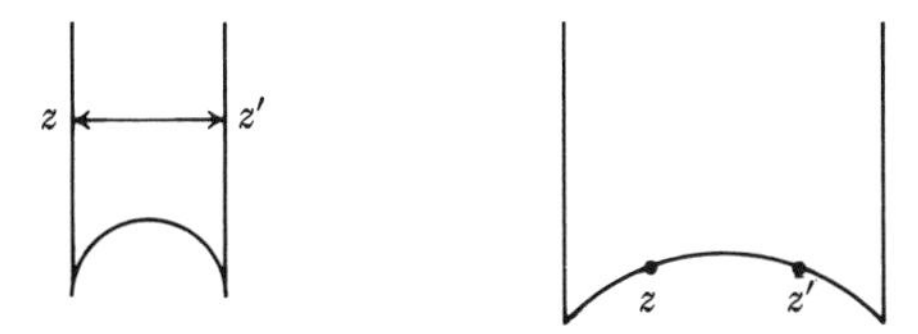

Fig. 3-2

If $\alpha(z) = z'$, the arguments will also determine α, which in particular will be seen to lie in Γ'. Say $\operatorname{Im} z' \geqq \operatorname{Im} z$, and $z' = \alpha(z)$ where

$$\alpha = \begin{pmatrix} a & b \\ c & d \end{pmatrix}.$$

Multiplying α by -1 if necessary, we may assume that $c \geqq 0$. From the formula for imaginary parts, we see that

$$|cz + d| \leqq 1.$$

Since $\operatorname{Im} z \geqq \sqrt{3}/2$, we must have $|c\sqrt{3}/2| \leqq 1$ so $c = 0$ or 1.

If $c = 0$, then

$$\alpha = \begin{pmatrix} 1 & b \\ 0 & 1 \end{pmatrix} = T^b,$$

and $\alpha z \in D$ implies that $b = \pm 1$, so we are in the obvious situation.

If $c = 1$, then $d = 0$ or $d = \pm 1$. If $d = 0$, then

$$\alpha = \begin{pmatrix} a & -1 \\ 1 & 0 \end{pmatrix} = T^a S, \quad \text{and} \quad \alpha(z) = a - \frac{1}{z}.$$

In this case $|z| = 1$, whence Sz also lies in D on the arc, and so z must be at the end points, i.e. $z = \rho$ or $z = S\rho$. It is then clear that $a = \pm 1$. If $d = \pm 1$, then $|z + d| \leqq 1$, and again obviously we have $z = \rho$ or $z = S\rho$. Say $z = \rho$. If $d = 1$, then

$$\alpha = \begin{pmatrix} a & a-1 \\ 1 & 1 \end{pmatrix},$$

and $\alpha(\rho) = \rho$ or $\alpha(\rho) = \rho + 1$. Say $\alpha(\rho) = \rho + 1$. Then

$$\alpha(\rho) = T^a S\rho = a - \frac{1}{\rho + 1} = \rho + 1.$$

But $-1/(\rho + 1) = \rho$, so that $a = 1$, and $\alpha = -TST$, so we are in one of the "obvious" cases. The other possible cases are treated similarly.

We have therefore shown that every orbit of the group generated by S, T has a representative in D, and also that if z, z' lie in D and $z' = \alpha z$ with $\alpha \in \Gamma$, then in fact $\alpha \in \Gamma'$, and the situation is an "obvious" one.

To show that S, T generate Γ, let $\alpha \in \Gamma$, and take an element z in the interior of D. There exists $\alpha' \in \Gamma'$ such that $\alpha'\alpha z \in D$. By the above, and since z is not on the boundary of D, it follows that $\alpha'\alpha z = z$. Again since z is not on the boundary, it follows that $\alpha'\alpha = \pm I$, whence α lies in Γ', and our theorem is proved.

Remark. We also have that $S^2 = (ST)^3 = I$, and that $\{S\}$, $\{ST\}$ are the isotropy groups of i and ρ, respectively. For all points which are not in an orbit of i or ρ, the isotropy group is $\pm I$. This follows at once from the arguments used to prove the theorem.

§2. AUTOMORPHIC FUNCTIONS OF DEGREE $2k$

Let $\mathfrak{H}$ be the upper half plane again, let $B > 0$, and let $\mathfrak{H}_B$ be the set of complex numbers z with $\operatorname{Im} z > B$. The map

$$z \mapsto e^{2\pi i z} = q_z$$

defines a holomorphic map from $\mathfrak{H}_B$ to the punctured disc of radius $e^{-2\pi B}$, i.e. the disc from which the origin is deleted. Furthermore, if $\mathfrak{H}_B/T$ denotes the quotient space of $\mathfrak{H}_B$ modulo translations by integers (essentially a cylinder), then q induces an analytic isomorphism between $\mathfrak{H}_B/T$ and this punctured disc (trivial verification, since for $z = x + iy$, we have

$$e^{2\pi i z} = e^{2\pi i x} e^{-2\pi y}.)$$

Consequently a meromorphic function f on $\mathfrak{H}_B$ which has period 1, i.e. is invariant under T, induces a meromorphic function f^* on the punctured disc.

A necessary and sufficient condition that f^* be also meromorphic at 0 is that there exist some positive integer N such that $f^*(q)q^N$ is bounded near 0. If this is the case, then f^* has a power series expansion

$$f^*(q) = \sum_{-N}^{\infty} c_n q^n.$$

We shall say that f is **meromorphic** (resp. **holomorphic**) **at infinity** if f^* is meromorphic (resp. holomorphic) at 0. By abuse of notation in this case, we also write

$$f = \sum_{-N}^{\infty} c_n q^n,$$

and call this the q-**expansion** of f **at infinity.** The coefficients c_n are called the **Fourier coefficients** of f. If $c_{-N} \neq 0$, we call $-N$ the **order** of f **at infinity,** and denote it by $v_\infty f$. For any $z \in \mathfrak{H}$ we let the order of f at z be denoted by $v_z f$.

Let $\mathfrak{M}$ be the field of meromorphic functions on $\mathfrak{H}$ and let

$$\alpha = \begin{pmatrix} a & b \\ c & d \end{pmatrix}$$

be in $\Gamma = SL_2(\mathbf{Z})$. For $f \in \mathfrak{M}$ and an integer $k \geqq 0$, define

$$(T_k(\alpha)f)(z) = f(\alpha(z))(cz + d)^{-2k}.$$

It is easily seen that this defines an operation of $SL_2(\mathbf{Z})$ on $\mathfrak{M}$. We say that f is **automorphic of weight** $2k$, or of **degree** $2k$, if $T_k(\alpha)f = f$ for all $\alpha \in \Gamma$, and if f is also meromorphic at infinity. Note that translation by 1 leaves f invariant, so our definition makes sense. The condition $T_k(\alpha)f = f$ also reads

$$f(\alpha(z)) = (cz + d)^{2k} f(z).$$

Remark. The literature is split on the convention whether to say of weight k or $2k$. The terminology of weight k is appropriate if one realizes that the condition can be interpreted to mean that the action of α leaves the differential form $f(z)(dz)^k$ invariant.

Theorem 2. *Let f be automorphic of weight $2k$, $f \neq 0$. Then*

$$v_\infty(f) + \frac{1}{3}v_\rho(f) + \frac{1}{2}v_i(f) + \sum_{P \neq i,\rho} v_P(f) = \frac{k}{6}.$$

The sum is taken over all points P of the upper half plane mod Γ, *not in the orbit of ρ or i.*

Proof. We integrate f'/f along the contour of Fig. 3(a), but modified by taking small arcs around the possible poles on the boundary, as on Fig. 3(b). For simplicity we phrase the proof under the assumption that f has no pole or zero on the edges other than at i or ρ, which are the most subtle possibilities. We have

$$\frac{1}{2\pi i}\int f'/f\,dz = \frac{1}{2\pi i}\int d\log f = \sum \text{Residues}$$
$$= \sum_{P\neq i,\rho} v_P(f).$$

We shall now compute the integral over the top, sides, arcs around the corners, arc around i, and the main arcs on the bottom circle.

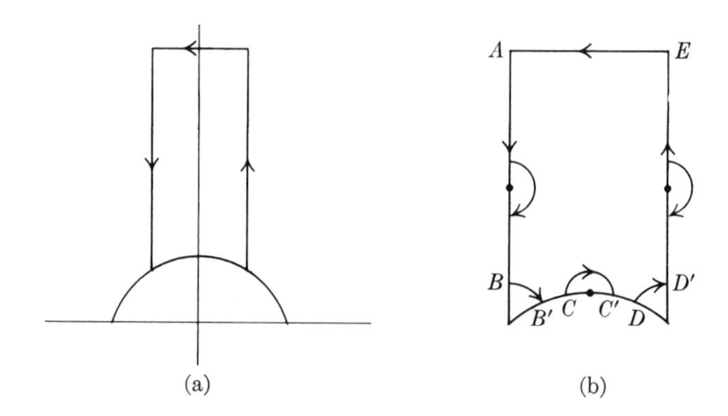

Fig. 3-3

Under the q-change of variables, the top segment between E and A transforms into the circle centered around the origin, clockwise. The integral over the top therefore gives

$$-v_\infty(f).$$

The integral over the left vertical side downward, plus the integral over the right vertical side upward yields 0 by the periodicity of f.

The integral around ρ over the small arc is equal to

$$\frac{1}{2\pi i}\int_B^{B'} d\log f.$$

We make the translation of ρ to 0, and thus suppose we consider a function also denoted by f near the origin, with power series expansion

$$f(z) = cz^m(1 + \cdots).$$

Then

$$\frac{f'(z)}{f(z)} = \frac{m}{z} + \text{holomorphic terms.}$$

As the radius of the small circle tends to 0, the integral of the holomorphic terms tend to 0. Integrating over an arc tending to $\pi/3$ in the clockwise direction, and taking the limit as the radius tends to 0 yields the value $-m/6$. We get a similar contribution on the small circle around $-\rho$, whence the contributions from these two small circles yield

$$-\tfrac{1}{3}v_\rho(f).$$

The same argument for the small arc around i shows that we get a contribution of

$$-\tfrac{1}{2}v_i(f).$$

There remains to compute the integrals over the main arcs

$$\int_{B'}^{C} + \int_{C'}^{D}.$$

The map S transforms the arc $B'C$ to the arc DC'. By definition,

$$f(Sz) = z^{2k}f(z),$$

and

$$\frac{df(Sz)}{dz} = f'(Sz)\frac{1}{z^2} = z^{2k}f'(z) + 2kz^{2k-1}f(z).$$

Since

$$\int_{C'}^{D} \frac{f'(w)}{f(w)}\,dw = \int_{C}^{B'} \frac{f'(Sz)}{f(Sz)}\,dz,$$

and

$$\frac{1}{z^2}\frac{f'(Sz)}{f(Sz)} = \frac{f'(z)}{f(z)} + \frac{2k}{z},$$

we see that the integral over the second arc has one term which cancels the integral over the first arc, plus another term which is

$$\frac{1}{2\pi i}\int_{B'}^{C} \frac{2k}{z}\,dz$$

and approaches $2k/12 = k/6$.

Putting all these contributions together proves our theorem.

Examples. They are constructed by using the following remark.

There is a bijection between functions of lattices, homogeneous of degree $-2k$, i.e. satisfying

$$G(\lambda L) = \lambda^{-2k}G(L), \qquad \lambda \in \mathbf{C},\ \lambda \neq 0,$$

and functions g on $\mathfrak{H}$ satisfying the condition

$$g(\alpha(z)) = (cz + d)^{2k}g(z).$$

The bijection is obtained as follows. Given a function G homogeneous of degree $-2k$, we let

$$g(z) = G(z, 1) = G\begin{pmatrix} z \\ 1 \end{pmatrix},$$

where by $G(z, 1)$ we mean the function G evaluated at the lattice $[z, 1]$. It then follows at once that

$$g(\alpha(z)) = (cz + d)^{2k} g(z).$$

Conversely, given a function g satisfying this condition, define

$$G(z, 1) = G\begin{pmatrix} z \\ 1 \end{pmatrix} = g(z),$$

and for any lattice $L = [\omega_1, \omega_2]$ define

$$G(L) = \omega_2^{-2k} g(\omega_1/\omega_2).$$

Then again it follows at once that $G(\lambda L) = \lambda^{-2k} G(L)$.

The fact that G is a function of lattices can be written in our vertical notation as

$$G\begin{pmatrix} \omega_1 \\ \omega_2 \end{pmatrix} = G\left(\alpha\begin{pmatrix} \omega_1 \\ \omega_2 \end{pmatrix}\right)$$

for any $\alpha \in SL_2(\mathbf{Z})$.

It is convenient to use the same symbol for the function of two variables and one variable, so that we shall also write

$$g(z) = g(z, 1) = g\begin{pmatrix} z \\ 1 \end{pmatrix}.$$

An automorphic function of weight $2k$ is called an **automorphic form** (of weight $2k$) if it is holomorphic on $\mathfrak{H}$ and at infinity. The special examples we now give will be of this type. In the next section, we construct an automorphic function of weight 0, holomorphic on $\mathfrak{H}$ but not at infinity.

Consider the functions

$$s_{2k}(L) = s_{2k} = \sum_{\omega \neq 0} \frac{1}{\omega^{2k}}.$$

Then the function

$$G_k(z) = \sum_{(m,n) \neq (0,0)} \frac{1}{(mz + n)^{2k}}$$

is obviously holomorphic on $\mathfrak{H}$, and substituting $z = \infty$ formally gives

$$G_k(\infty) = \sum_{n \neq 0} \frac{1}{n^{2k}} = 2\zeta(2k).$$

We shall actually get the q-expansion for G_k later, and see that G_k is holomorphic at infinity, with the above value. Hence G_k is an automorphic form of weight $2k$, and non-vanishing at infinity.

Let M_k be the set of automorphic forms of weight $2k$. Then M_k is a vector space over $\mathbf{C}$. It is clear that

$$M_k M_l \subset M_{k+l}.$$

The direct sum

$$\coprod_{k=0}^{\infty} M_k$$

can therefore be viewed as a graded algebra, whose structure is given by the next theorem.

Theorem 3. *The functions* $g_2 = 60s_4$ *and* $g_3 = 140s_6$ *are algebraically independent, and*

$$\coprod_{k=0}^{\infty} M_k = \mathbf{C}[g_2, g_3].$$

Proof. Note that g_2, g_3 generate a subalgebra of our graded algebra. To analyse M_k we shall apply the formula of Theorem 2, and observe that for $f \in M_k$, $f \neq 0$, all the orders on the left-hand side are $\geqq 0$. We now proceed systematically.

k = 0. The right-hand side is 0, so all the terms on the left are 0. If $f \in M_0$ and f is not identically 0, then f has no zero on $\mathfrak{H}$ or at infinity. The constants are contained in M_0. Let $c = f(\infty)$. Then $g = f - c$ vanishes at infinity, hence is identically 0, so $M_0 = \mathbf{C}$.

k = 1. The right-hand side is 1/6. The left-hand side shows that this is possible if and only if $f = 0$, so $M_1 = 0$.

k = 2. We prove that $M_2 = (g_2)$ is the 1-dimensional vector space generated by g_2. Let $f \in M_2$, $f \neq 0$. The right-hand side of the basic formula is 1/3. The only time this is compatible with the left-hand side is when all the terms on the left are 0 except for $\frac{1}{3}v_\rho(f)$, and we must have $v_\rho(f) = 1$, while f has no other zero. In particular, we have also proved:

g_2 has a zero only at ρ, and it is of order 1.

For some constant c, $f - cg_2$ has zero at infinity, and lies in M_2, hence is identically zero, and $f = cg_2$, thus proving what we wanted.

k = 3. We prove that $M_3 = (g_3)$. The right-hand side of the basic formula is 1/2, for f in M_3, $f \neq 0$. The only way this is possible is that $v_i(f) = 1$, and f has no other zero. In particular,

g_3 has a zero only at i, and it is of order 1.

The same argument as before shows that $f = cg_3$ for some constant c.

k = 4. We prove that $M_4 = (g_2^2)$. The right-hand side of the formula for $f \in M_4$, $f \neq 0$ is 2/3, and hence $v_\rho(f) = 2$, and f has no other zero. It follows that $f = cg_2^2$ as before.

k = 5. We prove that $M_5 = (g_2 g_3)$. In this case, the same arguments as before show that $f \in M_5$, $f \neq 0$ has a zero of order 1 at i and ρ, and no other zero, and also that $f = cg_3 g_3$.

k $\geqq$ 6. We recall that $\Delta = g_2^3 - 27g_3^2$ is nowhere zero on $\mathfrak{H}$, and Δ lies in M_6. The right-hand side of the formula for $k = 6$ is equal to 1, and shows that $v_\infty(\Delta) = 1$, i.e. Δ has a zero of order 1 at infinity.

Now $G_6 \in M_6$ and $G_6(\infty) \neq 0$. If $f \in M_6$, then there exists a constant c such that $f - cG_6$ vanishes at infinity. Then

$$\frac{f - cG_6}{\Delta} \in M_0 = \mathbf{C},$$

and we see that $f = b\Delta + cG_6$ for some constant b. Inductively, the same technique shows that for $k \geqq 6$,

$$\boxed{M_k = \Delta M_{k-6} \oplus (G_k).}$$

We can prove by induction that any $f \in M_k$ is a polynomial in g_2 and g_3. This has already been shown for $k \leqq 5$. If $k \geqq 6$, we write $k = 2r$ or $k = 2r + 1$, and we can subtract cg_2^r or $cg_2^{r-1}g_3$ from f, with a suitable constant c, to get a function vanishing at infinity, so that

$$\frac{f - cg_2^r}{\Delta} \quad \text{or} \quad \frac{f - cg_2^{r-1}g_3}{\Delta}$$

lies in M_{k-6}, and our proof is complete, by induction.

There remains to prove that g_2 and g_3 are algebraically independent, to be sure we get the formal polynomial ring. First it is clear from the homogeneity property that a non-trivial linear relation among elements of distinct M_k's cannot exist, i.e. if $f_1, \ldots, f_m$ are of distinct weights, then they are linearly independent over the complex numbers. If we had an algebraic relation among g_2, g_3, then we could assume that the monomials in it have the same weight. In such a relation, if a pure power of g_2 occurs, then the relation is of the form

$$g_2^m + g_3 P(g_2, g_3) = 0$$

where P is some polynomial. Evaluating this at i shows that it is impossible because $g_3(i) = 0$ and $g_2(i) \neq 0$. Similarly, no pure power of g_3 can occur. Hence g_2 divides each monomial, and cancelling g_2 yields a relation of lower degree, so the proof is finished by induction.

The exposition in this section follows Serre [B10].

§3. THE MODULAR FUNCTION j

We define the **modular function**

$$J = g_2^3/\Delta \qquad \text{and} \qquad j = 1728g_2^3/\Delta.$$

The reason for the 1728 is that certain power series expansions later will have integral coefficients. Note that $1728 = 2^6 3^3$.

From the properties of g_2, g_3 proved in the preceding section, we see that j is an automorphic function of weight 0, and since it is holomorphic, non-zero on $\mathfrak{H}$, we see that j has a pole of order 1 at infinity. We shall prove later that the residue is 1, in the q-expansion.

Theorem 4. *The map $j: \Gamma\backslash\mathfrak{H} \to \mathbf{C}$ is a bijection.*

Proof. We apply the basic relation of Theorem 2 with $k = 0$, so the right-hand side is 0, to the function $j - c$ for $c \in \mathbf{C}$. Then $j - c$ has a simple pole at infinity, and

$$\tfrac{1}{3}v_\rho + \tfrac{1}{2}v_i + \sum v_P = 1.$$

The terms on the left are all $\geqq 0$. This is possible if and only if the order of $j - c$ at some unique z in $\Gamma\backslash\mathfrak{H}$ is $\neq 0$. The multiplicity is 1 if z is not in the orbit of ρ, i and otherwise, it is 2 at i and 3 at ρ. In any case, our theorem is proved.

We can view j as a function of lattices according to our general scheme transforming functions of two variables into functions of one variable by homogeneity. But since j is of weight 0, we see that for a lattice $L = [\omega_1, \omega_2]$ we can write

$$j(L) = j(\tau)$$

if ω_1, ω_2 are selected such that $\omega_1/\omega_2 = \tau$ lies in $\mathfrak{H}$. If $L = \lambda M$ for some complex $\lambda \neq 0$ then $j(L) = j(M)$. Conversely, the fact that j gives a bijection of $\Gamma\backslash\mathfrak{H}$ with $\mathbf{C}$ can be stated in the homogeneous form, namely that the converse holds, i.e.:

Corollary 1. *Let L, M be two lattices in $\mathbf{C}$. Then $j(M) = j(L)$ if and only if M, L are equivalent.*

By Theorem 6 of Chapter 1, §4 we also see that the condition of the corollary is equivalent with the property that $\mathbf{C}/L$ is isomorphic to $\mathbf{C}/M$. Thus j gives us the desired analytic expression parametrizing isomorphism classes of elliptic curves (complex toruses).

Corollary 2. *Let c_2, c_3 be complex numbers such that*

$$c_2^3 - 27c_3^2 \neq 0.$$

Then there exists a lattice L such that

$$c_2 = g_2(L) \qquad \text{and} \qquad c_3 = g_3(L).$$

Proof. By the theorem, there exists $\tau \in \mathfrak{H}$ such that

$$j(\tau) = 1728\frac{c_2^3}{c_2^3 - 27c_3^2}.$$

Let $M = [\tau, 1]$. If $c_2 = 0$, then $j(\tau) = 0$ and $\tau = \rho$. Let $w \in \mathbf{C}^*$ be such that $w^{-6}g_3(L) = c_3 \neq 0$. Let $L = wM$. Then

$$g_2(L) = w^{-4}g_2(M) = w^{-4}g_2(\rho) = c_2 = 0,$$

and $g_3(L) = c_3$, so we are done.

If $c_2 \neq 0$, choose $w \in \mathbf{C}^*$ such that $w^{-4}g_2(M) = c_2$ and let $L = wM$ again. Then $g_2(L) = c_2$. Hence

$$\frac{c_2^3}{c_2^3 - 27c_3^2} = J(\tau) = J(M) = J(L) = \frac{g_2^3(L)}{g_2^3(L) - 27g_3^2(L)}$$
$$= \frac{c_2^3}{c_2^3 - 27g_3^2(L)}.$$

This shows that

$$g_3^2(L) = c_3^2, \qquad \text{whence} \quad g_3(L) = \pm c_3.$$

If necessary, replace w by iw. This does not change g_2 and changes g_3 by -1. Then L is a lattice whose g_2, g_3 have the desired values, thus concluding the proof of the corollary.

The above result shows that an arbitrary elliptic curve

$$y^2 = 4x^3 - c_2x - c_3$$

with non-vanishing discriminant can always be parametrized by elliptic functions, i.e. we can select a lattice L such that

$$c_2 = g_2(L) \qquad \text{and} \qquad c_3 = g_3(L).$$

The associated Weierstrass $\wp$ and $\wp'$ parametrize the curve.

If A is an elliptic curve, we denote by j_A the value $j(L)$, for any lattice L such that $A_{\mathbf{C}}$ is isomorphic to $\mathbf{C}/L$. This value is independent of the choice of L, and is called the **j-invariant** of the curve. Note that it is defined rationally in terms of the coefficients of the equation defining A. We can reformulate Corollary 1 as follows.

Corollary 3. *Two elliptic curves A and B are isomorphic if and only if $j_A = j_B$.*

Remark. Let τ be such that $j(\tau)$ is transcendental over $\mathbf{Q}$. Then an elliptic curve with invariant $j(\tau)$ necessarily has a trivial ring of endomorphisms. Indeed, we know from Chapter 1, §5 that if the curve has non-trivial endomorphisms, then τ is imaginary quadratic, and there are only denumerably such τ, while there are non-denumerably many transcendental complex numbers over $\mathbf{Q}$.

If A_1 is an elliptic curve with transcendental invariant j_1, and A_1 is defined over a field K_1 finitely generated over $\mathbf{Q}$, and similarly A_2 has invariant j_2 transcendental over $\mathbf{Q}$, and is defined over K_2, we let $\sigma: \mathbf{Q}(j_1) \to \mathbf{Q}(j_2)$ be an isomorphism sending j_1 on j_2 and extend σ to K_1. Then A_1^σ has invariant $j_1^\sigma = j_2$, and A_1^σ is therefore isomorphic to A_2. Extending K_1 to a bigger field if necessary, we may assume that all endomorphisms of A_1 are defined over K_1. Then $\mathrm{End}(A_1^\sigma) = \mathrm{End}(A_1)^\sigma$, and thus A_2 and A_1 have isomorphic rings of endomorphisms. This proves our remark.

4 *Fourier Expansions*

§1. EXPANSION FOR G_k

In this section we derive the promised expansions at infinity for the G_k, whence for Δ and j.

We start with the product expansion for the sine,

$$\sin \pi z = \pi z \prod_{n=1}^{\infty} \left(1 - \frac{z}{n}\right)\left(1 + \frac{z}{n}\right).$$

Taking the logarithmic derivative yields

$$\pi \frac{\cos \pi z}{\sin \pi z} = \frac{1}{z} + \sum_{n=1}^{\infty} \left[\frac{1}{z - n} + \frac{1}{z + n}\right]. \tag{1}$$

But

$$\cos w = \frac{e^{iw} + e^{-iw}}{2} \quad \text{and} \quad \sin w = \frac{e^{iw} - e^{-iw}}{2}$$

whence

$$\cos \pi z = \frac{1}{2} e^{-i\pi z} (e^{2\pi i z} + 1),$$

$$\sin \pi z = \frac{1}{2i} e^{-i\pi z} (e^{2\pi i z} - 1).$$

We let

$$\boxed{q = q_\tau = e^{2\pi i \tau}.}$$

Then for τ in the upper half plane $\mathfrak{H}$ we get

$$\pi \frac{\cos \pi\tau}{\sin \pi\tau} = \pi i \frac{q + 1}{q - 1} = \pi i + \frac{2\pi i}{q - 1} = \pi i - 2\pi i \sum_{\nu=0}^{\infty} q^\nu. \tag{2}$$

Differentiating the expressions in (1) and (2) repeatedly yields

$$(3) \qquad (-1)^{k-1}(k-1)! \sum_{n=-\infty}^{\infty} \frac{1}{(\tau - n)^k} = -\sum_{\nu=1}^{\infty} (2\pi i)^k \nu^{k-1} q^\nu.$$

Consequently from the definition

$$G_k(\tau) = \sum_{m,n}{}' \frac{1}{(m\tau + n)^{2k}}$$

we get, summing separately for $m = 0$ and $m \neq 0$,

$$\begin{aligned} G_k(\tau) &= 2\zeta(2k) + 2 \sum_{m=1}^{\infty} \sum_{n=-\infty}^{\infty} \frac{1}{(m\tau + n)^{2k}} \\ &= 2\zeta(2k) + 2 \sum_{m=1}^{\infty} \sum_{\nu=1}^{\infty} \frac{(2\pi i)^{2k} \nu^{2k-1}}{(2k-1)!} q_\tau^{m\nu}, \end{aligned}$$

We let

$$\sigma_k(n) = \sum_{d|n} d^k.$$

Proposition 1. *We have*

$$(4) \qquad G_k(\tau) = 2\zeta(2k) + 2\frac{(2\pi i)^{2k}}{(2k-1)!} \sum_{n=1}^{\infty} \sigma_{2k-1}(n) q_\tau^n.$$

The most interesting special cases give us:

$$(5) \qquad g_2 = 60G_2 = (2\pi)^4 \frac{1}{2^2 3}(1 + 240X)$$

$$(6) \qquad g_3 = 140G_3 = (2\pi)^6 \frac{1}{2^3 3^3}(1 - 504Y)$$

where

$$X = \sum_{n=1}^{\infty} \sigma_3(n) q^n \quad \text{and} \quad Y = \sum_{n=1}^{\infty} \sigma_5(n) q^n.$$

We have also used the standard values

$$\zeta(4) = \frac{\pi^4}{90} \quad \text{and} \quad \zeta(6) = \frac{\pi^6}{945}.$$

We then get

$$(7) \qquad \Delta = (2\pi)^{12} \frac{1}{2^6 3^3}[(1 + 240X)^3 - (1 - 504Y)^2].$$

We contend that all the coefficients in the q-expansion of the expression in brackets are $\equiv 0 \bmod 2^6 3^3 = 1728$. This is a simple matter. We see at once that

$$[\cdots] \equiv 3^2 2^4 (5X + 7Y) \bmod 2^6 3^3.$$

We have to show that $5X + 7Y \equiv 0 \bmod 4$ and mod 3. For this it suffices that

$$\sum_{d|n} d^3 \equiv \sum_{d|n} d^5 \bmod 4 \text{ and } \bmod 3.$$

But for all d, we already have $d^3 \equiv d^5$, so our contention holds.

Therefore the q-expansion for Δ has the form

$$\Delta = (2\pi)^{12} q\left(1 + \sum_{n=1}^{\infty} d_n q^n\right), \tag{8}$$

where the coefficients d_n are integers. From this we now see that the expansion for j has integer coefficients, namely

$$j = 12^3 \frac{g_2^3}{\Delta} = \frac{1}{q} + \sum_{n=0}^{\infty} a_n q^n \tag{9}$$

with $a_n \in \mathbf{Z}$. The first two coefficients are

$$j = \frac{1}{q} + 744 + 196884q + \cdots.$$

§2. EXPANSION FOR THE WEIERSTRASS FUNCTION

If $L_\tau = [\tau, 1]$ we write

$$\wp(z, L_\tau) = \wp(z; \tau, 1) = \wp(z; \tau).$$

From (1) and (2) in the preceding section, we have

$$\sum_{n=-\infty}^{\infty} \frac{1}{(w + n)^2} = (2\pi i)^2 \sum_{n=1}^{\infty} n q_w^n = (2\pi i)^2 \frac{q_w}{(1 - q_w)^2} \tag{10}$$

where $q_w = e^{2\pi i w}$, From the definition of the $\wp$-function, we find

$$\wp(z; \tau) = \frac{1}{z^2} + \sum_{m,n} \left[\frac{1}{(z - m\tau + n)^2} - \frac{1}{(m\tau + n)^2}\right]$$

$$= \frac{1}{z^2} + \sum_{m=0} \sum_{n \neq 0} + \sum_{m \neq 0} \sum_{n \in \mathbf{Z}}$$

$$= (2\pi i)^2 \frac{q_z}{(1 - q_z)^2} - 2\zeta(2)$$

$$+ \sum_{m=1}^{\infty} \left[\sum_{n \in \mathbf{Z}} \left[\frac{1}{(z + m\tau + n)^2} + \frac{1}{(-z + m\tau + n)^2}\right] - 2 \sum_{n \in \mathbf{Z}} \frac{1}{(m\tau + n)^2}\right] \tag{11}$$

Recall that $\zeta(2) = \pi^2/6$. Also use the fact that

$$q_{m\tau+z} = q_\tau^m q_z.$$

For $\operatorname{Im} \tau > 0$ we have $|q_\tau| < 1$. In the range

$$|q_\tau| < |q_z| < \frac{1}{|q_\tau|}$$

we therefore find a **first q-expansion for the $\wp$-function,** namely:

Proposition 2.

$$\frac{1}{(2\pi i)^2}\wp(z;\tau) = \frac{1}{12} + \frac{q_z}{(1-q_z)^2} + \sum_{m=1}^{\infty}\sum_{n=1}^{\infty} n q_\tau^{mn}(q_z^n + q_z^{-n}) - 2\sum_{m=1}^{\infty}\sum_{n=1}^{\infty} n q_\tau^{mn}.$$

Except for the 1/12, all the coefficients are integers.

On the other hand, we can use the second formula on the right of (10). Applying these to formula (11), we see that one of the sums has the form

$$\sum_{m=1}^{\infty}\left[\frac{q_\tau^m q_z}{(1-q_\tau^m q_z)^2} + \frac{q_\tau^m/q_z}{(1-q_\tau^m/q_z)^2}\right]. \tag{12}$$

We multiply the second term by q_τ^{-2m} and q_z^2 in the numerator and denominator. We also make a similar easy transformation for the other double sum in (11), and we come up with a **second expression for the q-expansion of the $\wp$-function,** namely:

Proposition 3.

$$\frac{1}{(2\pi i)^2}\wp(z;\tau) = \frac{1}{12} + \sum_{m\in\mathbf{Z}}\frac{q_\tau^m q_z}{(1-q_\tau^m q_z)^2} - 2\sum_{n=1}^{\infty}\frac{n q_\tau^n}{1-q_\tau^n}.$$

Differentiating yields

$$\frac{1}{(2\pi i)^3}\wp'(z;\tau) = \sum_{m\in\mathbf{Z}}\frac{q_\tau^m q_z(1+q_\tau^m q_z)}{(1-q_\tau^m q_z)^3}.$$

Using the splitting as in (12) or looking at these again directly, one sees that these second formulas are valid for all $z \in \mathbf{C}$ once τ is fixed.

The formulas for g_2 and g_3 found in the preceding section can be put in a similar form, say abbreviating $q = q_\tau$.

Proposition 4.

$$\frac{1}{(2\pi i)^4} g_2(\tau) = \frac{1}{12}\left[1 + 240 \sum_{n=1}^{\infty} \frac{n^3 q^n}{(1 - q^n)}\right]$$

$$\frac{1}{(2\pi i)^6} g_3(\tau) = \frac{1}{6^3}\left[-1 + 504 \sum_{n=1}^{\infty} \frac{n^5 q^n}{(1 - q^n)}\right].$$

From the expansions for g_2, g_3 and the Weierstrass function, we get trivially the expansion for the Weber function.

Proposition 5. *Let*

$$f_0(z, \tau) = -2^7 3^5 \frac{g_2(\tau) g_3(\tau)}{\Delta(\tau)} \wp(z; \tau, 1).$$

Let $q = q_\tau$ *and* $w = q_z = e^{2\pi i z}$. *Then*

$$f_0 = P(q)\left[1 + \frac{12w}{(1 - w)^2} + 12 \sum_{m,n=1}^{\infty} n q^{mn}(w^n + w^{-n} - 2)\right]$$

where

$$P(q) = q + c_2 q^2 + \cdots$$

is a power series with integer coefficients starting with q.

Let $L = [2\pi i\tau, 2\pi i]$. Then from our knowledge how $\wp$, $\wp'$, g_2, g_3 transform under isomorphisms in Chapter 1, §4 we see that the above expressions in fact give

$$g_2(L), \quad g_3(L), \quad \wp(z, L) \quad \text{and} \quad \wp'(z, L).$$

Thus if we let

$$g_2 = g_2(L) \quad \text{and} \quad g_3 = g_3(L),$$

then the elliptic curve

$$y^2 = 4x^3 - g_2 x - g_3$$

is parametrized by the functions having the second expansions. Furthermore, since the map

$$z \mapsto (1, \wp(z), \wp'(z))$$

is a homomorphism of $\mathbf{C}$ into the elliptic curve (actually surjective), and since the formulas for $\wp(z, \tau)$ depend only on q_z, it follows that the formulas of Proposition 3 give us a **homomorphism from the multiplicative group of complex numbers onto the complex points of the elliptic curve.** For the algebraic implications of this fact, see the Tate parametrization in Chapter 15.

§3. BERNOULLI NUMBERS

This section will not be used in the sequel and is included only for the convenience of the reader in reading some other literature, e.g. concerning elliptic functions and L-series. In particular, the von Staudt theorem is frequently used in such contexts.

We define the **Bernoulli numbers** B_n by the power series expansion

$$\frac{z}{e^z - 1} = \sum_{n=0}^{\infty} \frac{B_n}{n!} z^n.$$

From the relation

$$z = \sum_{m=1}^{\infty} \frac{z^n}{m!} \sum_{n=0}^{\infty} \frac{B_n}{n!} z^n$$

we get a recursion formula for the Bernoulli numbers, namely

$$\frac{B_0}{n!0!} + \frac{B_1}{(n-1)!1!} + \cdots + \frac{B_{n-1}}{1!(n-1)!} = \begin{cases} 1 & \text{if } n = 1 \\ 0 & \text{if } n > 1. \end{cases}$$

We get $B_0 = 1$,

$$2B_1 + B_0 = 0, \qquad \text{whence } B_1 = -1/2,$$
$$3B_2 + 3B_1 + B_0 = 0, \qquad \text{whence } B_2 = 1/6,$$

and so forth.

From the identity

$$\frac{z}{e^z - 1} + \frac{z}{2} = \frac{z}{2}\left(\frac{e^z + 1}{e^z - 1}\right) = \frac{z}{2} \frac{e^{z/2} + e^{-z/2}}{e^{z/2} - e^{-z/2}},$$

we see that the above function is even, and hence has only even terms in its power series expansion. This implies that, except for B_1, the odd Bernoulli numbers are equal to 0. The first few Bernoulli numbers are then:

$$B_4 = -\frac{1}{30} \qquad B_6 = \frac{1}{42} \qquad B_8 = -\frac{1}{30}$$

$$B_{10} = \frac{5}{66} \qquad B_{12} = -\frac{691}{20730} \qquad B_{14} = \frac{7}{6}.$$

We have

$$\frac{z}{2} \frac{e^{z/2} + e^{-z/2}}{e^{z/2} - e^{-z/2}} = \sum_{n=0}^{\infty} \frac{B_{2n}}{(2n)!} z^{2n}.$$

Replace z by $2\pi iz$. We then see that

$$\pi z \cot \pi z = \sum_{n=0}^{\infty} (-1)^n \frac{(2\pi)^{2n}}{(2n)!} B_{2n} z^{2n}.$$

Comparing with our previous expansion for $\pi \cot \pi z$, we see that

$$\zeta(2n) = (-1)^{n-1} \frac{2^{n-1}\pi^{2n}}{(2n)!} B_{2n}.$$

Von Staudt's theorem. *We have*

$$B_{2n} \equiv \sum_{(p-1)|2n} -\frac{1}{p} \pmod{\mathbf{Z}}.$$

Proof. Let $D = d/dz$. Then

$$B_n = D^n\left(\frac{z}{e^z - 1}\right)\bigg|_{z=0}$$
$$= D^n\left(\frac{-\log(1 - (1 - e^z))}{1 - e^z}\right)\bigg|_{z=0}.$$

Using the power series for the log, and differentiating term by term, we find that

$$B_n = \sum_{k=1}^{n+1} \frac{1}{k} D^n(1 - e^z)^{k-1}\bigg|_{z=0} = \sum_{k=1}^{n+1} \frac{1}{k} A_k$$

where

$$A_k = D^n(1 - e^z)^{k-1}\bigg|_{z=0}.$$

We assert: If $k \neq 4$ and is not a prime, then $k | A_k$.

Proof. Let $k = ab$, $2 \leqq a \leqq \sqrt{k}$. Write

$$(1 - e^z)^{k-1} = (1 - e^z)^a(1 - e^z)^b(1 - e^z)^{ab-a-b-1}.$$

We must have $ab - a - b - 1 \geqq 0$. Indeed, $y = k - x - k/x - 1$ has a maximum at k. The minimum is at $x = 2$, with value $(k - 6)/2$, which is $\geqq 0$ if $k \geqq 6$. Taking the derivative of

$$(1 - e^z)^a(1 - e^z)^b(1 - e^z)^c,$$

we see that there will be a non-zero contribution when we substitute $z = 0$ only for those terms for which we differentiate at least once the factors $(1 - e^z)^a$ and $(1 - e^z)^b$, in other words, such terms will be divisible by $ab = k$. This proves our assertion.

To compute $B_n \pmod{\mathbf{Z}}$ we are reduced to considering A_k for those values of k not already eliminated.

First, if $k = 4$, then we find the value directly by expanding out

$$(1 - e^z)^3 = 1 - 3e^z + 3e^{2z} - e^{3z},$$

and differentiating. We get

$$A_k = -3 + 3 \cdot 2^n - 3^n \equiv 0 \pmod{4}$$

if $n = 1$ or if n is even, which are the cases we want. Again in this case, we get no contribution to $B_n \pmod{\mathbf{Z}}$.

Finally, suppose that $k = p$ is a prime $\leqq n + 1$. Write

$$n = (p - 1)q + r, \qquad 0 \leqq r < p - 1.$$

Then

$$A_p = \sum_{i=0}^{p-1} (-1)^i \binom{p-1}{i} i^n = \sum_{i=0}^{p-1} (-1)^i \binom{p-1}{i} (i^{p-1})^q \, i^r$$

whence

$$A_p \equiv \begin{cases} \displaystyle\sum_{i=0}^{p-1} (-1)^i \binom{p-1}{i} i^r & \text{if } \; r > 0 \\ \displaystyle\sum_{i=0}^{p-1} (-1)^i \binom{p-1}{i} - 1 & \text{if } \; r = 0. \end{cases}$$

If $r = 0$, we get the contribution $-1 \pmod{\mathbf{Z}}$. If $r > 0$, then our value for A_p is the same as

$$D^r(1 - e^z)^{p-1}\Big|_{z=0},$$

which yields 0. This proves von Staudt's theorem.

5 The Modular Equation

We are interested in studying the j-invariants of isogenous elliptic curves, which, as we shall see, amounts to studying $j \circ \alpha$ where α is a rational matrix. For this we need some algebraic lemmas concerning integral matrices with positive determinant.

§1. INTEGRAL MATRICES WITH POSITIVE DETERMINANT

Let $M_2^+(\mathbf{Q})$, $M_2^+(\mathbf{Z})$ denote the sets of 2×2 matrices with components in $\mathbf{Q}$ and $\mathbf{Z}$ respectively and positive determinant. We also write $M_2^+(\mathbf{Q}) = GL_2^+(\mathbf{Q})$. If

$$\alpha = \begin{pmatrix} a & b \\ c & d \end{pmatrix}$$

is in $M_2^+(\mathbf{Z})$, we shall say that α is **primitive** if $(a, b, c, d) = 1$, i.e. a, b, c, d are relatively prime. The set of integral matrices with determinant n is denoted by Δ_n, and the subset of those which are primitive is denoted by Δ_n^*. It is immediately clear that multiplication on the left or right by elements of $\Gamma = SL_2(\mathbf{Z})$ maps Δ_n^* into itself.

Since $j \circ \alpha = j \circ \gamma\alpha$ for all $\gamma \in \Gamma$, we are led to study the cosets $\Gamma\alpha$ for $\alpha \in \Delta_n^*$.

Theorem 1. *The group Γ operates left transitively on the right Γ-cosets, and also right transitively on the left Γ-cosets of Δ_n^*.*

Proof. Let α be a primitive integral matrix as above. Let $L = [\tau, 1]$ be a lattice. Then

$$M = [a\tau + b, c\tau + d]$$

is a sublattice, and by the elementary divisor theorem, there exists a basis $\{\omega_1, \omega_2\}$ of L and a basis $\{\omega'_1, \omega'_2\}$ of M such that

$$\omega'_1 = e_1\omega_1$$
$$\omega'_2 = e_2\omega_2,$$

and $e_1|e_2$. Since $(a, b, c, d) = 1$, it follows that $e_1 = 1$. This means that there exist elements $\gamma, \gamma' \in \Gamma$ such that

$$\gamma\alpha\gamma' = \begin{pmatrix} 1 & 0 \\ 0 & n \end{pmatrix},$$

and we see that $\Delta_n^* = \Gamma\alpha\Gamma$. This also proves that Γ operates transitively on the cosets as desired.

We now want to obtain a simple set of representatives for the left cosets of Γ in Δ_n^*. Given $\alpha \in \Delta_n^*$ as above, we can always find $\gamma \in \Gamma$ such that

$$\gamma\alpha = \begin{pmatrix} a_1 & b_1 \\ 0 & d_1 \end{pmatrix}.$$

For instance, select relatively prime integers z, w such that $za + wc = 0$, and then $x, y \in \mathbf{Z}$ such that $xw - zy = 1$. Then

$$\gamma = \begin{pmatrix} x & y \\ z & w \end{pmatrix}$$

works.

Suppose now that α is triangular, i.e.

$$\alpha = \begin{pmatrix} a & b \\ 0 & d \end{pmatrix}.$$

Since

$$\begin{pmatrix} 1 & k \\ 0 & 1 \end{pmatrix}\begin{pmatrix} a & b \\ 0 & d \end{pmatrix} = \begin{pmatrix} a & b + kd \\ 0 & d \end{pmatrix},$$

we see that a left coset contains a representative with $0 \leqq b < d$. Finally one verifies that the elements

$$\begin{pmatrix} a & b \\ 0 & d \end{pmatrix},$$

with $0 < a$, $0 \leqq b < d$, and $ad = n$ form distinct left coset representatives of Δ_n^*, i.e. that no two of them lie in the same coset.

We let $\psi(n)$ be the number of left cosets of Δ_n^*. If $n = p$ is a prime number, then we see that $\psi(p) = p + 1$, the coset representatives being the matrices

$$\begin{pmatrix} p & 0 \\ 0 & 1 \end{pmatrix} \quad \text{and} \quad \begin{pmatrix} 1 & i \\ 0 & p \end{pmatrix} \quad \text{with} \quad 0 \leqq i < p.$$

In general, we have

$$\boxed{\psi(n) = n \prod_{p|n} \left(1 + \frac{1}{p}\right).}$$

Although we won't use this fact, we give the simple proof.

We have to count the number of matrices in normalized form as above. For given d, $a = n/d$ is determined. Let $e = (a, d)$. There are then

$$\frac{d}{e} \varphi(e)$$

possible values for b, so

$$\psi(n) = \sum_{d|n} \frac{d}{e} \varphi(e)$$

where $e = (d, n/d)$.

The function ψ is multiplicative (in the sense of elementary number theory), i.e. if $n = n_1 n_2$ with n_1, n_2 relatively prime, then

$$\psi(n_1 n_2) = \psi(n_1)\psi(n_2).$$

Indeed, $d = d_1 d_2$, $e = e_1 e_2$, and hence

$$\psi(n_1 n_2) = \sum_{\substack{d_1|n_1 \\ d_2|n_2}} \frac{d_1 d_2}{e_1 e_2} \varphi(e_1)\varphi(e_2) = \psi(n_1)\psi(n_2).$$

This reduces our study of ψ to the case when $n = p^r$ is a prime power. For $d = 1$, $e = 1$, we get a contribution of 1 in the sum for $\psi(p^r)$. For $d = p^r$ and $e = 1$, we get a contribution of p^r. Hence

$$\begin{aligned} \psi(p^r) &= 1 + p^r + \sum_{\nu=1}^{r-1} \frac{p^\nu}{e} e\left(1 - \frac{1}{p}\right) \\ &= 1 + p^r + \sum_{\nu=1}^{r-1} (p^\nu - p^{\nu-1}) \\ &= p^r + p^{r-1} = p^r\left(1 + \frac{1}{p}\right), \end{aligned}$$

thereby proving that the value $\psi(n)$ is given by the desired formula.

§2. THE MODULAR EQUATION

By a **Γ-modular function,** or simply a **modular function** for this section, we mean an automorphic function on $\mathfrak{H}$ of weight 0, in other words a function meromorphic on $\mathfrak{H}$, invariant under Γ, and having a q-expansion at infinity.

Theorem 2. *Let f be a Γ-modular function which is holomorphic on $\mathfrak{H}$, and with a q-expansion*

$$f = \sum c_n q^n.$$

Then f is a polynomial in j with coefficients in the module over $\mathbf{Z}$ generated by the Fourier coefficients c_n.

Proof. Write

$$f = \frac{c_{-M}}{q^M} + \text{terms of higher degree},$$

so that $f - c_{-M}j^M$ is holomorphic on $\mathfrak{H}$ and has a q-expansion starting with at most a polar term of order $M - 1$. Repeating the procedure, we can subtract a polynomial in j whose coefficients lie in the module generated by all c_n over $\mathbf{Z}$, so as to get a modular function holomorphic on $\mathfrak{H}$, vanishing at infinity, and therefore identically zero, thus proving our assertion.

Let $\alpha \in M_2^+(\mathbf{Q})$. Let m be a positive integer such that $m\alpha$ is an integral matrix. By homogeneity, we have

$$j \circ m\alpha = j \circ \alpha.$$

Thus the study of $j \circ \alpha$ for rational matrices α is reduced to the study of $j \circ \alpha$ for integral α. Also, for any integral α we can factor out the greatest common divisor of its components, and therefore we can always consider primitive α.

Let

$$\{\alpha_i\} \qquad (i = 1, \ldots, \psi(n))$$

be representatives of the right cosets of Δ_n^* for Γ. Then the functions $j \circ \alpha_i$ are permuted transitively by the operation of Γ, where as usual, Γ operates on a function f by

$$f \mapsto f \circ \gamma.$$

Let

$$\Phi_n(X) = \prod_{i=1}^{\psi(n)} (X - j \circ \alpha_i),$$

where X is a variable. The coefficients of $\Phi_n(X)$ are the elementary symmetric functions of the $f \circ \alpha_i$, and are therefore holomorphic on $\mathfrak{H}$, invariant under Γ, and are meromorphic at infinity. To see this last property, one replaces τ by

$$\frac{a\tau + b}{d}$$

in the q_τ-expansion of j, and one sees that the resulting expansion is a power series in $q_\tau^{1/d}$, whence each function $f \circ \alpha_i$ grows at most like a power of q at infinity.

Furthermore, the coefficients of $q^{1/d}$ in the expansion of $j \circ \alpha_i$ lie in $\mathbf{Z}[\zeta_d]$, where $\zeta_d = e^{2\pi i/d}$. In fact, if

$$j = \frac{1}{q} + P(q),$$

where P is a power series with integer coefficients, and if

$$\alpha = \begin{pmatrix} a & b \\ 0 & d \end{pmatrix},$$

then

$$j \circ \alpha = \frac{1}{q^{a/d}\zeta_d^b} + P(q^{a/d}\zeta_d^b). \tag{1}$$

By Theorem 2 we conclude that the coefficients of $\Phi_n(X)$ are polynomials in j, whose coefficients are in $\mathbf{Z}[\zeta_n]$. Furthermore, we may view all these functions as embedded in the power series field

$$\mathbf{Q}(\zeta_n)((q^{1/n})).$$

If k is any field and X a variable, and if σ is an automorphism of k, then σ extends to an automorphism of the power series field $k((X))$ by

$$\sum c_m X^m \mapsto \sum c_m^\sigma X^m.$$

Let $r \in (\mathbf{Z}/N\mathbf{Z})^*$. The automorphism σ_r on $\mathbf{Q}(\zeta_n)$ such that

$$\sigma_r\colon \zeta_n \mapsto \zeta_n^r$$

extends to the power series field $\mathbf{Q}(\zeta_n)((q^{1/n}))$, and we see from (1) that this automorphism permutes the functions $j \circ \alpha_i$. Consequently the coefficients of $\Phi_n(X)$ are invariant under all such automorphisms σ_r, $r \in (\mathbf{Z}/N\mathbf{Z})^*$. Hence their q-expansions lie in $\mathbf{Z}((q))$.

By Theorem 2 we now conclude that the coefficients of $\Phi_n(X)$ are in $\mathbf{Z}[j]$, i.e. are polynomials in j with integer coefficients. Thus we may view $\Phi_n(X)$ as a polynomial in the two independent variables X and j, and we write it as

$$\Phi_n(X) = \Phi_n(X, j) \in \mathbf{Z}[X, j].$$

We call this the **modular polynomial of order** n.

Theorem 3.

i) *The polynomial $\Phi_n(X, j)$ is irreducible over $\mathbf{C}(j)$, and has degree $\psi(n)$.*

ii) *We have $\Phi_n(X, j) = \Phi_n(j, X)$.*

iii) *If n is not a square, then $\Phi_n(j, j)$ is a polynomial in j of degree > 1 and with leading coefficient 1.*

Proof. The first assertion comes from the fact that Γ permutes the functions $j \circ \alpha_i$ $(i = 1, \ldots, \psi(n))$ transitively, and acts as a group of automorphisms on the field $\mathbf{C}(j, j \circ \alpha_1, \ldots, j \circ \alpha_{\psi(n)})$.

Next, we prove the symmetry of (ii). One of the matrices α_i can be taken as

$$\begin{pmatrix} 1 & 0 \\ 0 & n \end{pmatrix}.$$

Hence $j \circ \frac{1}{n}$ is a root of $\Phi_n(X, j)$, i.e.

$$\Phi_n(j(\tau/n), j(\tau)) = 0, \qquad \text{for all } \tau.$$

Hence

$$\Phi_n(j(\tau), j(n\tau)) = 0, \qquad \text{for all } \tau,$$

or in other words,

$$\Phi_n(j, j \circ n) = 0.$$

So $j \circ n$ is a root of $\Phi_n(j, X)$, but it is also a root of $\Phi_n(X, j)$, corresponding to the matrix

$$\begin{pmatrix} n & 0 \\ 0 & 1 \end{pmatrix}.$$

Since $\Phi_n(X, j)$ is irreducible, we conclude that

$$\Phi_n(X, j) \text{ divides } \Phi_n(j, X),$$

i.e.

$$\Phi_n(j, X) = g(X, j)\Phi_n(X, j)$$

for some polynomial $g(t, j) \in \mathbf{Z}[t, j]$, by the Gauss lemma. It follows that

$$\Phi_n(j, X) = g(X, j)g(j, X)\Phi_n(j, X),$$

whence

$$g(X, j)g(j, X) = 1,$$

and $g(X, j)$ is constant, $= \pm 1$. If $g(X, j) = -1$, then

$$\Phi_n(j, j) = -\Phi_n(j, j),$$

and hence j must be a root of $\Phi_n(X)$. But $\Phi_n(X)$ is irreducible over $\mathbf{Q}(j)$, so this is impossible, and $g(X, j) = 1$. This proves (ii).

To prove (iii), assume that n is not a square, so that if

$$\alpha = \begin{pmatrix} a & b \\ 0 & d \end{pmatrix},$$

α is primitive and $ad = n$, then $a \neq d$. We have the q-expansion

$$j - j \circ \alpha = \frac{1}{q} + \cdots - \frac{1}{\zeta_d^b q^{a/d}} - \cdots.$$

Since $a \neq d$, there is no cancellation in the polar term, and the leading coefficient of this q-expansion is a root of unity. But $\Phi_n(j, j) \in \mathbf{Z}[j]$. Taking the product of the $j - j \circ \alpha_i$, we see that the q-expansion for $\Phi_n(j, j)$ starts with

$$\frac{c_m}{q^m} + \cdots$$

with $c_m = \pm 1$, because c_m has to be an integer and also a root of unity. Hence

$$\Phi_n(j, j) = c_m j^m + \cdots$$

is a polynomial in j with leading coefficient $c_m = \pm 1$, as was to be shown.

Corollary. *For any $\alpha \in M_2^+(\mathbf{Q})$, the function $j \circ \alpha$ is integral over $\mathbf{Z}[j]$.*

Proof. We may assume that α is integral, has determinant n, and then $j \circ \alpha$ is a root of $\Phi_n(X)$ which has leading coefficient 1, and lies in $\mathbf{Z}[j, X]$.

Theorem 4. *If $\tau \in \mathfrak{H}$ is imaginary quadratic, then $j(\tau)$ is an algebraic integer.*

Proof. Let $K = Q(\tau)$, and let $\mathfrak{o} = [z, 1]$ be the ring of algebraic integers in K. We can always find an element $\lambda \in \mathfrak{o}$ such that the norm of λ is square free. If $K = \mathbf{Q}(i)$, we take $\lambda = 1 + i$, and if $K = Q(\sqrt{-m})$ with $m > 1$ square free, we take $\lambda = \sqrt{-m}$. Then

$$\lambda z = az + b$$
$$\lambda = cz + d$$

with integers a, b, c, d and the norm of λ (over $\mathbf{Q}$) is the determinant $ad - bc$. Then

$$\alpha = \begin{pmatrix} a & b \\ c & d \end{pmatrix}$$

is primitive, and $z = \alpha z$. Hence $j(z)$ is a root of the polynomial $\Phi_n(X, X)$ which lies in $\mathbf{Z}[X]$ and has leading coefficient 1 according to Theorem 3, whence $j(z)$ is an algebraic integer. We have $\mathbf{Q}(z) = \mathbf{Q}(\tau)$, and $\tau = uz + v$ with rational u, v, i.e. $\tau = \beta z$ with some primitive $\beta \in M_2^+(\mathbf{Z})$. Since $j \circ \beta$ is integral over $\mathbf{Z}[j]$ by Theorem 3, it follows that $j(\beta z) = j(\tau)$ is integral over $\mathbf{Z}[j(z)]$, and therefore $j(\tau)$ is also an algebraic integer, as was to be shown.

It will be proved in the complex multiplication that $j(\tau)$ generates an abelian extension of $\mathbf{Q}(\tau)$.

The proofs which we have given here are very classical, going back to Kronecker and Weber. So far, these proofs for integrality are the simplest ones, through the q-expansions. Algebraically, one could give proofs which are fairly complicated. This is one reason why in the higher dimensional theory, integrality statements like the above are completely lacking.

For a finer analysis of the factorization of the polynomial $\Phi_m(X, X)$, we refer the reader to the appendix of Chapter 10.

We shall now see how the above techniques also give **the Kronecker congruence relation**

$$\Phi_p(X, j) \equiv (X - j^p)(X^p - j) \pmod{p},$$

for any prime number p. Stronger results will be derived later by other techniques and the reader can skip the present arguments.

For a prime p, representatives for the primitive matrices of determinant p are given by

$$\alpha_i = \begin{pmatrix} 1 & i \\ 0 & p \end{pmatrix}, \qquad i = 0, \ldots, p-1$$

and

$$\alpha_p = \begin{pmatrix} p & 0 \\ 0 & 1 \end{pmatrix}.$$

For a modular function f, we shall write $f^*(q)$ for its q-expansion, and similarly for a $q^{1/N}$-expansion. Such an expansion is a power series in $q^{1/N}$. If it has coefficients in a ring $\mathbf{Z}[\zeta_p]$ where ζ_p is a primitive p-th root of unity, we shall write congruences

$$f^*(q) \equiv g^*(q) \pmod{1 - \zeta}$$

to mean that all the coefficients of $f^*(q) - g^*(q)$ in the $q^{1/N}$-expansion lie in the ideal generated by $1 - \zeta$ in $\mathbf{Z}[\zeta]$.

Making the given substitutions in the q-expansion for $j \circ \alpha_i$, we find at once that

$$(j \circ \alpha_p)^*(q) \equiv j^*(q)^p \qquad \pmod{p}$$

and

$$(j \circ \alpha_i)^*(q) \equiv j^*(q)^{1/p} \pmod{1 - \zeta}.$$

Observe that $1 - \zeta$ is a prime element at the prime dividing p in $\mathbf{Z}[\zeta_p]$. Therefore we conclude that

$$\Phi_p(X, j^*(q)) \equiv (X - j^*(q)^p)(X^p - j^*(q)) \pmod{1 - \zeta},$$

in the sense that the power series in q which are the coefficients of the polynomials in t on both sides of this congruence satisfy the desired congruence. Let

$$\Phi_p(X, j) - (X - j^p)(X^p - j) = \sum \psi_\nu(j) X^\nu$$

where $\psi_\nu(j) \in \mathbf{Z}[j]$. Then $\psi_\nu(j^*(q))$ has coefficients divisibly by $1 - \zeta$, hence by p because these coefficients are ordinary integers. This proves the desired congruence relation.

§3. RELATIONS WITH ISOGENIES

Let A, B be elliptic curves over the complex numbers. If $A_{\mathbf{C}} \approx \mathbf{C}/L$ and $M \subset L$ is a sublattice such that $B_{\mathbf{C}} \approx \mathbf{C}/M$, then we have an isogeny $\lambda\colon B \to A$ and a commutative diagram

$$\begin{array}{ccc} \mathbf{C}/M & \longrightarrow & \mathbf{C}/L \\ \downarrow & & \downarrow \\ B_{\mathbf{C}} & \xrightarrow[\lambda]{} & A_{\mathbf{C}} \end{array}$$

where the top homomorphism is the canonical one. Its kernel is the finite group L/M. Let $L = [\tau, 1]$. Then

$$M = [a\tau + b, c\tau + d]$$

with some matrix

$$\alpha = \begin{pmatrix} a & b \\ c & d \end{pmatrix}$$

in $M_2^+(\mathbf{Z})$. Hence

$$j_B = j(\alpha\tau) = j(M) \qquad \text{and} \qquad j_A = j(\tau) = j(L).$$

In particular, we see that $j(\alpha\tau)$ is a root of the polynomial

$$\Phi_n(X, j(\tau)) \in \mathbf{Z}[j(\tau), X].$$

Evaluating functions at τ in fact shows that for any special value of $\tau \in \mathfrak{H}$, the roots of $\Phi_n(X, j(\tau))$ are precisely the values

$$j(\alpha_i\tau), \qquad i = 1, \ldots, \psi(n).$$

A sublattice M of L is called **primitive** if when we express a $\mathbf{Z}$-basis of M in terms of $\mathbf{Z}$-basis of L, by a matrix α in $M_2(\mathbf{Z})$, then α is primitive. It is immediately verified that M is primitive in L if and only if the factor group L/M is cyclic (using the elementary divisor theorem). Thus the primitive sublattices of L correspond to the isogenies with a cyclic kernel, whose order is precisely the determinant of α, or equivalently the index $(L : M)$.

For any given value of $\tau \in \mathfrak{H}$, we see that the roots of

$$\Phi_n(X, j(\tau))$$

are exactly the j-invariants of all the elliptic curves B which admit a cyclic isogeny

$$\lambda\colon B \to A$$

of degree n. In other words:

Theorem 5. *Let A, B be elliptic curves over the complex. There exists an isogeny $\lambda\colon B \to A$ with cyclic kernel of degree n if and only j_B is a root of the equation*

$$\Phi_n(X, j_A) = 0.$$

The theorem is true in characteristic 0 simply by embedding any field of characteristic 0 in the complex numbers. Igusa [22] has shown how it is valid in characteristic p, for $p \nmid n$. In a later paper, he analyses the situation when n is a power of p [24].

6 Higher Levels

§1. CONGRUENCE SUBGROUPS

Let $\Gamma = SL_2(\mathbf{Z})$ again. We define Γ_N (or $\Gamma(N)$) for each positive integer N to be the subgroup of Γ consisting of those matrices satisfying the condition

$$\begin{pmatrix} a & b \\ c & d \end{pmatrix} \equiv 1 \pmod{N},$$

in other words

$$a \equiv d \equiv 1 \pmod{N} \qquad \text{and} \qquad c \equiv b \equiv 0 \pmod{N}.$$

We call Γ_N the **congruence subgroup of level** N. By $SL_2(\mathbf{Z}/N\mathbf{Z})$ we shall mean the group of matrices with components in the ring $\mathbf{Z}/N\mathbf{Z}$ having determinant 1 in $\mathbf{Z}/N\mathbf{Z}$. Reducing $SL_2(\mathbf{Z})$ mod N maps $SL_2(\mathbf{Z})$ into $SL_2(\mathbf{Z}/N\mathbf{Z})$, and the kernel by definition is Γ_N. Actually one has an exact sequence

$$0 \to \Gamma_N \to SL_2(\mathbf{Z}) \to SL_2(\mathbf{Z}/N\mathbf{Z}) \to 0,$$

and the surjectivity on the right is proved as follows.

Let

$$\alpha = \begin{pmatrix} a & b \\ c & d \end{pmatrix}$$

be an integral matrix representing an element of $SL_2(\mathbf{Z}/N\mathbf{Z})$, so that

$$ad - bc \equiv 1 \pmod{N}.$$

By elementary divisor theory, there exist elements $\gamma, \gamma' \in SL_2(\mathbf{Z})$ such that $\gamma\alpha\gamma'$ is diagonal, and if we can find $\beta \in SL_2(\mathbf{Z})$ such that

$$\beta \equiv \gamma\alpha\gamma' \pmod{N},$$

then $\gamma^{-1}\beta\gamma'^{-1}$ solves our problem. Without loss of generality we may therefore assume that α is diagonal, say

$$\alpha = \begin{pmatrix} a & 0 \\ 0 & d \end{pmatrix}.$$

It will suffice to find integers x, y such that

$$\begin{pmatrix} a + xN & yN \\ N & d \end{pmatrix}$$

has determinant 1. Let $ad = 1 + rN$. Our problem amounts to solving

$$r + dx - yN = 0,$$

which we can do since $(d, N) = 1$. This proves the surjectivity.

By a simple counting argument, one sees that the order of $SL_2(\mathbf{Z}/N\mathbf{Z})$ is

$$N^3 \prod_{p|N} \left(1 - \frac{1}{p^2}\right).$$

This general fact will not be used in this book.

By $GL_2(\mathbf{Z}/N\mathbf{Z})$, we shall mean the group of matrices with components in $\mathbf{Z}/N\mathbf{Z}$ whose determinant is a unit in $\mathbf{Z}/N\mathbf{Z}$. Thus $SL_2(\mathbf{Z}/N\mathbf{Z})$ is a subgroup of $GL_2(\mathbf{Z}/N\mathbf{Z})$. In fact, let G_N be the group of matrices

$$\begin{pmatrix} 1 & 0 \\ 0 & d \end{pmatrix}$$

with $d \in (\mathbf{Z}/N\mathbf{Z})^*$. Thus $G_N \approx (\mathbf{Z}/N\mathbf{Z})^*$. Then

$$GL_2(\mathbf{Z}/N\mathbf{Z}) = G_N \cdot SL_2(\mathbf{Z}/N\mathbf{Z}) = SL_2(\mathbf{Z}/N\mathbf{Z}) \cdot G_N.$$

Indeed, any matrix in $GL_2(\mathbf{Z}/N\mathbf{Z})$ can be multiplied, say on the left, by a suitable element of G_N, so that the product has determinant 1 in $\mathbf{Z}/N\mathbf{Z}$. The product decomposition is clearly unique. Furthermore, we have an exact sequence

$$0 \to SL_2(\mathbf{Z}/N\mathbf{Z}) \to GL_2(\mathbf{Z}/N\mathbf{Z}) \xrightarrow{\det} (\mathbf{Z}/N\mathbf{Z})^* \to 0.$$

§2. THE FIELD OF MODULAR FUNCTIONS OVER C

Let f be a function on the upper half plane $\mathfrak{H}$, meromorphic and invariant by Γ_N, i.e. such that

$$f(\gamma\tau) = f(\tau), \qquad \tau \in \mathfrak{H},\ \gamma \in \Gamma_N.$$

Let $q = e^{2\pi i\tau}$ and $q^{1/N} = e^{2\pi i\tau/N}$. The map

$$\tau \mapsto q^{1/N}$$

defines a holomorphic map from $\mathfrak{H}_B$ (the set of $\tau \in \mathfrak{H}$ with $\operatorname{Im}\tau > B$) onto a punctured disc, and is defined on $\mathfrak{H}$ modulo the translation by N. Since the matrix

$$\begin{pmatrix} 1 & N \\ 0 & 1 \end{pmatrix}$$

lies in Γ_N and acts as translation by N on $\mathfrak{H}$, it follows that f induces a meromorphic function f^* on this punctured disc. If there exists a positive power q^M such that $|f^*(q)q^M|$ is bounded near 0, then in fact f^* is also meromorphic on the disc and has a power series expansion in the parameter $q^{1/N}$, with at most a finite number of negative terms. If for every $\gamma \in SL_2(\mathbf{Z})$ the function $(f \circ \gamma)^*$ also has such a power series expansion in $q^{1/N}$, then f is called **modular of level N on $\mathfrak{H}$.**

We denote by $F_{N,\mathbf{C}}$ the field of modular functions of level N. The group Γ operates as a group of automorphisms of $F_{N,\mathbf{C}}$ by $f \mapsto f \circ \gamma$. Indeed, let $\gamma \in \Gamma$, and $\alpha \in \Gamma_N$. Since Γ_N is normal in Γ, it follows that $\gamma\alpha = \alpha'\gamma$ for some $\alpha' \in \Gamma_N$. If $f \in F_{N,\mathbf{C}}$ then

$$f(\gamma\alpha\tau) = f(\alpha'\gamma\tau) = f(\gamma\tau),$$

so that $f \circ \gamma$ is invariant under Γ_N. Clearly, $f \circ \gamma$ is meromorphic on $\mathfrak{H}$. The last condition about q-expansions is immediate from the definition, so we see that $f \circ \gamma$ is modular of level N, and Γ operates by composition.

By definition, $F_{1,\mathbf{C}}$ is the field of automorphic functions of weight 0, defined in Chapter 3. We let $F_{\mathbf{C}}$ be the union of all fields $F_{N,\mathbf{C}}$, and call $F_{\mathbf{C}}$ the **modular function field over the complex numbers.**

Theorem 1. $F_{1,\mathbf{C}} = \mathbf{C}(j)$.

Proof. Let $f \in F_{1,\mathbf{C}}$. For some polynomial $P(j)$ the function $fP(j)$ is holomorphic on $\mathfrak{H}$. (For instance, if f has a pole at z_0, then $f(j - j(z_0))^m$ has no pole at z_0 for high m, and the number of possible poles in a fundamental domain is bounded since f is meromorphic at infinity.) Suppose that f has no pole on $\mathfrak{H}$, and has a pole of order n at infinity. Using the fact that j has a pole of order 1 at infinity, we see that there exists a constant c such that $f - cj^n$ has a pole of order $\leqq n - 1$ at infinity. Consequently by induction, we can find a polynomial in j such that $f - \mathrm{Pol}(j)$ has no pole on $\mathfrak{H}$ and no pole at infinity. Then $f - \mathrm{Pol}(j)$ lies in the space of automorphic functions of weight 0, i.e. the constants, and this concludes the proof that $f \in \mathbf{C}(j)$.

We shall now find generators for $F_{N,\mathbf{C}}$. Let

$$f_0(w; \tau) = -2^7 3^5 \frac{g_2(\tau)g_3(\tau)}{\Delta(\tau)} \wp(w; \tau, 1)$$

so that $w \in \mathbf{C}$ and $\tau \in \mathfrak{H}$. This is called the **first Weber function.** Having fixed the integer $N > 1$, for $r, s \in \mathbf{Z}$ and not both divisible by N, let

$$f_{r,s}(\tau) = f_0\left(\frac{r\tau + s}{N}; \tau\right).$$

The point of the factors involving g_2, g_3, Δ in front of $\wp$ is to make the resulting function homogeneous of degree 0 in the vector $(\tau, 1)$. Because of this homogeneity, we sometimes also write

$$f_{r,s}(\tau) = f_0\left(\frac{r\omega_1 + s\omega_2}{N}; \omega_1, \omega_2\right)$$

if $\tau = \omega_1/\omega_2$. For a fixed τ, the above functions give the normalized x-coordinates of the points of period N on the corresponding elliptic curve. If $(r, s, N) = 1$, the function $f_{r,s}$ is said to be **primitive of level** N. In view of the periodicity property of the $\wp$-function, it follows that $f_{r,s}$ depends only on the residue classes of r, s mod N. Thus it is appropriate to use a notation exhibiting this property. If

$$a = (a_1, a_2) \in \mathbf{Q}^2 \quad \text{but} \quad a \notin \mathbf{Z}^2,$$

we shall write

$$f_a(\tau) = f(a; \tau) = f_0(a_1\tau + a_2; \tau).$$

Then each function f_a is holomorphic on $\mathfrak{H}$, and f_a depends only on the residue class of a (mod $\mathbf{Z}^2$). We call the functions f_a the **Fricke functions.**

It is also sometimes useful to use vertical notation, and write

$$f_a(\tau) = f_0\left(a\begin{pmatrix}\tau \\ 1\end{pmatrix}; \tau\right).$$

If $\alpha \in SL_2(\mathbf{Z})$, this notation makes the following relation obvious:

$$\boxed{f_{a\alpha}(\tau) = f_a(\alpha\tau).}$$

If we look at the q-expansions of Chapter 4, Proposition 5, then we see that the Fricke functions have a $q_\tau^{1/N}$ expansion with only a finite number of negative terms. Furthermore the powers of $2\pi i$ cancel in the definition of f_0, and all the coefficients of $q_\tau^{1/N}$ lie in the field of N-th roots of unity over $\mathbf{Q}$, because for $w = \dfrac{r\tau + s}{N}$ we have

$$q_w = q_{\tau/N}^r q_{s/N},$$

and $q_{s/N} = \zeta_N^s$ where $\zeta_N = e^{2\pi i/N}$ is a primitive N-th root of unity. For the moment we disregard this special nature of the coefficients since we first do the theory over $\mathbf{C}$.

In any case, we have proved that the Fricke functions are modular functions of level N, because if $\alpha \equiv 1 \pmod N$, then $a\alpha \equiv a \pmod N$ and hence $f_{a\alpha} = f_a$ and $f_a(\alpha\tau) = f_a(\tau)$.

The relation $f_a(\alpha\tau) = f_{a\alpha}(\tau)$ also shows that the modular group operates as a group of permutations of the functions f_a. Furthermore, if $a = \left(\dfrac{r}{N}, \dfrac{s}{N}\right)$ has exact denominator N (i.e. $(r, s, N) = 1$), then $a\alpha$ also has exact denominator N, and thus $SL_2(\mathbf{Z})$ permutes the primitive Fricke functions of level N among themselves.

Of course, $\Gamma = SL_2(\mathbf{Z})$ operates as a group of analytic automorphisms of $\mathfrak{H}$, and hence operates on $F_{N,\mathbf{C}}$ by composition,

$$f \mapsto f \circ \alpha.$$

Since Γ_N operates trivially, we may view the finite group Γ/Γ_N as operating on $F_{N,\mathbf{C}}$, the kernel containing ± 1.

We are essentially in a situation of Galois theory, with a group Γ/Γ_N operating on the field $F_{N,\mathbf{C}}$, with fixed field $F_{N,\mathbf{C}}$.

Theorem 2. *We have*

$$F_{N,\mathbf{C}} = F_{1,\mathbf{C}}(f_{r,s})_{\text{all } r,s} = \mathbf{C}(j, f_{r,s})_{\text{all } r,s}.$$

Furthermore, the Galois group of $F_{N,\mathbf{C}}$ over $F_{1,\mathbf{C}}$ is precisely

$$\Gamma/\pm\Gamma_N = SL_2(\mathbf{Z}/N\mathbf{Z})/\pm 1.$$

Proof. Let E be the subfield of $F_{N,\mathbf{C}}$ generated over $\mathbf{C}(j)$ by all $f_{r,s}$. Since Γ permutes the $f_{r,s}$ it follows that Γ/Γ_N acts as a finite group of automorphisms of E. Note that ± 1 acts trivially, because the $\wp$-function is an even function. We shall now prove that any element $\gamma \in \Gamma$ which acts trivially on E must lie in $\pm\Gamma_N$. We consider the effect of γ on the two functions $f_{1,0}$ and $f_{0,1}$. Since $\wp(u) = \wp(v)$ if and only if $u \equiv v \pmod{L}$, we see that if γ leaves $f_{(1,0)}$ and $f_{(0,1)}$ fixed, then

$$f_{(1,0)} \circ \gamma = f_{(\pm 1,0)} \quad \text{and} \quad f_{(0,1)} \circ \gamma = f_{(0,\pm 1)}.$$

From this one sees at once that

$$\gamma \equiv \begin{pmatrix} \pm 1 & 0 \\ 0 & \pm 1 \end{pmatrix} \pmod{N}.$$

Since $\gamma \in SL_2(\mathbf{Z})$, it follows that $\gamma \equiv \pm 1 \pmod{N}$. Hence we have an injection

$$\Gamma/\pm\Gamma_N \to \mathrm{Gal}(E/\mathbf{C}(j)),$$

and the fixed field is $\mathbf{C}(j)$. Since we have a fortiori an injection of $\Gamma/\pm\Gamma_N$ in $\mathrm{Gal}(F_{N,\mathbf{C}}/\mathbf{C}(j))$, it follows that $F_{N,\mathbf{C}} = E$ and that the Galois group is that stated in the theorem.

§3. THE FIELD OF MODULAR FUNCTIONS OVER Q

Let f be a modular function (of level 1). We shall say that f is **defined over a field** k if $f \in k(j)$.

Fix an integer $N > 1$ as before. Form the polynomial

$$\prod (X - f_{r,s}),$$

the product being taken over all (r, s) mod N (we could also take the product over those (r, s) such that $(r, s, N) = 1$). We obtain a polynomial in X, whose coefficients are invariant under Γ, because Γ permutes the $f_{r,s}$. Hence these coefficients are modular functions of level 1, holomorphic on $\mathfrak{H}$. Furthermore, their Fourier coefficients are in the field $\mathbf{Q}(\zeta_N)$.

By Theorem 2 of Chapter 5, it follows that these coefficients are polynomials in j with coefficients in $\mathbf{Q}(\zeta_N)$, and hence the functions $f_{r,s}$ are algebraic over $\mathbf{Q}(j)$.

Let $\mathbf{Q}_N = \mathbf{Q}(\zeta_N)$, and let

$$F_N = \mathbf{Q}(j, f_{r,s})_{\text{all } r,s}.$$

We shall call F_N the **modular function field of level** N over $\mathbf{Q}$, and omit the reference to $\mathbf{Q}$ in a discussion when the context makes it clear.

From the function theory of the preceding section, we already know that its Galois group contains

$$SL_2(\mathbf{Z}/N\mathbf{Z})/\pm 1 = \Gamma/\pm\Gamma_N.$$

Theorem 3. *The Galois group of $F_N/\mathbf{Q}(j)$ is precisely*

$$GL_2(\mathbf{Z}/N\mathbf{Z})/\pm 1.$$

The algebraic closure of $\mathbf{Q}$ in F_N is $\mathbf{Q}_N = \mathbf{Q}(\zeta_N)$. If $\alpha \in GL_2(\mathbf{Z}/N\mathbf{Z})$, then the automorphism induced by α on $\mathbf{Q}_N$ is given by the determinant, i.e. *if $\sigma(\alpha)$ is the automorphism given by α on F_N, then*

$$\sigma(\alpha)\zeta = \zeta^{\det \alpha}.$$

The Galois group of F_N over $\mathbf{Q}_N(j)$ is $SL_2(\mathbf{Z}/N\mathbf{Z})/\pm 1$.

Proof. We shall prove $\mathrm{Gal}(G_N/F_1)$ contains the group

$$G_N = \left\{\begin{pmatrix} 1 & 0 \\ 0 & d \end{pmatrix}, \quad d \in (\mathbf{Z}/N\mathbf{Z})^*\right\}.$$

We consider the q-expansion given for the Weber function in Chapter 4. At

$$\frac{r\tau + s}{N}$$

it will be of the form of a power series in q_τ with integer coefficients, times the power series

$$1 + \frac{12q^{r/N}\zeta^s}{(1 - q^{r/N}\zeta^s)^2} + 12 \sum_{m,n=1}^{\infty} nq^{mn}(q^{nr/N}\zeta^s + q^{-nr/N}\zeta^{-s} - 2),$$

with $q = q_\tau$. This power series is therefore contained in the power series field

$$\mathbf{Q}_N((q^{1/N})).$$

If k is any field and X a variable, then any automorphism σ of k extends to the power series field $k((X))$ by the mapping

$$\sum c_n X^n \mapsto \sum c_n^\sigma X^n, \qquad c_n \in k.$$

If $d \in (\mathbf{Z}/N\mathbf{Z})^*$ we let σ_d be the automorphism of $\mathbf{Q}_N((q^{1/N}))$ obtained in the above manner, from the automorphism of $\mathbf{Q}_N$ such that $\zeta_N \mapsto \zeta_N^d$. Then g_2, g_3, j are fixed since their q-expansions are in $\mathbf{Q}((q))$. On the other hand, we see from the q-expansions of the Weber function that

$$\sigma_d: f_{r,s}(\tau) \mapsto f_{r,sd}(\tau).$$

In other words σ_d defines an element of $\mathrm{Gal}(F_N/\mathbf{Q}(j))$, and σ_d is represented by the matrix

$$\begin{pmatrix} 1 & 0 \\ 0 & d \end{pmatrix}.$$

Hence G_N is contained in $\mathrm{Gal}(F_N/\mathbf{Q}(j))$. It follows now at once that

$$\mathrm{Gal}(F_N/\mathbf{Q}(j)) = GL_2(\mathbf{Z}/N\mathbf{Z})/\pm 1.$$

Furthermore, from the way we defined σ_d, and the decomposition of an element in $GL_2(\mathbf{Z}/N\mathbf{Z})$ as a product from an element in G_N and an element in $SL_2(\mathbf{Z}/N\mathbf{Z})$, we see that the effect of an element in $GL_2(\mathbf{Z}/N\mathbf{Z})$ on the roots of unity is given by the determinant of the matrix.

Finally, let k be the algebraic closure of $\mathbf{Q}$ in F_N, so that $k = \mathbf{C} \cap F_N$. Then

$$\mathrm{Gal}(F_N/k(j)) \approx \mathrm{Gal}(F_{N,\mathbf{C}}/\mathbf{C}(j)) \approx SL_2(\mathbf{Z}/N\mathbf{Z})/\pm 1.$$

Hence

$$[k(j) : \mathbf{Q}(j)] = [k : \mathbf{Q}] = \text{order of } (\mathbf{Z}/N\mathbf{Z})^* = [\mathbf{Q}(\zeta_N) : \mathbf{Q}].$$

Since $F_N \subset \mathbf{Q}_N((q^{1/N}))$ it follows that $k \subset \mathbf{Q}_N$, and we get equality by the fact that k and $\mathbf{Q}_N$ have the same dimension over $\mathbf{Q}$. This settles the Galois group of $F_N/\mathbf{Q}(j)$.

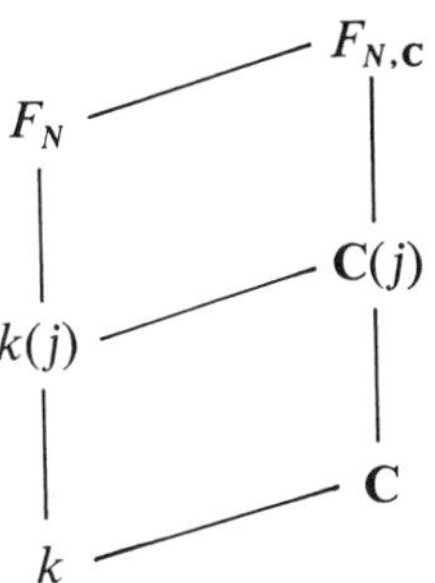

We shall now give the formulation of Theorems 2 and 3 in terms of points of finite order on a "generic" elliptic curve.

Let $\tau \in \mathfrak{H}$ be such that $j(\tau)$ is transcendental over $\mathbf{Q}$. Then the map

$$f \mapsto f(\tau)$$

gives an isomorphism of F_N (which is an algebraic extension of $\mathbf{Q}(j)$) on a field which we denote by $F_N(\tau)$. Let A^τ be an elliptic curve defined over $\mathbf{Q}(j(\tau))$ whose

j-invariant is $j(\tau)$, say in Weierstrass form with coordinates (x, y). Let h be the first Weber function, so that

$$h(x, y) = -2^7 3^5 \frac{g_2 g_3}{\Delta} x,$$

and let $\varphi\colon \mathbf{C}/L \to A^\tau_{\mathbf{C}}$ be the analytic parametrization given by the Weierstrass functions. Let $P_1 = \varphi(\omega_1/N)$ and $P_2 = \varphi(\omega_2/N)$. Then

$$h(P_1) = f_{(1,0)}(\tau) \qquad \text{and} \qquad h(P_2) = f_{(0,1)}(\tau).$$

In general,

$$f_{r,s}(\tau) = h(rP_1 + sP_2).$$

Therefore the field $F_N(\tau)$ is none other than the field

$$\mathbf{Q}(j(\tau), h(A^\tau_N))$$

of x-coordinates of division points of order N on A^τ. Its Galois group is a subgroup of $GL_2(\mathbf{Z}/N\mathbf{Z})/\pm 1$, as we saw in Chapter 2.

Corollary 1. *Let j be transcendental over $\mathbf{Q}$. Let A be an elliptic curve with invariant j, defined over $\mathbf{Q}(j)$. Let*

$$K_N = \mathbf{Q}(j, A_N).$$

i) *The Galois group of K_N over $\mathbf{Q}(j)$ is isomorphic to the full group $GL_2(\mathbf{Z}/N\mathbf{Z})$ in its representation on $A_N \approx (\mathbf{Z}/N\mathbf{Z})^2$.*
ii) *The algebraic closure of $\mathbf{Q}$ in K_N is $\mathbf{Q}(\zeta_N)$.*
iii) *The Galois group of K_N over $\mathbf{Q}(\zeta_N, j)$ is $SL_2(\mathbf{Z}/N\mathbf{Z})$.*

Proof. Let $G = \mathrm{Gal}(K_N/\mathbf{Q}(j))$. By the result for F_N we see that

$$G \cdot \{\pm 1\} = GL_2(\mathbf{Z}/N\mathbf{Z}).$$

Let

$$\gamma = \begin{pmatrix} 0 & -1 \\ 1 & 0 \end{pmatrix},$$

so that $\gamma \in SL_2(\mathbf{Z}/N\mathbf{Z})$ and $\gamma^2 = -1$. Then γ or $-\gamma$ lies in G, and hence $-1 \in G$, whence $G = GL_2(\mathbf{Z}/N\mathbf{Z})$. This proves the first assertion, and the argument also proves the following lemma.

Lemma. *Let G be a subgroup of $GL_2(\mathbf{Z}/N\mathbf{Z})$* [resp. $SL_2(\mathbf{Z}/N\mathbf{Z})$] *which maps onto $GL_2(\mathbf{Z}/N\mathbf{Z})/\pm 1$* [resp. onto $SL_2(Z/NZ)/\pm 1$] *under the canonical homomorphism. Then $G = GL_2(\mathbf{Z}/N\mathbf{Z})$* [resp. $G = SL_2(\mathbf{Z}/N\mathbf{Z})$].

If $\alpha \in GL_2(\mathbf{Z}/N\mathbf{Z})$, we denote by σ_α the corresponding automorphism of K_N over $\mathbf{Q}(j)$, relative to a fixed basis of A_N over $\mathbf{Z}/N\mathbf{Z}$. Let k be the algebraic closure of $\mathbf{Q}$ in K_N. We know from Theorem 3 that k contains ζ_N, and that

$$\sigma_\alpha \zeta_N = \zeta_N^{\det \alpha}.$$

Let G_1 be the Galois group of K_N over $k(j)$. If $\sigma_\alpha \in G_1$, then σ_α leaves the N-th roots of unity fixed, and hence $\det \alpha = 1$. Hence $G_1 \subset SL_2(\mathbf{Z}/N\mathbf{Z})$, and G_1 is naturally isomorphic with the Galois group of $\mathbf{C}(j, A_N)$ over $\mathbf{C}(j)$ (assuming that j is transcendental over $\mathbf{C}$, i.e. making the constant field extension to $\mathbf{C}$ from k). Using Theorem 2 of the preceding section we conclude that

$$G_1 \cdot \{\pm 1\} = SL_2(\mathbf{Z}/N\mathbf{Z}).$$

By the lemma, it follows that $G_1 = SL_2(\mathbf{Z}/N\mathbf{Z})$. Hence the order of the Galois group of $k(j)$ over $\mathbf{Q}(j)$ is exactly the order of $(\mathbf{Z}/N\mathbf{Z})^*$, by the exact sequence at the end of §1. This implies that

$$[k : \mathbf{Q}] = \text{order of } (\mathbf{Z}/N\mathbf{Z})^*,$$

and since k contains the N-th roots of unity, we conclude that $k = \mathbf{Q}(\zeta_N)$, thereby proving both (ii) and (iii), and concluding the proof of the corollary.

Let k be an algebraically closed field of characteristic 0 and let j_0 be transcendental over k. Let us assume that the cardinality of k is at most that of $\mathbf{C}$. We can then embed k into $\mathbf{C}$, and even in such a way that $\mathbf{C}$ has infinite degree of transcendence over k. Let A be an elliptic curve defined over $k(j_0)$, with invariant j_0. Taking a suitable isomorphism of $k(j_0)$ over k, we may assume that j_0 is transcendental over $\mathbf{C}$. Select $\tau \in \mathfrak{H}$ such that $j(\tau)$ is transcendental over k. Let $F_{N,k} = kF_N$ be the compositum of the modular function field over $\mathbf{Q}$ with k. The map $f \mapsto f(\tau)$ induces an isomorphism of $F_{N,k}$ with a subfield $F_{N,k}(\tau)$ of $\mathbf{C}$. There is also an isomorphism of $k(j_0)$ with $k(j(\tau))$, sending j_0 on $j(\tau)$, and transforming A on an elliptic curve A^τ defined over $k(j(\tau))$, having invariant $j(\tau)$. Thus we have isomorphisms

$$k(j_0, A_N) \approx k(j(\tau), A_N^\tau),$$

and

$$k(j_0, h(A_N)) \approx k(j(\tau), h(A_N^\tau)) \approx F_{N,k}.$$

Having assumed that j_0 is transcendental over $\mathbf{C}$, it follows that $\mathbf{C}$ is linearly disjoint from the algebraic closure of $k(j_0)$ over k. Making the constant field extension from k to $\mathbf{C}$, we see that

$$\mathbf{C}(j_0, h(A_N)) \approx F_{N,\mathbf{C}}.$$

Corollary 2. *Let k be an algebraically closed field of characteristic* 0 *and let j be transcendental over k. Let A be an elliptic curve with invariant j, defined over $k(j)$. The Galois group of $k(j, A_N)$ over $k(j)$ is isomorphic to $SL_2(\mathbf{Z}/N\mathbf{Z})$ in its representation on $A_N \approx (\mathbf{Z}/N\mathbf{Z})^2$.*

Proof. There exists a subfield k_0 of k which is finitely generated over the rationals, such that A is defined over $k_0(j)$, and such that k_0 is algebraically closed in $k_0(j, A_N)$, i.e. k_0 is the constant field of $k_0(j, A_N)$. We may then replace k by the algebraic closure of k_0, and therefore we may assume that k has finite transcendence degree over $\mathbf{Q}$. We may then also assume that k is contained in the complex numbers, and we may identify j with $j(\tau)$ for some value τ such that $j(\tau)$ is transcendental over k. Letting $\varphi: \mathbf{C}/L \to A_{\mathbf{C}}$ be an analytic parametrization, we let $P_1 = \varphi(\omega_1/N)$ and $P_2 = \varphi(\omega_2/N)$ as usual. Let G be the Galois groups of $\mathbf{C}(j, A_N)$ over $\mathbf{C}(j)$. We can represent an element $\sigma \in G$ by a matrix $\alpha \in GL_2(\mathbf{Z}/N\mathbf{Z})$ with respect to the basis $\{P_1, P_2\}$. We may identify the subfield $\mathbf{C}(j, h(A_N))$ with $F_{N,\mathbf{C}}$, and σ induces an automorphism of $F_{N,\mathbf{C}}$ over $\mathbf{C}(j)$, also induced by an element $\beta \in SL_2(\mathbf{Z}/N\mathbf{Z})$. We shall prove first that $\alpha = \pm\beta$. Let

$$P_{r,s} = rP_1 + sP_2, \qquad r, s \in \mathbf{Z}/N\mathbf{Z}.$$

For any $P_{r,s}$ we have

$$h(P_{(r,s)\beta}) = \sigma h(P_{r,s}) = h(\sigma P_{r,s}) = h(P_{(r,s)\alpha}).$$

Let

$$\alpha = \begin{pmatrix} a & b \\ c & d \end{pmatrix}.$$

For each (r, s) we therefore have $(r, s)\alpha = \pm(r, s)\beta$. Taking (r, s) to be $(1, 0)$ and $(0, 1)$, respectively, shows that $\beta = \pm\alpha$ or

$$\beta = \pm\begin{pmatrix} -a & -b \\ c & d \end{pmatrix}.$$

Say $\beta = \begin{pmatrix} -a & -b \\ c & d \end{pmatrix}$. Take $(r, s) = (1, 1)$. We see that

$$(a + c, b + d) \equiv (-a + c, -b + d) \pmod{N},$$

whence $2a \equiv 0 \pmod{N}$ and $2b \equiv 0 \pmod{N}$. If $N = 2$, then $1 \equiv -1 \pmod{2}$ and $GL_2(\mathbf{Z}/N\mathbf{Z}) = SL_2(\mathbf{Z}/N\mathbf{Z})$, so we may assume $N > 2$. If N is odd, then $a \equiv b \equiv 0 \pmod{N}$, which is impossible. If N is even, then $a \equiv b \equiv 0 \pmod{N/2}$, which is also impossible. Hence $\beta = \pm\alpha$, and we have proved that

$$G \subset SL_2(\mathbf{Z}/N\mathbf{Z}).$$

The lemma shows that $G = SL_2(\mathbf{Z}/N\mathbf{Z})$, and proves our corollary.

Remark. Some sort of argument is needed to prove Corollary 2, beyond Corollary 1. Indeed, let A, B be two elliptic curves defined over $\mathbf{C}(j)$, where j is transcendental over $\mathbf{C}$, and suppose that they are isomorphic, but not over $\mathbf{C}(j)$ (i.e. over some finite extension of $\mathbf{C}(j)$). The fields $\mathbf{C}(j, h(A_N))$ and $\mathbf{C}(j, h(B_N))$ are then equal, but as far as I know, it is not known if the fields $\mathbf{C}(j, A_N)$ and $\mathbf{C}(j, B_N)$ are distinct if $N > 2$. The problem lies with the extra quadratic extension, and the answer may depend on the parity of N. In any case, this shows that to prove Corollary 2, we cannot use the model of Corollary 1, defined over $\mathbf{Q}(j)$, without some additional considerations.

The main part of the argument was to show that Galois group of $\mathbf{C}(j, A_N)$ over $\mathbf{C}(j)$ is contained in $SL_2(\mathbf{Z}/N\mathbf{Z})$. One can use a quite different approach, based on a canonical skew-symmetric non-degenerate pairing

$$A_N \times A_N \to \mu_N,$$

where μ_N is the group of N-th roots of unity, due to Weil on abelian varieties. Cf. my book on abelian varieties, and Shimura's book [B12], where Shimura actually selects this approach to the question. Hence it seemed worthwhile to describe the other way in the present book. An analytic description of this pairing will be given in Chapter 18. The pairing is compatible with the action of the Galois group, i.e.

$$\langle \sigma P, \sigma Q \rangle = \langle P, Q \rangle^{\sigma}.$$

From this it is immediate that if α is the matrix representing σ in its action on A_N relative to a basis of A_N over $\mathbf{Z}/N\mathbf{Z}$, then

$$\zeta_N^{\sigma} = \zeta_N^{\det \alpha}.$$

Consequently, over the complex numbers, we see right away that the image of the Galois group in $GL_2(\mathbf{Z}/N\mathbf{Z})$ is in fact contained in $SL_2(\mathbf{Z}/N\mathbf{Z})$.

The proofs in this section are classical. Weber [B16], §63, knew the structure of the Galois group of the division points of order N, both over the complex numbers and over the rationals, especially that the roots of unity came up as the new constants. Fricke [B2], Vol. Two, I.4, gave precisely the same arguments we have chosen here, through the automorphism on roots of unity acting on the coefficients of the q-expansion.

Shimura in [38] gave new birth to these questions, and to the study of the modular function field, using these arguments. It was of considerable help for the present-day generations to have Shimura's paper available, rather than plow through Weber or Fricke, whom we had to learn to read all over again.

The analogous results in characteristic p were given by Igusa [22], [25], who even works integrally over $\mathbf{Z}[j]$. He gives different arguments, based on ramification theory, and finds the unipotent elements in the Galois group over the complex numbers to see that it is all of $SL_2(\mathbf{Z}/N\mathbf{Z})$. We shall recover this ramification theory later, when we discuss the Tate parametrization.

One of the reasons why it is still hard to read Weber is that he uses extensively the Jacobi elliptic functions, rather than the Weierstrass function more or less exclusively, as we have done.

Actually, there is some point in using the same functions Weber uses, or similar ones, constructed out of theta functions, because their values are special algebraic numbers, which are units when suitably normalized, and in this sense Weber knew perfectly well what he was doing (cf. [B16], §157). We shall consider this type of question in the last part of the book, since it is much more subtle than the general question of generating class fields any old way by values of modular functions of some level.

In this book we are exclusively concerned with congruence subgroups of $\Gamma = SL_2(\mathbf{Z})$, i.e. subgroups which contain some Γ_N. It is known that there are infinitely many subgroups of finite index which are not congruence subgroups. One can factor the upper half plane $\mathfrak{H}$ by these to obtain coverings of the projective line, ramified at 0, 1, and ∞. The pullback of any one of these to a model of some modular curve of suitable level yields an unramified covering of such a curve, and conversely, any unramified covering of a modular curve of any level belongs to a subgroup of finite index of Γ. Very little is known about the curves obtained from non-congruence subgroups. A very deep conjecture was made by Ihara [B6], who considers their reduction mod p, and conjectures roughly that "supersingular" values of j cannot split completely in these coverings, unless they arise from congruence subgroups. The beginnings of computational data have been provided by Atkin and Swinnerton-Dyer for "non-congruence" coverings (AMS Proceedings of Symposia on Pure Mathematics, XIX, (1971) pp. 1–26).

§4. SUBFIELDS OF THE MODULAR FUNCTION FIELD

By the **modular function field** F we mean the union of all the fields F_N. Similarly, $F_{\mathbf{C}}$ is the union of all fields $F_{N,\mathbf{C}}$. We shall deal mainly with F.

We denote by $M_2^+(\mathbf{Z})$ the set of 2×2 matrices with components in $\mathbf{Z}$, and positive determinant. Similarly for $M_2^+(\mathbf{Q}) = GL_2^+(\mathbf{Q})$.

Theorem 4. *If* $\alpha \in M_2^+(\mathbf{Z})$ *and* $\det \alpha = N$, *then* $j \circ \alpha$ *is a modular function of level* N. *For any* $\alpha \in M_2^+(\mathbf{Z})$, *the map*

$$f \mapsto f \circ \alpha$$

is an automorphism of F (*or* $F_{\mathbf{C}}$) *leaving the constants fixed.*

Proof. Let $\gamma \in \Gamma_N$, and write $\gamma = I + N\beta$. Then

$$\gamma' = \alpha\gamma\alpha^{-1} = I + N\alpha\beta\alpha^{-1}$$

has integral components and determinant 1, so lies in $SL_2(\mathbf{Z})$. Since

$$j \circ \alpha \circ \gamma = j \circ \gamma' \circ \alpha = j \circ \alpha,$$

it follows that $j \circ \alpha$ is invariant under Γ_N. The other conditions for $j \circ \alpha$ to be modular are immediately verified, so the first assertion is proved. The second assertion is proved similarly. Observe that if $\alpha \in M_2(\mathbf{Q})$ and m is integer such that $m\alpha \in M_2(\mathbf{Z})$, then for any function on the upper half plane, we have

$$f \circ \alpha = f \circ (m\alpha)$$

(the m cancels in the fractional transformation). Thus the inverse automorphism of

$$f \mapsto f \circ \alpha$$

is

$$f \mapsto f \circ \alpha^{-1}.$$

Although Fricke [B2], Vol. Two, I.4 also gives some discussion of subfields of the modular function fields, his discussion is not so clear (to me), and I follow Shimura [38], [B12].

In selecting τ such that $j(\tau)$ is transcendental, we could always pick τ transcendental itself (for trivial cardinality reasons, the set of algebraic values of j on $\mathfrak{H}$ is denumerable). In particular, an elliptic curve A^τ with transcendental $j(\tau)$ always has a trivial ring of endomorphisms, i.e. $\mathrm{End}(A^\tau) \approx \mathbf{Z}$.

The first case we consider is that of $j(N\tau)$, which is the invariant of an elliptic curve with lattice

$$[N\tau, 1] \sim \left[\tau, \frac{1}{N}\right].$$

Let $\tau = \omega_1/\omega_2$ and let $L = [\omega_1, \omega_2]$ be the lattice of A^τ. Put as before

$$P_1 = \varphi\left(\frac{\omega_1}{N}\right) \quad \text{and} \quad P_2 = \varphi\left(\frac{\omega_2}{N}\right),$$

where $\varphi\colon \mathbf{C}/L \to A^\tau_{\mathbf{C}}$ is an analytic representation of A^τ. Then

$$A^{N\tau} \approx A^\tau/(P_2),$$

as one sees at once from the nature of its associated lattice. From Proposition 3 of Chapter 2, §2, we know that $A/\mathfrak{g}_1 \approx A/\mathfrak{g}_2$ if and only $\mathfrak{g}_1 = \mathfrak{g}_2$ (whenever $\mathfrak{g}_1, \mathfrak{g}_2$ are finite subgroups of the same order, and A has a trivial ring of endomorphisms). Consequently we conclude that a matrix

$$\begin{pmatrix} a & b \\ c & d \end{pmatrix}$$

leaves $j(N\tau)$ fixed if and only if it maps (P_2) into itself. But

$$\begin{pmatrix} a & b \\ c & d \end{pmatrix}\begin{pmatrix} P_1 \\ P_2 \end{pmatrix} = \begin{pmatrix} aP_1 + bP_2 \\ cP_1 + dP_2 \end{pmatrix}.$$

Hence this happens if and only if $c \equiv 0 \pmod{N}$. From this we conclude:

Theorem 5. *The Galois group of F_N over $\mathbf{Q}(j, j \circ N)$ is the group*

$$\left\{\begin{pmatrix} a & b \\ 0 & d \end{pmatrix} \in GL_2(\mathbf{Z}/N\mathbf{Z})\right\} \Big/ \pm 1.$$

Corollary 1. *The fixed field of F_N under the group G_N consisting of all matrices*

$$\begin{pmatrix} 1 & 0 \\ 0 & d \end{pmatrix}, \quad d \in (\mathbf{Z}/N\mathbf{Z})^*$$

is the field

$$\mathbf{Q}(j, j \circ N, f_{1,0}).$$

Proof. The elements of the Galois group in Theorem 5 which leave $f_{1,0}$ fixed are represented by those matrices

$$\begin{pmatrix} a & b \\ 0 & d \end{pmatrix}$$

such that

$$(1, 0)\begin{pmatrix} a & b \\ 0 & d \end{pmatrix} = (\pm 1, 0).$$

This immediately implies the corollary.

Corollary 2. *The field of Corollary* 1 *is a maximal subfield of* F_N *consisting of functions whose Fourier coefficients in the* $q^{1/N}$*-expansion are rational.*

Proof. Clear.

Theorem 6. *The Galois group of* F_N *over the field* $\mathbf{Q}(j, j \circ \alpha)_{\text{all } \alpha}$, *with* $\alpha \in M_2^+(\mathbf{Z})$ *and* $\det \alpha = N$, *is the diagonal group*

$$\left\{\begin{pmatrix} e & 0 \\ 0 & e \end{pmatrix}\right\} \mod \pm 1, \qquad e \in (\mathbf{Z}/N\mathbf{Z})^*.$$

Proof. A diagonal matrix eI has the effect $P \mapsto eP$ on a point of finite order P, and hence maps every subgroup of A_N into itself. Consequently, since $j(\alpha\tau)$ is the invariant of some factor curve $A/\mathfrak{g}$ where $\mathfrak{g} \subset A_N$, it follows that $j \circ \alpha$ is fixed under such a diagonal matrix. Conversely, if an automorphism represented by

$$\begin{pmatrix} k & l \\ m & n \end{pmatrix}$$

leaves $j \circ \alpha$ fixed for all α, then it leaves $j \circ \alpha$ fixed for the special α corresponding to the factor curves $A/(P_1)$, $A/(P_2)$ and $A/(P_1 + P_2)$. The matrix

$$\begin{pmatrix} k & l \\ m & n \end{pmatrix}$$

must map each one of the vectors (1, 0), (0, 1), (1, 1) into a scalar multiple of itself, and from this one sees at once that the m..trix must be diagonal, thus proving the theorem.

One usually denotes by $\Gamma_0(N)$ the group of elements $\gamma \in \Gamma = SL_2(\mathbf{Z})$ consisting of matrices

$$\gamma = \begin{pmatrix} a & b \\ c & d \end{pmatrix}$$

with $c \equiv 0 \pmod{N}$.

Theorem 7. *The fixed field of* F_N *by* $\Gamma_0(N)$ *is the field* $\mathbf{Q}(j, j \circ N, \zeta_N)$.

Proof. This is immediate from Theorem 5, the fact that elements of $SL_2(\mathbf{Z})$ leave the constants fixed, and that the group of Theorem 5 is the product

$$\Gamma_0(N)G_N,$$

where G_N consists of the matrices

$$\begin{pmatrix} 1 & 0 \\ 0 & d \end{pmatrix}, \qquad d \in (\mathbf{Z}/N\mathbf{Z})^*.$$

7 Automorphisms of the Modular Function Field

§1. RATIONAL ADELES OF GL_2

If N, M are positive integers, and $N|M$, then we have a canonical homomorphism

$$GL_2(\mathbf{Z}/M\mathbf{Z}) \to GL_2(\mathbf{Z}/N\mathbf{Z}),$$

and we can take the projective limit. By the Chinese remainder theorem, if $N = \prod p_i^{r_i}$ is the prime factorization, then

$$GL_2(\mathbf{Z}/N\mathbf{Z}) \approx \prod_i GL_2(\mathbf{Z}/p_i^{r_i}\mathbf{Z}),$$

and so taking the projective limit can be done "componentwise" with respect to the primes. The projective limit of the rings $\mathbf{Z}/p^r\mathbf{Z}$ as $r \to \infty$ is simply the ring of p-adic integers $\mathbf{Z}_p$. Let $\mathbf{Z}_p^*$ be the group of p-adic units (invertible elements in $\mathbf{Z}_p$). Then we see that

$$\varprojlim_N GL_2(\mathbf{Z}/N\mathbf{Z}) = \prod_p GL_2(\mathbf{Z}_p),$$

where $GL_2(\mathbf{Z}_p)$ is the group of matrices with components in $\mathbf{Z}_p$, having their determinants in $\mathbf{Z}_p^*$. We abbreviate

$$GL_2(\mathbf{Z}_p) = U_p$$

and let

$$U = \prod_p U_p = \prod_p GL_2(\mathbf{Z}_p).$$

We let the finite adelic group of GL_2 be

$$GL_2(\mathbf{A}_f) = \prod_p{}' GL_2(\mathbf{Q}_p),$$

where the prime on the product means restricted product: For almost all p the

p-component of an element of $GL_2(\mathbf{A}_f)$ lies in $GL_2(\mathbf{Z}_p)$. We let $GL_2^+(\mathbf{Q})$ denote the group of rational 2×2 matrices with positive determinant.

Of course we can also form the usual ideles

$$\mathbf{A}_{\mathbf{Q}}^* = \mathbf{R}^* \times \prod_p{}' \mathbf{Q}_p^*,$$

with p-adic component in $\mathbf{Q}_p^*$, and almost all components in $\mathbf{Z}_p^*$. Using the prime factorization of an integer, one sees at once that

$$\mathbf{A}_{\mathbf{Q}}^+ = \mathbf{Q}^+(\mathbf{R}^+ \times \prod_p \mathbf{Z}_p^*),$$

where $\mathbf{A}_{\mathbf{Q}}^+$ denotes the subgroup of ideles with positive component in $\mathbf{R}$. We shall next prove the analogous result for GL_2 and SL_2.

Theorem 1. *We have*

$$GL_2(\mathbf{A}_f) = GL_2^+(\mathbf{Q})U$$
$$SL_2(\mathbf{A}_f) = SL_2(\mathbf{Q}) \prod_p SL_2(\mathbf{Z}_p).$$

Proof. We shall first prove the second equality.

For any field k it is easy to see that $SL_2(k)$ is generated by the elements

$$X(b) = \begin{pmatrix} 1 & b \\ 0 & 1 \end{pmatrix} \quad \text{and} \quad \begin{pmatrix} 1 & 0 \\ c & 1 \end{pmatrix} = Y(c)$$

with $b, c \in k$. Indeed, multiplying an arbitrary element of $SL_2(k)$ by matrices of the above type on the right and on the left corresponds to elementary row and column operations (e.g. adding a scalar multiple of a row to the other, etc.). Thus the given matrix can always be brought into a form

$$\begin{pmatrix} a & 0 \\ 0 & a^{-1} \end{pmatrix}$$

by such multiplications. Letting $W(a) = X(a)Y(-a^{-1})X(a)$ we get

$$W(a)W(-1) = \begin{pmatrix} a & 0 \\ 0 & a^{-1} \end{pmatrix},$$

thereby proving our assertion about $SL_2(k)$.

Now given $\alpha \in SL_2(\mathbf{A}_f)$, let p be a prime where α_p is not p-integral. Write α as a product

$$\alpha = Z(b_1) \cdots Z(b_m)$$

where $Z(b_i)$ is either $X(b_i)$ or $Y(b_i)$, and $b_i \in \mathbf{Q}_p$. For each i, select a rational number r_i with only powers of p in the denominator, and approximating b_i very closely at p. Let $x_p = Z(r_1) \cdots Z(r_m)$. Then $x_p \in SL_2(\mathbf{Q})$, and $x_p^{-1}\alpha$ is very close to the unit matrix in $SL_2(\mathbf{Q}_p)$, whence lies in $SL_2(\mathbf{Z}_p)$. Furthermore, x_p is ℓ-integral for any prime $\ell \neq p$. We can now repeat the procedure successively for the finite number of primes where α is not integral, and thus obtain an element $x \in SL_2(\mathbf{Q})$ such that

$$x\alpha \in \prod_p SL_2(\mathbf{Z}_p),$$

as desired.

To handle GL_2 we multiply an element $\alpha \in GL_2(\mathbf{A}_f)$ by an element β of the form

$$\beta_p = \begin{pmatrix} 1 & 0 \\ 0 & s_p \end{pmatrix}, \qquad s_p \in \mathbf{Q}_p^*$$

so that $\beta\alpha \in SL_2(\mathbf{A}_f)$. Approximating the idele $s = (\ldots, s_p \ldots)$ at a finite number of p by a positive rational number, we can find a rational matrix

$$y = \begin{pmatrix} 1 & 0 \\ 0 & r \end{pmatrix}, \qquad r \in \mathbf{Q}^+$$

such that $y\alpha \in SL_2(\mathbf{A}_f)U$. This reduces our problem to the preceding one, and proves our theorem.

We view $\mathbf{Q}^2 = \mathbf{Q} \times \mathbf{Q}$ as a space of row vectors, and let 2×2 matrices operate on the right, so that $GL_2(\mathbf{Q})$ operates on $\mathbf{Q}^2$. Similarly, $GL_2(\mathbf{Q}_p)$ operates on the right of $\mathbf{Q}_p^2$.

We have a natural isomorphism

$$\mathbf{Q}^2/\mathbf{Z}^2 \approx \coprod_p \mathbf{Q}_p^2/\mathbf{Z}_p^2,$$

which corresponds to the primary decomposition of the torsion group $(\mathbf{Q}/\mathbf{Z})^2$. An element $u_p \in GL_2(\mathbf{Z}_p)$, operates on $\mathbf{Q}_p^2/\mathbf{Z}_p^2$ and hence if

$$u = (u_p) \in U,$$

then u operates on $\mathbf{Q}^2/\mathbf{Z}^2$, according to the above prime decomposition.

§2. OPERATION OF THE RATIONAL ADELES ON THE MODULAR FUNCTION FIELD

Let $A^\tau = A$ be an elliptic curve with invariant $j(\tau)$, $\tau \in \mathfrak{H}$, and assume that A is defined over $\mathbf{Q}(j(\tau))$. We let

$$L_\tau = [\tau, 1].$$

We have an analytic representation

$$\varphi = \varphi_\tau \colon \mathbf{C}/L_\tau \to A_{\mathbf{C}}.$$

For $a = (a_1, a_2) \in \mathbf{Q}^2$ we get an element of $\mathbf{Q}L_\tau$ by taking the dot product

$$a\begin{pmatrix} \tau \\ 1 \end{pmatrix} = a_1\tau + a_2,$$

whence an isomorphism

$$\mathbf{Q}^2/\mathbf{Z}^2 \to \mathbf{Q}L_\tau/L_\tau.$$

The group $\mathbf{Q}L_\tau/L_\tau$ is the torsion subgroup of $\mathbf{C}/L_\tau$, and its image under φ_τ consists of the points of finite order on A. We shall also denote by φ the homomorphism of $\mathbf{Q}^2/\mathbf{Z}^2 \to A$ obtained by the composition of mappings

$$\mathbf{Q}^2/\mathbf{Z}^2 \to \mathbf{Q}L_\tau/L_\tau \to A.$$

We see that our analytic representation gives us a coordinate system for the points of finite order on A. If $a \in \mathbf{Q}^2$ and $\bar{a}$ denotes the class of a in $\mathbf{Q}^2/\mathbf{Z}^2$, we also write

$$\varphi(a) = \varphi(\bar{a}).$$

Thus we also view φ as giving a homomorphism

$$\varphi: \mathbf{Q}^2 \to \mathbf{Q}L_\tau/L_\tau \to A.$$

Let us assume that $\mathrm{End}(A) \approx \mathbf{Z}$. Then any other analytic parametrization

$$\psi: \mathbf{C}/L_\tau \to A_{\mathbf{C}}$$

must be such that $\psi = \pm\varphi$, because $\psi \circ \varphi^{-1}$ is an automorphism of A. Let us assume that A is in Weierstrass form, and let h be the Weber function such that

$$h(x, y) = -2^7 3^5 \frac{g_2 g_3}{\Delta} x,$$

so that h is an isomorphism invariant. Then we have

$$h_\tau \circ \varphi_\tau(a) = f_a(\tau), \qquad a \in \mathbf{Q}^2,$$

where f_a is the Fricke function.

Theorem 2. *Let F be the modular function field, and let f_a ($a \in \mathbf{Q}^2/\mathbf{Z}^2$, $a \neq 0$) be the Fricke functions. For each $u \in U$ there is an automorphism $\sigma(u)$ of F over $\mathbf{Q}(j)$ such that*

$$f_a^{\sigma(u)} = f_{au},$$

and the map

$$u \mapsto \sigma(u)$$

is a homomorphism of U onto $\mathrm{Gal}(F/\mathbf{Q}(j))$ whose kernel is ± 1.

Proof. This is but a reformulation of the results of the preceding chapter, taking into account the projective limit

$$U = \lim_{\leftarrow} GL_2(\mathbf{Z}/N\mathbf{Z}).$$

Theorem 3. *Let $\tau \in \mathfrak{H}$ be such that $j(\tau)$ is transcendental over $\mathbf{Q}$, and let A be an elliptic curve such that $j_A = j(\tau)$, and defined over $\mathbf{Q}(j(\tau))$. Let $\varphi: \mathbf{C}/L_\tau \to A_{\mathbf{C}}$ be an analytic parametrization of A. Let U be as in §1. Then* *for each $u \in U$ there is an automorphism $\sigma(u)$ of the field of all division points on A such that*

$$\varphi(a)^{\sigma(u)} = \varphi(au),$$

and the map $u \mapsto \sigma(u)$ is an isomorphism of U onto the Galois group of the field of all division points over $\mathbf{Q}(j(\tau))$.

Proof. This is a reformulation of Theorem 2, and Theorem 3 of the preceding chapter, taking into account the projective limits.

In particular we get the formula

$$\boxed{h_\tau(\varphi_\tau(a))^{\sigma(u)} = h_\tau(\varphi_\tau(au)).}$$

There is another type of automorphism. For any $\alpha \in GL_2^+(\mathbf{Q})$ we let $\sigma(\alpha)$ be the automorphism such that for any $f \in F$ we have

$$f^{\sigma(\alpha)} = f \circ \alpha.$$

In other words,

$$f^{\sigma(\alpha)}(\tau) = f(\alpha\tau).$$

This yields a homomorphism of $GL_2^+(\mathbf{Q})$ into $\mathrm{Aut}(F)$, whose kernel is the subgroup of matrices

$$\begin{pmatrix} a & 0 \\ 0 & a \end{pmatrix}, \qquad a \in \mathbf{Q}^*.$$

Remark 1. Note that $U \cap GL_2^+(\mathbf{Q}) = SL_2(\mathbf{Z})$. If $\alpha \in SL_2(\mathbf{Z})$, then the definition of $\sigma(\alpha)$ viewing α as an element of U or as an element of $GL_2^+(\mathbf{Q})$ is the same.

Indeed, we have the obvious relation

$$f_a(\alpha\tau) = f_{a\alpha}(\tau)$$

for the Fricke functions, and for any $\alpha \in SL_2(\mathbf{Z})$, viewed as an element of U, the corresponding automorphism leaves j fixed because $j(\alpha\tau) = j(\tau)$.

Remark 2. Suppose $u \in U$ and in addition $u_p \in SL_2(\mathbf{Z}_p)$ for all primes p. Let f be a modular function of level N. Then there exists an element $\alpha \in SL_2(\mathbf{Z})$ such that if $n(p)$ is the order of N at p, then

$$\alpha \equiv u_p \pmod{p^{n(p)}}$$

for all $p|N$. We then see that

$$f^{\sigma(u)} = f^{\sigma(\alpha)} = f \circ \alpha,$$

first for the Fricke functions f_a, where a has exact denominator N, and then for any $f \in F_N$ since the functions f_a generate F_N.

If σ, σ' are two automorphisms of F, then to have associativity in the exponential notation, we make their composite act so that

$$f^{\sigma\sigma'} = (f^\sigma)^{\sigma'}.$$

There is another important consistency relation.

Theorem 4. (**Shimura**) *Let $\alpha, \beta \in GL_2^+(\mathbf{Q})$ and let $u, v \in U$ be such that $\alpha u = v\beta$. Then $\sigma(\alpha)\sigma(u) = \sigma(v)\sigma(\beta)$.*

Proof. For the proof, we have to look into the meaning of this relation, and on its interpretation in terms of isogenies.

Let $\gamma \in M_2(\mathbf{Z})$ be a 2×2 integral matrix. Then γ operates on $\mathbf{Q}^2/\mathbf{Z}^2$ and its kernel is represented by those elements $a \in \mathbf{Q}^2$ such that

$$a\gamma \in \mathbf{Z}^2,$$

i.e. its kernel is

$$\mathbf{Z}^2\gamma^{-1}/\mathbf{Z}^2.$$

The next lemma is a basic formal tool for the study of isogenies of elliptic curves and their points of finite order.

Lemma. *Let $\alpha \in GL_2^+(\mathbf{Q})$. Let A^τ and $A^{\alpha(\tau)}$ be elliptic curves with invariants $j(\tau)$ and $j(\alpha(\tau))$ respectively, and let*

$$\varphi: \mathbf{C}/L_\tau \to A^\tau_{\mathbf{C}} \quad \text{and} \quad \psi: \mathbf{C}/L_{\alpha(\tau)} \to A^{\alpha(\tau)}_{\mathbf{C}}$$

be corresponding analytic representations of these curves. Assume that $\alpha^{-1} \in M_2(\mathbf{Z})$ has integral coefficients. Let

$$\alpha = \begin{pmatrix} a & b \\ c & d \end{pmatrix},$$

and let $\mu = c\tau + d$. Then there exists a unique isogeny

$$\lambda = \lambda_\alpha: A^\tau \to A^{\alpha(\tau)}$$

such that the following diagram is commutative.

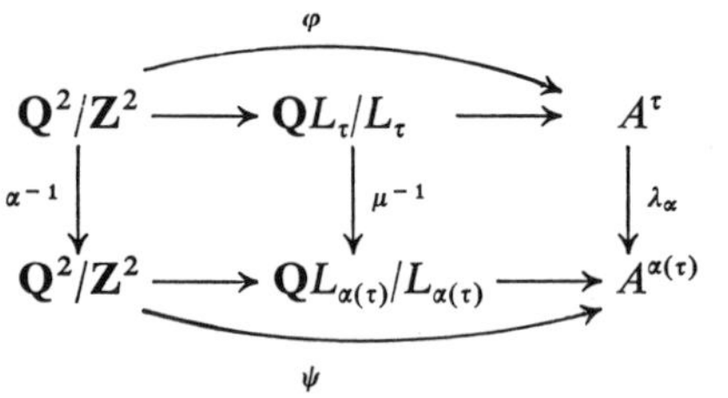

$$\begin{array}{ccccc} \mathbf{Q}^2/\mathbf{Z}^2 & \longrightarrow & \mathbf{Q}L_\tau/L_\tau & \longrightarrow & A^\tau \\ \downarrow{\scriptstyle \alpha^{-1}} & & \downarrow{\scriptstyle \mu^{-1}} & & \downarrow{\scriptstyle \lambda_\alpha} \\ \mathbf{Q}^2/\mathbf{Z}^2 & \longrightarrow & \mathbf{Q}L_{\alpha(\tau)}/L_{\alpha(\tau)} & \longrightarrow & A^{\alpha(\tau)} \end{array}$$

(The top row composite is φ, the bottom row composite is ψ.)

The middle arrow is multiplication by μ^{-1}.

Proof. We have

$$\alpha\begin{pmatrix} \tau \\ 1 \end{pmatrix} = \mu\begin{pmatrix} \alpha(\tau) \\ 1 \end{pmatrix},$$

whence

$$\mu^{-1}\begin{pmatrix} \tau \\ 1 \end{pmatrix} = \alpha^{-1}\begin{pmatrix} \alpha(\tau) \\ 1 \end{pmatrix}.$$

Since $\alpha^{-1} \in M_2(\mathbf{Z})$ by assumption, we see that multiplication by μ^{-1} maps $\mathbf{C}/L_\tau$ into $\mathbf{C}/L_{\alpha(\tau)}$. There exists a unique isogeny λ_α which makes the following diagram commutative.

$$\begin{array}{ccc} \mathbf{C}/L_\tau & \xrightarrow{\varphi} & A^\tau_{\mathbf{C}} \\ \downarrow{\scriptstyle \mu^{-1}} & & \downarrow{\scriptstyle \lambda_\alpha} \\ \mathbf{C}/L_{\alpha(\tau)} & \xrightarrow[\psi]{} & A^{\alpha(\tau)}_{\mathbf{C}} \end{array}$$

Then

$$\mu^{-1}(a_1, a_2)\begin{pmatrix}\tau \\ 1\end{pmatrix} = (a_1, a_2)\mu^{-1}\begin{pmatrix}\tau \\ 1\end{pmatrix} = (a_1, a_2)\alpha^{-1}\begin{pmatrix}\alpha(\tau) \\ 1\end{pmatrix},$$

and therefore the square on the left is commutative. This proves the lemma.

Since $A^{\alpha(\tau)}$ has invariant $j(\alpha(\tau))$, we can always select $A^{\alpha(\tau)}$ defined over $\mathbf{Q}(j(\alpha(\tau)))$. A way of doing this is to take the elliptic curve with transcendental invariant j to be defined by

$$y^2 = 4x^3 - gx - g,$$

such that $g/(g - 27) = j/12^3$. If we select $A^{\alpha(\tau)}$ defined over $\mathbf{Q}(j(\alpha(\tau)))$, then any automorphism of $F(\tau)$ over $F_1(\tau)$, for instance $\sigma(u)$, can be applied to $A^{\alpha(\tau)}$.

Theorem* 5.** (Shimura***) *Let $u, v \in U$ and let $\alpha, \beta \in GL_2^+(\mathbf{Q})$ be such that $\alpha u = v\beta$. Assume that $j(\tau)$ is transcendental over* $\mathbf{Q}$, *and that A^τ* (resp. *$A^{\alpha(\tau)}$*) *is defined over* $\mathbf{Q}(j(\tau))$ [resp. over $\mathbf{Q}(j(\alpha(\tau)))$]. *Then $\sigma(u)A^{\alpha(\tau)}$ has invariant $j(\beta(\tau))$. Select $A^{\beta(\tau)} = \sigma(u)A^{\alpha(\tau)}$. Let $\lambda_\alpha, \lambda_\beta$ be the isogenies which make the diagram in the lemma commutative. Then*

$$\lambda_\alpha^{\sigma(u)} = \pm \lambda_\beta.$$

Proof. We first prove that independently of how we choose $A^{\beta(\tau)}$, the two isogenies

$$\lambda_\alpha^{\sigma(u)} \quad \text{and} \quad \lambda_\beta$$

have the same kernel.

The kernel of λ_α is $\varphi(\mathbf{Z}^2\alpha/\mathbf{Z}^2)$. Hence

$$\begin{aligned} \operatorname{Ker} \lambda_\alpha^{\sigma(u)} = (\operatorname{Ker} \lambda_\alpha)^{\sigma(u)} &= \varphi(\mathbf{Z}^2\alpha/\mathbf{Z}^2)^{\sigma(u)} \\ &= \varphi(\mathbf{Z}^2\alpha u/\mathbf{Z}^2) \quad \text{(see below)} \\ &= \varphi(\mathbf{Z}^2 v\beta/\mathbf{Z}^2) \\ &= \varphi(\mathbf{Z}^2\beta/\mathbf{Z}^2) \\ &= \operatorname{Ker} \lambda_\beta. \end{aligned}$$

This proves the first assertion, except that we must explain the notation

$$\mathbf{Z}^2\alpha u/\mathbf{Z}^2.$$

We recall that

$$\mathbf{Q}^2/\mathbf{Z}^2 = \coprod_p \mathbf{Q}_p^2/\mathbf{Z}_p^2,$$

and an element $u_p \in GL_2(\mathbf{Z}_p)$ acts on the p-component $\mathbf{Q}_p^2/\mathbf{Z}_p^2$. What we mean by $\mathbf{Z}^2\alpha u/\mathbf{Z}^2$ is the direct sum

$$\mathbf{Z}^2\alpha u/\mathbf{Z}^2 = \coprod_p \mathbf{Z}_p^2\alpha u_p/\mathbf{Z}_p^2,$$

and since $\alpha u_p = v_p\beta$, we have $\mathbf{Z}_p^2\alpha u_p = \mathbf{Z}_p^2 v\beta_p = \mathbf{Z}_p^2\beta$. From these remarks, the notation makes sense, and the equalities in the above proof are valid.

The two isogenies $\lambda_\alpha^{\sigma(u)}$ and λ_β having the same kernel shows that their images are isomorphic, and hence have the same j-invariant, so that the first assertion of our theorem is valid. We may then choose $A^{\beta(\tau)} = \sigma(u)A^{\alpha(\tau)}$. Both $\lambda_\alpha^{\sigma(u)}$ and λ_β then map A^τ on the same image, and have the same kernel, so they differ by an automorphism of the image. Since we selected τ such that $j(\tau)$ is transcendental, we know that the only possible automorphisms are ± 1. This proves Theorem 5.

We can now return to Theorem 4, and verify the relation of Theorem 4 for the functions j and f_a.

First, we have

$$j(\tau)^{\sigma(\alpha)\sigma(u)} = j(\alpha(\tau))^{\sigma(u)}$$

and

$$j(\tau)^{\sigma(v)\sigma(\beta)} = j(\tau)^{\sigma(\beta)} = j(\beta(\tau)).$$

The two expressions on the right are equal by Theorem 5, so our relation is proved for the j function.

Next, we consider $a \in \mathbf{Q}^2$ and $b = a\alpha^{-1}$. Then:

$$\begin{aligned}
\varphi_\tau(b)^{\sigma(\alpha)\sigma(u)} &= \varphi_{\alpha(\tau)}(b)^{\sigma(u)} \\
&= \varphi_{\alpha(\tau)}(a\alpha^{-1})^{\sigma(u)} \\
&= (\lambda_\alpha(\varphi_\tau(a)))^{\sigma(u)} \\
&= \lambda_\alpha^{\sigma(u)}(\varphi_\tau(a))^{\sigma(u)} \\
&= \pm\lambda_\beta \circ \varphi_\tau(au) \\
&= \pm\varphi_{\beta(\tau)}(au\beta^{-1}) \\
&= \pm\varphi_{\beta(\tau)}(a\alpha^{-1}v) \\
&= \pm\varphi_{\beta(\tau)}(bv).
\end{aligned}$$

Taking the h-coordinate yields

$$f_b(\alpha(\tau))^{\sigma(u)} = f_{bv}(\beta(\tau))$$

which means that

$$f_b^{\sigma(\alpha)\sigma(u)} = f_b^{\sigma(v)\sigma(\beta)},$$

and proves our theorem.

§3. THE SHIMURA EXACT SEQUENCE

For an arbitrary finite adele $x \in GL_2(\mathbf{A}_f)$ we write

$$x = \alpha u \qquad \text{or} \qquad x = v\beta,$$

with $u, v \in U$ and $\alpha, \beta \in GL_2^+(\mathbf{Q})$. Then Theorem 4 shows that we can define the automorphism $\sigma(x)$ on F by

$$\sigma(x) = \sigma(\alpha)\sigma(u) = \sigma(v)\sigma(\beta).$$

This is well defined by Theorem 4, and a trivial computation shows that the association

$$x \mapsto \sigma(x)$$

gives a homomorphism of $GL_2(\mathbf{A}_f)$ into $\mathrm{Aut}(F)$. It is easily proved that the kernel is precisely the group of diagonal matrices

$$\begin{pmatrix} a & 0 \\ 0 & a \end{pmatrix}, \qquad a \in \mathbf{Q}^*,$$

simply by using the results of the preceding chapter. We leave this as an exercise.

Theorem 6. (***Shimura***) *The Sequence*

$$0 \to \mathbf{Q}^* \to GL_2(\mathbf{A}_f) \to \mathrm{Aut}(F) \to 0$$

is exact, in other words, every automorphism of F is of the form αu (i.e. $\sigma(\alpha)\sigma(u)$) *for some $\alpha \in GL_2^+(\mathbf{Q})$ and $u \in U$.*

Proof. The proof which we shall give for the surjectivity now differs from Shimura's arguments, and is based on a different principle.

Let σ be an automorphism of F. If $\sigma j = j$, then $\sigma \in \sigma(U)$ and we are done. We shall reduce our proof to this case.

First we may assume that σ leaves the roots of unity fixed, because we can compose σ with some $\sigma(u)$ to achieve this. It then suffices to prove that we can compose σ with some $\sigma(\alpha)$ so as to fix j. Since σ is now assumed to leave the roots of unity fixed, it may be extended to an automorphism of the modular function field $F_{\mathbf{C}}$ over $\mathbf{C}$, leaving the constants fixed.

Let A be an elliptic curve having invariant j, defined over $\mathbf{C}(j)$, say by the standard Weierstrass equation. We identify the modular function field of level N over $\mathbf{C}$ with $\mathbf{C}(j, h(A_N))$. The field

$$F_{\mathbf{C},A}^{(p)} = \mathbf{C}(j, h(A^{(p)}))$$

is the subfield of $F_{\mathbf{C}}$ obtained from the points of p-power order on A. It is a p-extension of $\mathbf{C}(j, A_p)$, and $\sigma F_{\mathbf{C},A}^{(p)}$ is the corresponding p-tower over $\mathbf{C}(j^\sigma, A_p^\sigma)$.

Let $E = \mathbf{C}(j, j^\sigma, A_p, A_p^\sigma)$. Then

$$E(h(A^{(p)})) \qquad \text{and} \qquad E(h(A^{\sigma(p)}))$$

are p-towers over E. We shall now prove that there exists a finite extension K of E such that

$$K(h(A^{(p)})) = K(h(A^{\sigma(p)})).$$

The Galois group of $F_{\mathbf{C}}$ over E contains an open subgroup of the form

$$W = \prod_{\ell \in S} W_\ell \times \prod_{\ell \notin S} SL_2(\mathbf{Z}_\ell),$$

where S is a finite set of primes, and W_ℓ is such a small open neighborhood of 1 in $SL_2(\mathbf{Z}_\ell)$ for all $\ell \in S$, that W_ℓ is an ℓ-group without torsion. We select S so large as to contain 2, 3 and p. Let K be the fixed field of W. Let H_p be the Galois group of $K(h(A^{\sigma(p)})$ over K. Then we have a surjective homomorphism

$$\psi \colon W \to H_p$$

of Galois theory, corresponding to the inclusion of fields

$$K \subset K(h(A^{\sigma(p)})) \subset F_{\mathbf{C}}.$$

Each factor W_ℓ for $\ell \in S$, $\ell \neq p$, maps onto 1 under this homomorphism, because an ℓ-group can only map trivially into a p-group. If $\ell \notin S$, then the subgroup of $SL_2(\mathbf{Z}_\ell)$ projecting on 1 in $SL_2(\mathbf{Z}/\ell\mathbf{Z})$ is an ℓ-group, and the same reasoning applies, to see that this subgroup maps onto 1 under ψ. Finally, any homomorphic image

$$SL_2(\mathbf{Z}/\ell\mathbf{Z}) \to H_p$$

must be trivial, because ± 1 maps into 1 (since H_p has no torsion), and $SL_2(\mathbf{Z}/\ell\mathbf{Z})/\pm 1$ is simple for $\ell \geqq 5$.

Therefore H_p is in fact a homomorphic image of W_p, and in terms of field extensions, this means that

$$K(h(A^{\sigma(p)})) \subset K(h(A^{(p)})).$$

Replacing K be a finite extension if necessary and using a symmetry argument, we conclude that in fact these two fields are the same. (Alternatively, one could also use the fact that since the Lie algebra of $SL_2(\mathbf{Z}_p)$ is simple, the above extension is finite, and hence of degree 1 since W_p is assumed without torsion.)

It now follows from a theorem to be proved by entirely different methods later (Chapter 16, §5, Theorem 7, and §1, Corollary of Theorem 1), that A and A^σ must be isogeneous. Consequently there exists an integral matrix α such that $j^\sigma = j \circ \alpha$. Thus finally $\sigma(\alpha)^{-1}\sigma$ is an automorphism of F leaving j fixed, as was to be shown.

Groups of automorphisms of infinite modular function fields were considered by Shafarevič and Piateckii-Shapiro [31] and [32]. The latter considers the field of all functions $j \circ \alpha$, with rational matrices α. The section of the paper dealing with the automorphisms is not entirely clear. For instance, what we gave here as Theorem 5, due to Shimura, seems to be completely overlooked by Piateckii-Shapiro. On the other hand, the rest of the paper deals with the reduction mod p of the modular function field, and has results related to the Shimura reciprocity law, proved in Chapter 11.

Part Two

Complex Multiplication Elliptic Curves with Singular Invariants

In this part we study special curves whose rings of endomorphisms are strictly bigger than **Z**. This involves both elliptic curves whose j-invariant $j(z)$ is such that z is an imaginary quadratic number over **Q**, giving rise to the theory of complex multiplication, and elliptic curves over finite fields. We shall also relate this special theory with the generic theory of the preceding part, and show how the various mappings of an arithmetic nature which we obtain are related at all three levels: generic, number fields, and finite fields, specializing from one level to the next.

The term **complex multiplication** arises because the algebras of endomorphisms of elliptic curves which are bigger than **Z** must be complex, i.e. cannot have real embeddings. Over the complex numbers, complex multiplication arises from the endomorphisms induced by multiplication in **C** with a complex number α sending the given lattice into itself.

The main development of the theory will be carried out by the Deuring reduction method. However, it is illuminating to see some of the results derived by the older analytic method of Kronecker, Weber and Hasse, so we have done this on a selective basis. For instance, you may find it useful to look right away at the analytic derivation of the congruence relation reproduced in Chapter 12, §3, and also the factorization results of Chapter 12, §2 which are self-contained, before, or simultaneously with, the algebraic arguments using reduction mod p.

8 *Results from Algebraic Number Theory*

In this chapter we assume that the reader is acquainted with the ordinary ideal theory in number fields. Cf. for instance [B7]. The first two sections should be read as technical background for Chapter 10, §2. On the other hand, although we strive for some completeness, once the reader sees the first results that the proper $\mathfrak{o}$-lattices form a multiplicative group, he can wait to read the other results until he needs them, as they are slightly technical. They are all classical, known to Dedekind, except possibly for the fact that a proper $\mathfrak{o}$-lattice is locally principal, which seems to have been first pointed out by Ihara [26]. The localization technique will be used heavily for the idelic formulation of the complex multiplication, as in Shimura [B12].

§1. LATTICES IN QUADRATIC FIELDS

Proper $\mathfrak{o}$-ideals

Let k be a number field, i.e. a finite extension of the rationals. We denote by $\mathfrak{o}_k$ the ring of algebraic integers of k. By an **order** $\mathfrak{o}$ in k we mean a subring of $\mathfrak{o}_k$ whose dimension over $\mathbf{Z}$ is equal to the degree $[k : \mathbf{Q}]$. By a **lattice** in k we mean an additive subgroup of k which is free of dimension $[k : \mathbf{Q}]$ over $\mathbf{Z}$. If L is a lattice in k, we define the **order of** L to be the set of elements $\lambda \in k$ such that $\lambda L \subset L$. By one of the definitions of algebraic integers, it follows that the order of L is contained in $\mathfrak{o}_k$, and it is easily verified that it is in fact an order, i.e. has rank $[k : \mathbf{Q}]$ over $\mathbf{Z}$.

For the rest of this section, we assume that τ *is quadratic over* $\mathbf{Q}$ *and we let*

$k = \mathbf{Q}(\tau)$. *We let* $\lambda \mapsto \lambda'$ *be the non-trivial automorphism of* k. Let τ satisfy the quadratic equation

$$A\tau^2 + B\tau + C = 0$$

with integers A, B, C which are relatively prime and $A > 0$. Let the discriminant be

$$D = B^2 - 4AC,$$

so that

$$\tau = \frac{-B + \sqrt{D}}{2A}.$$

We clearly have

$$B \equiv D \pmod{2}. \tag{1}$$

Theorem 1. *Notation as above, let*

$$\mathfrak{o} = \left[1, \frac{D + \sqrt{D}}{2}\right] = \left[1, \frac{B + \sqrt{D}}{2}\right].$$

Then $\mathfrak{o}$ *is the order of the lattice* $[\tau, 1]$.

Proof. The congruence (1) shows that the equality on the right is true. By a straightforward multiplication, one sees that $1 \cdot L \subset L$, and that

$$\frac{B + \sqrt{D}}{2}\tau = -C \in \mathbf{Z} \subset L,$$

$$\frac{B + \sqrt{D}}{2} = A\tau + B \in L.$$

Hence $\left[1, \frac{B + \sqrt{D}}{2}\right]$ is contained in the order of $[\tau, 1]$. To prove the converse, we prove another basic result first.

Theorem 2. *Let* $L' = [\tau', 1]$ *where* τ' *is the conjugate of* τ, *and let* $\mathfrak{o}$ *be as in Theorem* 1. *Then*

$$LL' = \frac{1}{A}\mathfrak{o}.$$

Proof. We have

$$\begin{aligned} LL' = [\tau\tau', \tau, \tau', 1] &= \left[\frac{B^2 - D}{4A^2}, \frac{-B + \sqrt{D}}{2A}, \frac{-B - \sqrt{D}}{2A}, 1\right] \\ &= \frac{1}{A}\left[C, B, A, \frac{B + \sqrt{D}}{2}\right] \\ &= \frac{1}{A}\mathfrak{o}, \end{aligned}$$

as was to be shown.

In particular, we see that L is invertible (with respect to $\mathfrak{o}$), in other words

$$L^{-1} = AL'.$$

To finish the proof of Theorem 1, suppose that $\lambda L \subset L$. Then

$$\lambda LL^{-1} \subset LL^{-1} = \mathfrak{o},$$

so $\lambda\mathfrak{o} \subset \mathfrak{o}$, and since $\mathfrak{o}$ contains 1, we get $\lambda \in \mathfrak{o}$, thus proving that

$$\mathfrak{o} = \{\lambda \in k, \lambda L \subset L\}.$$

Given an order $\mathfrak{o}$ in k, we shall say that a lattice L **belongs** to $\mathfrak{o}$, or is a **proper** $\mathfrak{o}$-lattice, if

$$\mathfrak{o} = \{\lambda \in k, \lambda L \subset L\}.$$

By an $\mathfrak{o}$-ideal we mean an ordinary ideal $\mathfrak{a} \subset \mathfrak{o}$, which is a lattice.

Corollary. *Let $\mathfrak{o}$ be an order in the quadratic field k. Every proper $\mathfrak{o}$-lattice in k is $\mathfrak{o}$-invertible, and conversely any lattice which is $\mathfrak{o}$-invertible is a proper $\mathfrak{o}$-lattice. The set of proper $\mathfrak{o}$-lattices is a multiplicative group.*

If $\mathfrak{a}$, $\mathfrak{c}$ are proper $\mathfrak{o}$-ideals, we define $\mathfrak{c}|\mathfrak{a}$ to mean that there exists an $\mathfrak{o}$-ideal $\mathfrak{b}$ such that $\mathfrak{b}\mathfrak{c} = \mathfrak{a}$. Multiplying by $\mathfrak{c}^{-1}$ shows that $\mathfrak{b}$ is necessarily a proper $\mathfrak{o}$-ideal. Furthermore, as usual, one sees that this condition is equivalent with the condition $\mathfrak{a} \subset \mathfrak{c}$. An **irreducible** proper $\mathfrak{o}$-ideal $\mathfrak{p}$ is a proper $\mathfrak{o}$-ideal $\neq \mathfrak{o}$ which cannot be factored $\mathfrak{p} = \mathfrak{a}\mathfrak{b}$, with proper $\mathfrak{o}$-ideals $\mathfrak{a}$, $\mathfrak{b}$ such that $\mathfrak{a} \neq \mathfrak{p}$ and $\mathfrak{b} \neq \mathfrak{p}$. [We shall see later as a result of Theorem 4, that an irreducible proper $\mathfrak{o}$-ideal $\mathfrak{p}$ prime to the conductor, is a prime ideal.]

The conductor and ideals prime to the conductor

Theorem 3. *Let $\mathfrak{o}$ be an order in k, and let $\mathfrak{o}_k = [z, 1]$. There exists a unique positive integer c such that*

$$\mathfrak{o} = [cz, 1] = \mathbf{Z} + c\mathfrak{o}_k.$$

Proof. Note that $\mathfrak{o}$ is a sublattice of $\mathfrak{o}_k$, whence of finite index. Let $c > 0$ be the unique positive integer such that

$$\mathfrak{o} \cap \mathbf{Z}z = \mathbf{Z}cz.$$

We contend this c does it. Indeed, let $\lambda \in \mathfrak{o}$, $\lambda = m + nz$. Then

$$nz = \lambda - m \in \mathfrak{o} \cap \mathbf{Z}z,$$

whence $c|n$, and $\lambda \in \mathbf{Z} + \mathbf{Z}cz$. This proves the theorem.

The number c in Theorem 3 is called the **conductor** of $\mathfrak{o}$.

Let $\mathfrak{o}$ be an order and $\mathfrak{a}$ an $\mathfrak{o}$-ideal. Let c be the conductor of $\mathfrak{o}$. We shall say that $\mathfrak{a}$ is **prime** to c if either $\mathfrak{a} + c\mathfrak{o} = \mathfrak{o}$ or $\mathfrak{a} + c\mathfrak{o}_k = \mathfrak{o}$. The two conditions are actually equivalent, for suppose $\mathfrak{a} + c\mathfrak{o} = \mathfrak{o}$. If $\mathfrak{a} + c\mathfrak{o}_k \neq \mathfrak{o}$, then $\mathfrak{a} + c\mathfrak{o}_k$ is contained in a maximal ideal $\mathfrak{p}$ which also contains $\mathfrak{a} + c\mathfrak{o}$, impossible. Conversely, suppose $\mathfrak{a} + c\mathfrak{o}_k = \mathfrak{o}$. If $\mathfrak{a} + c\mathfrak{o} \neq \mathfrak{o}$ then $\mathfrak{a} + c\mathfrak{o}$ is contained in a maximal ideal $\mathfrak{p}$, and since $\mathfrak{o}_k$ is integral over $\mathfrak{o}$, there is a maximal ideal of $\mathfrak{o}_k$ lying above $\mathfrak{p}$. This contradicts $\mathfrak{a} + c\mathfrak{o}_k = \mathfrak{o}$.

We let $I_k(c)$ be the set of $\mathfrak{o}_k$-ideals prime to c, and we let $I_{\mathfrak{o}}(c)$ be the set of $\mathfrak{o}$-ideals prime to c.

Theorem 4. *There is a multiplicative bijection between the monoid of ideals of $\mathfrak{o}_k$ prime to c and the monoid of $\mathfrak{o}$-ideals prime to c, given by the two inverse mappings*

$$\mathfrak{a} \mapsto \mathfrak{a} \cap \mathfrak{o}, \qquad \mathfrak{a} \in I_k(c)$$

$$\mathfrak{a} \mapsto \mathfrak{a}\mathfrak{o}_k, \qquad \mathfrak{a} \in I_{\mathfrak{o}}(c).$$

An ideal of $\mathfrak{o}$ prime to c is a proper $\mathfrak{o}$-ideal.

Proof. i) Let $\mathfrak{a}$ be an $\mathfrak{o}$-ideal and $\mathfrak{a} + c\mathfrak{o}_k = \mathfrak{o}$. We shall prove that $\mathfrak{a} = \mathfrak{a}\mathfrak{o}_k \cap \mathfrak{o}$. The inclusion $\subset$ is clear. Conversely,

$$\begin{aligned} \mathfrak{a}\mathfrak{o}_k \cap \mathfrak{o} = (\mathfrak{a}\mathfrak{o}_k \cap \mathfrak{o})\mathfrak{o} &= (\mathfrak{a}\mathfrak{o}_k \cap \mathfrak{o})(\mathfrak{a} + c\mathfrak{o}_k) \\ &\subset \mathfrak{a} + \mathfrak{a}\mathfrak{o}_k c \\ &\subset \mathfrak{a} + \mathfrak{a}\mathfrak{o} \subset \mathfrak{a}. \end{aligned}$$

This proves our first assertion.

ii) Let $\mathfrak{a}$ be an $\mathfrak{o}_k$-ideal such that $\mathfrak{a} + c\mathfrak{o}_k = \mathfrak{o}_k$. Then we prove that $(\mathfrak{a} \cap \mathfrak{o})\mathfrak{o}_k = \mathfrak{a}$. We have:

$$\begin{aligned} \mathfrak{o} = \mathfrak{o}_k \cap \mathfrak{o} &= (\mathfrak{a} + c\mathfrak{o}_k) \cap \mathfrak{o} \\ &\subset (\mathfrak{a} \cap \mathfrak{o}) + c\mathfrak{o}_k \subset \mathfrak{o}. \end{aligned}$$

Hence $\mathfrak{a} \cap \mathfrak{o}$ is prime to c. Now

$$\mathfrak{a} = \mathfrak{a}\mathfrak{o} = \mathfrak{a}((\mathfrak{a} \cap \mathfrak{o}) + c\mathfrak{o}_k) \subset \mathfrak{o}_k(\mathfrak{a} \cap \mathfrak{o}) + c\mathfrak{a}.$$

But $\mathfrak{a}c \subset \mathfrak{a} \cap \mathfrak{o}$, so $\mathfrak{a} \subset (\mathfrak{a} \cap \mathfrak{o})\mathfrak{o}_k$. The converse inclusion is obvious, thus proving (ii).

iii) We prove that an $\mathfrak{o}$-ideal $\mathfrak{a}$ prime to c is proper. Suppose $\lambda \in k$ and $\lambda\mathfrak{a} \subset \mathfrak{a}$. Then

$$\lambda\mathfrak{o} = \lambda(\mathfrak{a} + c\mathfrak{o}_k) = \lambda\mathfrak{a} + \lambda c\mathfrak{o}_k \subset \mathfrak{a} + c\mathfrak{o}_k = \mathfrak{o}.$$

Since $1 \in \mathfrak{o}$, we get $\lambda \in \mathfrak{o}$.

iv) In (i) and (ii) we got the desired bijection. It preserves multiplication, for let $\mathfrak{a}_{\mathfrak{o}}$, $\mathfrak{b}_{\mathfrak{o}}$ be $\mathfrak{o}$-ideals prime to c, where

$$\mathfrak{a}_{\mathfrak{o}} = \mathfrak{a} \cap \mathfrak{o} \quad \text{and} \quad b_{\mathfrak{o}} = b \cap \mathfrak{o},$$

with $\mathfrak{o}_k$-ideals $\mathfrak{a}$, $\mathfrak{b}$ prime to c. Then $\mathfrak{a}_{\mathfrak{o}}\mathfrak{b}_{\mathfrak{o}}$ is prime to c, and

$$\mathfrak{a}_{\mathfrak{o}}\mathfrak{b}_{\mathfrak{o}} = (\mathfrak{a}_{\mathfrak{o}}\mathfrak{b}_{\mathfrak{o}}\mathfrak{o}_k) \cap \mathfrak{o} = (\mathfrak{a}\mathfrak{b}) \cap \mathfrak{o}.$$

This proves our theorem.

Remark. The above arguments work for any number field, with an order $\mathfrak{o}$, defining the conductor $\mathfrak{c}$ to be the largest ideal of $\mathfrak{o}$ which is also an $\mathfrak{o}_k$-ideal.

Theorem 5. *Let L be a proper $\mathfrak{o}$-lattice and m a positive integer. Then there exists an element $\lambda \in k$ such that $\lambda L \subset \mathfrak{o}$ and*

$$\lambda L + m\mathfrak{o} = \mathfrak{o}.$$

In other words, in the equivalence class of L, there exists a lattice which is prime to m, and is integral.

Proof. Suppose that we start with a lattice of the form $L = [\tau, 1]$, such that τ satisfies the equation

$$A\tau^2 + B\tau + C = 0,$$

with integers A, B, C relatively prime, and $A > 0$. Then

$$L = \frac{1}{A}\left[A, \frac{-B + \sqrt{D}}{2}\right] \sim \left[A, \frac{-B + \sqrt{D}}{2}\right].$$

Without loss of generality, we may assume that L is the $\mathfrak{o}$-ideal

$$\mathfrak{a} = \left[A, \frac{-B + \sqrt{D}}{2}\right].$$

Then $\mathfrak{a}\mathfrak{a}' = A\mathfrak{o}$. Finally, we could also change τ by an element of $SL_2(\mathbf{Z})$, i.e. prove our assertion for the lattice $L_1 = [\tau_1, 1]$ where

$$\tau = \frac{a\tau_1 + b}{c\tau_1 + d}.$$

The equation for such τ_1 is

$$\begin{aligned} 0 &= A(a\tau_1 + b)^2 + B(a\tau_1 + b)(c\tau_1 + d) + C(c\tau_1 + d)^2 \\ &= A_1\tau_1^2 + B_1\tau_1 + C_1, \end{aligned}$$

where $A_1 = Aa^2 + Bac + Cc^2$. It will therefore suffice to prove that we can select a, c relatively prime such that A_1 is prime to m. We take a, c to be products of primes p dividing m as follows. If $p \nmid A$, select a prime to p and p divides c. If $p \mid A$ but $p \nmid C$, take c prime to p but p divides a. If $p \mid A$ and $p \mid C$, then necessarily $p \nmid B$. Take both a, c prime to p. This yields the desired integers a and c, and proves our theorem.

The proper $\mathfrak{o}$-ideal classes

Let $I_{\mathfrak{o}}$ be the multiplicative monoid of proper $\mathfrak{o}$-ideals, and $P_{\mathfrak{o}}$ the submonoid of principal $\mathfrak{o}$-ideals (automatically proper). We let

$$G_{\mathfrak{o}} = I_{\mathfrak{o}}/P_{\mathfrak{o}},$$

and call $G_{\mathfrak{o}}$ the group of **proper $\mathfrak{o}$-ideal classes.** Let $I_{\mathfrak{o}}(c)$ be the monoid of proper $\mathfrak{o}$-ideals prime to the conductor c, and let $P_{\mathfrak{o}}(c)$ be the submonoid of principal $\mathfrak{o}$-ideals prime to c. Then by Theorem 5, we have an isomorphism

$$G_{\mathfrak{o}} \approx I_{\mathfrak{o}}(c)/P_{\mathfrak{o}}(c).$$

We shall express $G_{\mathfrak{o}}$ as a factor group of a generalized ideal class group of $\mathfrak{o}_k$. We let

$$P_{\mathbf{Z}}(c)$$

be the monoid of $\mathfrak{o}_k$-ideals $\mathfrak{a}$ which are principal, of the form

$$\mathfrak{a} = \mathfrak{o}_k \alpha,$$

where

$$\alpha \equiv a \pmod{c\mathfrak{o}_k}$$

for some $a \in \mathbf{Z}$, $(a, c) = 1$.

Lemma 1. Let $\mathfrak{a} \in P_{\mathbf{Z}}(c)$ be as above. Then

$$\mathfrak{a} \cap \mathfrak{o} = \mathfrak{o}\alpha.$$

Proof. Since $\alpha \in \mathfrak{o}$ we get $\mathfrak{o}\alpha \subset \mathfrak{a} \cap \mathfrak{o}$. Conversely, if $x \in \mathfrak{o}_k$ and $x\alpha \in \mathfrak{o}$, let us write

$$x = m + nz \qquad \text{and} \qquad \alpha = a + cbz$$

with integers m, n, a, b such that $(a, c) = 1$. Then

$$x\alpha \equiv ma + nza \pmod{c\mathfrak{o}_k}.$$

Hence na is divisible by c, so that $c \mid n$. Hence $x \in \mathfrak{o}$, proving our Lemma.

Theorem 6. Consider the homomorphism

$$I_k(c) \to I_{\mathfrak{o}}(c)$$

such that $\mathfrak{a} \mapsto \mathfrak{a} \cap \mathfrak{o}$. The inverse image of $P_{\mathfrak{o}}(c)$ is $P_{\mathbf{Z}}(c)$.

Proof. The lemma shows that $P_{\mathbf{Z}}(c)$ is contained in the inverse image. Conversely, suppose that $\mathfrak{a} \cap \mathfrak{o} = \mathfrak{o}\alpha$ with $\alpha \equiv a \pmod{c\mathfrak{o}_k}$ and $a \in \mathbf{Z}$. Then $\mathfrak{a} = \mathfrak{o}_k\alpha$ so $\mathfrak{a} \in P_{\mathbf{Z}}(c)$.

It follows from Theorem 6 that we have an isomorphism

$$\boxed{G_{\mathfrak{o}} \approx I_k(c)/P_{\mathbf{Z}}(c).}$$

Note that $P_{\mathbf{Z}}(c)$ contains the ideals which are principal and generated by an element $\equiv 1 \pmod{c}$, the monoid of such ideals being denoted by $P_1(c)$. So we have a tower

$$I_k(c) \supset P_{\mathbf{Z}}(c) \supset P_1(c).$$

From this we can easily determine the order of the group $G_{\mathfrak{o}}$.

Theorem 7. *The order of the group $G_{\mathfrak{o}}$ is equal to*

$$h_{\mathfrak{o}} = h\frac{c}{(\mathfrak{o}_k^* : \mathfrak{o}^*)}\prod_{p|c}\left(1 - \left(\frac{k}{p}\right)\frac{1}{p}\right),$$

where h is the class number of k, c is the conductor of $\mathfrak{o}$, $\mathfrak{o}_k^*$ and $\mathfrak{o}^*$ are the groups of units in $\mathfrak{o}_k$ and $\mathfrak{o}$, respectively, and $\left(\frac{k}{p}\right)$ is the usual symbol, equal to 1 if p splits completely in k, -1 if p remains prime, and 0 if p ramifies in k.

Proof. We shall give the same argument as in Fueter and Weber, §98. The theorem is very classical. We know from general algebraic number theory that the order of the generalized ideal class group $I_k(c)/P_1(c)$ is given by

$$h_c = \frac{h\varphi(c\mathfrak{o}_k)}{(\mathfrak{o}_k^* : U_c)},$$

where φ is the Euler function, and U_c consists of those units in $\mathfrak{o}_k$ which are congruent to 1 mod $c\mathfrak{o}_k$. See for instance my *Algebraic Number Theory*, Chapter VI, §1, Theorem 1. It follows that

$$h_{\mathfrak{o}} = \frac{h_c}{(P_{\mathbf{Z}}(c) : P_1(c))}.$$

Suppose first for simplicity that ± 1 are the only units of $\mathfrak{o}_k$. We have a map

$$(\mathbf{Z}/c\mathbf{Z})^* \to P_{\mathbf{Z}}(c)/P_1(c),$$

given by $a \mapsto$ class of $a\mathfrak{o}_k$ modulo $P_1(c)$, whose kernel is ± 1, of order 2 if $c > 2$. Suppose that p is a prime number and p^m divides c exactly. The p-contribution to $(\mathbf{Z}/c\mathbf{Z})^*$ is $p^m\left(1 - \frac{1}{p}\right)$. Suppose that p splits completely in k. Then $p\mathfrak{o}_k = \mathfrak{p}\mathfrak{p}'$ and $\mathfrak{o}_k/\mathfrak{p}$, $\mathfrak{o}_k/\mathfrak{p}'$ have order p. Hence the p-contribution to $\varphi(c\mathfrak{o}_k)$ is the

$$\text{order of } \mathfrak{o}_k/(\mathfrak{p}\mathfrak{p}')^m = p^{2m}\left(1 - \frac{1}{p}\right)^2.$$

Dividing these p-contributions gives the proper factor in the product. On the other hand, if $-1 \equiv 1 \pmod{c}$, then $\mathfrak{o}_k^* = \mathfrak{o}^*$, and $c = 1$ or 2. The unit contribution is then precisely the right one. If $-1 \not\equiv 1$, then it is also clear that the unit contribution is the correct one. If p remains prime in $\mathfrak{o}_k$, then the p-contribution to $\varphi(c\mathfrak{o}_k)$ is the order of the multiplicative group of $\mathfrak{o}_k/p^m\mathfrak{o}_k$ which is $p^{2m}\left(1 - \frac{1}{p^2}\right)$. Dividing by the p-contribution to $(\mathbf{Z}/c\mathbf{Z})^*$ yields precisely

$$p^m\left(1 + \frac{1}{p}\right),$$

which is the desired factor. If $p\mathfrak{o}_k = \mathfrak{p}^2$, then the same type of argument again shows that we get the right contribution to our factor. Finally, when $\mathfrak{o}_k$ contains i or ρ, one argues the same way, which we safely leave to the reader.

Remark. We worked above with ideals of $\mathfrak{o}$, i.e. contained in $\mathfrak{o}$. Of course, one can also work with the group of proper $\mathfrak{o}$-lattices, with respect to the usual equivalence, $L \sim M$ if and only if there exists $\lambda \in k$ such that $\lambda L = M$. If $\alpha \in k$ is such that $\alpha \equiv 1 \pmod{^* c}$, meaning that

$$\mathrm{ord}_{\mathfrak{p}}(\alpha - 1) \geqq \mathrm{ord}_{\mathfrak{p}}\, c$$

for all primes $\mathfrak{p}$ of $\mathfrak{o}_k$ such that $\mathfrak{p}|c$, then we can write $\alpha = \beta/\gamma$, where

$$\beta, \gamma \equiv 1 \quad (\mathrm{mod}^*\ c\mathfrak{o}_k), \quad \beta, \gamma \in \mathfrak{o}_k.$$

[If d is a positive rational denominator for α, prime to c, we can select d_1 having the same divisibility as d for $p \nmid c$, and $d_1 \equiv 1 \pmod{c}$ by the Chinese remainder theorem. Then $d_1\alpha \in \mathfrak{o}_k$ and $d_1\alpha \equiv 1 \pmod{c\mathfrak{o}_k}$.] If $\mathfrak{a}, \mathfrak{b}$ are proper $\mathfrak{o}$-ideals such that $\alpha\mathfrak{a} = \mathfrak{b}$, then $\beta\mathfrak{a} = \gamma\mathfrak{b}$.

Corollary. *There is only a finite number of imaginary quadratic $\tau \in \mathfrak{H}$ inequivalent under the modular group, such that $j(\tau)$ lies in a given number field K.*

Proof. One knows that the class number of a quadratic imaginary field k goes to infinity with the discriminant, in fact

$$\log h(D) \sim \log |D|^{\frac{1}{2}}$$

by a theorem of Siegel. Therefore $j(\mathfrak{o}_k)$ has degree tending to infinity as $|D| \to \infty$. For any order $\mathfrak{o}$ of $\mathfrak{o}_k$, we see from Theorem 7 that the class number of $\mathfrak{o}$ also tends to infinity with the conductor, and $j(\mathfrak{o})$ has degree equal to this class number over k (proved later, complex multiplication). This proves our corollary.

Note that Theorem 7 gives very explicitly the rate at which the degree of $j(\mathfrak{o})$ goes to infinity as a function of the conductor, once the absolute class number is known. The Riemann Hypothesis would give an explicit and very good inequality for the absolute class number in terms of the discriminant, but at the moment, one has to go through various contorsions to prove Siegel's theorem because of the lack of a proof for RH. See for instance [B7], Chapter 13, §4, and Chapter 16, and [B15].

Localization

Finally we consider localization at a prime number p. Let S_p be the set of positive integers not divisible by p. We define the localization of a lattice L at p to be

$$L_{(p)} = S_p^{-1}L.$$

For the rest of this section, in order to simplify the notation, we shall write L_p instead of $L_{(p)}$. When we consider completions in the next section, we shall use L_p to denote the completion of $L_{(p)}$.

If L, M are lattices, then we have trivially

$$S_p^{-1}(L \cap M) = S_p^{-1}L \cap S_p^{-1}M.$$

The inclusion $\subset$ is clear. Conversely, if an element can be written as $x/m = y/n$ with $x \in L$, $y \in M$, and mn not divisible by p, then $my = nx$ lies in $L \cap M$, and our element is equal to nx/mn, as desired.

As usual, LM consists of all sums

$$\sum x_i y_i$$

with $x_i \in L$ and $y_i \in M$. It is an additive subgroup of k, finitely generated over $\mathbf{Z}$, whence it is a lattice. We have

$$(LM)_p = L_p M_p.$$

Theorem 8. *Let L, M be lattices in k. If $L_p \subset M_p$ for all primes p, then $L \subset M$.*

Proof. Let $x \in L$. Then we can write $x = y_p/n_p$ with $y_p \in M$ and an integer n_p prime to p, so that $n_p x \in M$. The family of n_p's is relatively prime, so there exists $m_p \in \mathbf{Z}$ such that

$$\sum m_p n_p = 1.$$

It follows that

$$x = \sum m_p n_p x \in M,$$

as was to be shown.

The theorem shows that to prove that two lattices are equal, it suffices to do so locally for each p. A similar argument as in the theorem shows that

$$L = \bigcap_p L_p.$$

We note that if p does not divide the conductor c of $\mathfrak{o}$, then

$$\mathfrak{o}_p = (\mathfrak{o}_k)_p,$$

and in particular, if L is a proper $\mathfrak{o}$-lattice, then by ordinary ideal theory we find that L_p is locally principal, i.e. there exists an element $\alpha \in k$ such that

$$L_p = \mathfrak{o}_p \alpha.$$

For quadratic fields, this property remains true even if $p|c$, as was pointed out by Ihara [25].

Theorem 9. *Let L be a lattice in k belonging to the order $\mathfrak{o}$. Then there exists $\alpha \in k$ such that $L_p = \mathfrak{o}_p \alpha$,* i.e. *$L$ is locally principal.*

Proof. Let $\mathfrak{o}_k = [z, 1]$ again. We may assume that $p|c$. Since $L_p L_p^{-1} = \mathfrak{o}_p$, there exists $y \in L_p^{-1}$ and $x \in L$ such that $yx = m + ncz$, with integers m, n and $(m, p) = 1$. Dividing by m, we conclude that

$$1 \in yL_p + cz\mathfrak{o}_p.$$

Multiplying by cz, which lies in $\mathfrak{o}$ so that $czL_p \subset L_p$, we get

$$cz \in yL_p + c^2z^2\mathfrak{o}_p,$$

and substituting back, using induction, we get

$$1 \in yL_p + (cz)^\nu \mathfrak{o}_p \subset yL_p + p^\nu \mathfrak{o}_p$$

for all positive integers ν. Hence $\mathfrak{o}_p \subset yL_p + p^\nu \mathfrak{o}_p$ for all ν. Since the index $(\mathfrak{o}_p : yL_p)$ is a power of p, we have $\mathfrak{o}_p \subset yL_p$, and $\mathfrak{o}_p = yL_p$. This proves our theorem.

The next lemma is sometimes useful to find a local generator for a proper $\mathfrak{o}$-ideal.

Lemma 2. *Let $\mathfrak{a} \in I_k(c)$, and for a prime p suppose that $\mathfrak{a}_p = \mathfrak{o}_{k,p}\alpha$, with $\alpha \in \mathfrak{a}$. Let $x, y \in \mathfrak{o}_{k,p}$ be such that*

$$x\alpha + yc = 1.$$

Then

$$\mathfrak{a}_p \cap \mathfrak{o}_p = \mathfrak{o}_p x\alpha.$$

Proof. Note that $1, yc \in \mathfrak{o}_p$, so $x\alpha \in \mathfrak{o}_p$, whence the inclusion $\supset$ follows. The converse inclusion is proved by a jacking up argument similar to that of Theorem 9.

§2. COMPLETIONS

Let k be a number field, and let L be a lattice in k. For a prime number p we let $\mathbf{Z}_p$ be the ring of p-adic integers, and

$$L_p = \mathbf{Z}_p \otimes L, \quad \text{also written } \mathbf{Z}_p L.$$

We let $L_{(p)} = S_p^{-1}L$ be the localization of L at p as defined in the previous section. Then L_p can be viewed as the completion of $L_{(p)}$, and there is a natural injection

$$L_{(p)} \to L_p,$$

which we treat as an inclusion.

Let $\mathbf{Q}_p$ be the p-adic numbers. We define

$$k_p = \mathbf{Q}_p \otimes k, \quad \text{also written } \mathbf{Q}_p k.$$

By a **$\mathbf{Z}_p$-lattice** in k_p we mean a $\mathbf{Z}_p$-submodule of k_p of dimension $[k:\mathbf{Q}]$ over $\mathbf{Z}_p$. Note that

$$k_p = \mathbf{Q}_p \otimes \mathfrak{o}_k,$$

so that the powers of p appearing in the denominators of elements of a $\mathbf{Z}_p$-lattice are bounded. Consequently, if M_p is a $\mathbf{Z}_p$-lattice there exists a power p^r of p such that

$$p^r M_p \subset \mathfrak{o}_{k,p} = \mathbf{Z}_p \otimes \mathfrak{o}_k.$$

On the other hand, there exists a power p^s such that

$$p^s \mathfrak{o}_{k,p} \subset M_p,$$

because $\mathfrak{o}_{k,p}$ and M_p have the same dimension over $\mathbf{Z}_p$.

By a $\mathbf{Z}_{(p)}$-lattice in k we mean a $\mathbf{Z}_{(p)}$ submodule of k of dimension $[k:\mathbf{Q}]$ over $\mathbf{Z}_{(p)}$. If M_p is a $\mathbf{Z}_p$-lattice in k_p, then

$$M_p \cap k$$

is a $\mathbf{Z}_{(p)}$-lattice in k. The intersection is taken by viewing k as embedded naturally in k_p. The assertion is easily seen, because $M_p \cap k$ is a module over $\mathbf{Z}_{(p)}$, it contains $p^s \mathfrak{o}_{k,(p)}$ for some integer s, and it is contained in $p^r \mathfrak{o}_{k,(p)}$ so that it has the correct dimension, over $\mathbf{Z}_{(p)}$.

Theorem 10. i) *Given for each p a $\mathbf{Z}_{(p)}$-lattice $M_{(p)}$ in k such that $M_{(p)} = \mathfrak{o}_{k,(p)}$ for almost all p, there exists a unique lattice L in k such that $L_{(p)} = M_{(p)}$ for all p.*

ii) *Given a $\mathbf{Z}_p$-lattice M_p in k_p such that $M_p = \mathfrak{o}_{k,p}$ for almost all p, there exists a unique lattice L in k such that $L_p = M_p$ for all p.*

Proof. We let $L = \bigcap M_{(p)}$ to prove (i). It is immediately verified that L is a lattice, and that $L_{(p)} = M_{(p)}$ for all p. The second part follows from the first by the remarks we have made, relating $\mathbf{Z}_{(p)}$-lattices and $\mathbf{Z}_p$-lattices.

Let L be a lattice in k. For each p we have a natural isomorphism

$$k/L_{(p)} \approx k_p/L_p,$$

because $L_{(p)} = L_p \cap k$. Since $\bigcap_p L_{(p)} = L$, we get a canonical isomorphism

$$k/L \approx \coprod_p k/L_{(p)} \approx \coprod_p k_p/L_p.$$

Indeed, any $x \in k$ lies in $\mathfrak{o}_{k,(p)}$ for almost all p, so we have a map of k into the direct sum of the k_p/L_p. The above isomorphism essentially gives the p-primary decomposition of the torsion group k/L.

Recall that the ideles J_k of k can be defined as the restricted product

$$k_{\mathbf{R}}^* \times \prod_p{}' k_p^*$$

where $k_{\mathbf{R}} = \mathbf{R} \otimes k$, and the * indicates the group of invertible elements. If

$$s = (\ldots, s_p, \ldots)$$

is an idele with $s_p \in k_p^*$, then for any $\mathbf{Z}_p$-lattice L_p we can take the product

$$s_p L_p,$$

and it is again a $\mathbf{Z}_p$-lattice.

If L is a lattice in k, then $s_p L_p = \mathfrak{o}_{k,p}$ for almost all p, and consequently there exists a unique lattice M such that

$$M_p = s_p L_p.$$

This lattice M is denoted by sL. Observe that s is an idele, and that there is no multiplication defined directly between s and L. The notation sL is merely symbolic.

Multiplication by s_p induces an isomorphism also denoted s_p,

$$s_p \colon k_p/L_p \to k_p/s_p L_p,$$

given by $x_p \mapsto s_p x_p$. From the decomposition

$$k/L \approx \coprod_p k_p/L_p,$$

we can define an isomorphism

$$s \colon k/L \to k/sL$$

by letting s operate componentwise, i.e. s_p operates by multiplication on each component k_p/L_p.

If $L = \mathfrak{a}$ is a fractional ideal of the ring of algebraic integers $\mathfrak{o}_k$, then we can work with the prime components in k. If we denote by $\mathfrak{a}_\mathfrak{p}$ the closure of $\mathfrak{a}$ in the local field $k_\mathfrak{p}$, and if s is an idele with $\mathfrak{p}$-component $s_\mathfrak{p}$, then $s_\mathfrak{p} a_\mathfrak{p}$ is defined, and

$$k/\mathfrak{a} \approx \coprod_\mathfrak{p} k_\mathfrak{p}/a_\mathfrak{p}$$

so that

$$k/s\mathfrak{a} \approx \coprod_\mathfrak{p} k_\mathfrak{p}/s_\mathfrak{p} a_\mathfrak{p}.$$

Let $\mathfrak{c}$ be an ideal of $\mathfrak{o}_k$ (and so contained in $\mathfrak{o}_k$). Then $\mathfrak{c}^{-1}\mathfrak{a} \supset \mathfrak{a}$, and

$$\mathfrak{c}^{-1}\mathfrak{a}/\mathfrak{a}$$

is a finite subgroup of $k/\mathfrak{a}$. Furthermore, $k/\mathfrak{a}$ is the union of such finite subgroups taken for $\mathfrak{c}$ tending to infinity (ideals being ordered by divisibility). Let $\mathfrak{p}_1, \ldots, \mathfrak{p}_m$ be the prime ideals dividing $\mathfrak{c}$ or entering in the factorization of $\mathfrak{a}$. By localizing $\mathfrak{o}_k$ at these primes, we obtain a Dedekind ring $\mathfrak{o}'$ having only a finite number of prime ideals, and hence a principal ring. Then $\mathfrak{c}\mathfrak{o}' = (c)$ for some element c and $\mathfrak{a}\mathfrak{o}' = (a)$. We have an isomorphism

$$\mathfrak{c}^{-1}\mathfrak{a}/\mathfrak{a} \approx (c^{-1}a)/(a)$$

where (x) is the ideal generated by x in $\mathfrak{o}'$. Let $x \in \mathfrak{o}'$ be such that $x\mathfrak{a}\mathfrak{o}' = \mathfrak{a}\mathfrak{o}'$. Let $u \in \mathfrak{c}^{-1}\mathfrak{a}/\mathfrak{a}$. Then xu is defined in the natural way by multiplication by x, and we have $xu = u$ for all $u \in \mathfrak{c}^{-1}\mathfrak{a}/\mathfrak{a}$ if and only if

$$x \equiv 1 \pmod{^* \mathfrak{c}}.$$

This follows at once from the definitions. The congruence mod* means ordinary congruence in the local ring for each prime power component of $\mathfrak{c}$.

§3. THE DECOMPOSITION GROUP AND FROBENIUS AUTOMORPHISM

In this section we summarize pertinent facts about the decomposition group of a prime ideal in a Galois extension. The results are basic, but we emphasize that although they are sometimes stated only for Dedekind rings, e.g. in number fields, they are valid more generally, and this is important when we consider an elliptic curve over the ring $\mathbf{Z}[j]$.

Throughout this section, **ring** means ring without divisor of zero and commutative.

Proposition 1. *Let R be a ring, integrally closed in its quotient field K. Let L be a finite Galois extension of K with group G. Let $\mathfrak{p}$ be a maximal ideal of R, and let $\mathfrak{P}$, $\mathfrak{Q}$ be prime ideals of the integral closure of R in L lying above $\mathfrak{p}$. Then there exists $\sigma \in G$ such that $\sigma\mathfrak{P} = \mathfrak{Q}$.*

Proof. Suppose that $\mathfrak{P} \neq \sigma\mathfrak{Q}$ for any $\sigma \in G$. There exists an element $x \in S$ such that

$$x \equiv 0 \pmod{\mathfrak{P}}$$
$$x \equiv 1 \pmod{\sigma\mathfrak{Q}}, \qquad \text{all } \sigma \in G$$

(use the Chinese remainder theorem). The norm

$$N_K^L(x) = \prod_{\sigma \in G} \sigma x$$

lies in $B \cap K = R$ (because R is integrally closed), and lies in $\mathfrak{P} \cap R = \mathfrak{p}$. But $x \notin \sigma\mathfrak{Q}$ for all $\sigma \in G$, so that $\sigma x \notin \mathfrak{Q}$ for all $\sigma \in G$. This contradicts the fact that the norm of x lies in $\mathfrak{p} = \mathfrak{Q} \cap R$.

Corollary. *Let R be a ring, integrally closed in its quotient field K. Let E be a finite separable extension of K, and S the integral closure of R in E. Let $\mathfrak{p}$ be a maximal ideal of R. Then there exists only a finite number of prime ideals of S lying above $\mathfrak{p}$.*

Proof. Let L be the smallest Galois extension of K containing E. If $\mathfrak{Q}_1, \mathfrak{Q}_2$ are two distinct prime ideals of S lying above $\mathfrak{p}$, and $\mathfrak{P}_1, \mathfrak{P}_2$ are two prime ideals of the integral closure of R in L lying above $\mathfrak{Q}_1$ and $\mathfrak{Q}_2$ respectively, then $\mathfrak{P}_1 \neq \mathfrak{P}_2$. This argument reduces our assertion to the case that E is Galois over K, and it then becomes an immediate consequence of the proposition.

Let R be integrally closed in its quotient field K, and let S be its integral closure in a finite Galois extension L, with group G. Then $\sigma S = S$ for every $\sigma \in G$. Let $\mathfrak{p}$ be a maximal ideal of R, and $\mathfrak{P}$ a maximal ideal of S lying above $\mathfrak{p}$. We denote by $G_{\mathfrak{P}}$ the subgroup of G consisting of those automorphisms such that $\sigma\mathfrak{P} = \mathfrak{P}$. Then $G_{\mathfrak{P}}$ operates in a natural way on the residue class field $S/\mathfrak{P}$, and leaves $R/\mathfrak{p}$ fixed. To each $\sigma \in G_{\mathfrak{P}}$ we can associate an automorphism $\bar{\sigma}$ of $S/\mathfrak{P}$ over $R/\mathfrak{p}$, and the map given by

$$\sigma \mapsto \bar{\sigma}$$

induces a homomorphism of $G_{\mathfrak{P}}$ into the group of automorphisms of $S/\mathfrak{P}$ over $R/\mathfrak{p}$.

The group $G_{\mathfrak{P}}$ will be called the **decomposition group** of $\mathfrak{P}$. Its fixed field will be denoted by L^d, and will be called the **decomposition field** of $\mathfrak{P}$. Let S^d be the integral closure of R in L^d, and let $\mathfrak{Q} = \mathfrak{P} \cap S^d$. By Proposition 1, we know that $\mathfrak{P}$ is the only prime of S lying above $\mathfrak{Q}$.

Let $G = \bigcup \sigma_j G_{\mathfrak{P}}$ be a coset decomposition of $G_{\mathfrak{P}}$ in G. Then the prime ideals $\sigma_j\mathfrak{P}$ are precisely the distinct primes of S lying above $\mathfrak{p}$. Indeed, for two elements $\sigma, \tau \in G$ we have $\sigma\mathfrak{P} = \tau\mathfrak{P}$ if and only if $\tau^{-1}\sigma\mathfrak{P} = \mathfrak{P}$, i.e. $\tau^{-1}\sigma$ lies in $G_{\mathfrak{P}}$. Thus τ, σ lie in the same coset mod $G_{\mathfrak{P}}$.

It is then immediately clear that the decomposition group of a prime $\sigma\mathfrak{P}$ is $\sigma G_{\mathfrak{P}}\sigma^{-1}$.

Proposition 2. *The field L^d is the smallest subfield E of L containing K such that $\mathfrak{P}$ is the only prime of S lying above $\mathfrak{P} \cap E$ (which is prime in $S \cap E$).*

Proof. Let E be as above, and let H be the Galois group of L over E. Let $\mathfrak{q} = \mathfrak{P} \cap E$. By Proposition 1, all primes of S lying above $\mathfrak{q}$ are conjugate by elements of H. Since there is only one prime, namely $\mathfrak{P}$, it means that H leaves $\mathfrak{P}$ invariant. Hence $H \subset G_{\mathfrak{P}}$ and $E \supset L^d$. We have already observed that L^d has the required property.

Proposition 3. *Notation being as above, we have $R/\mathfrak{p} = S^d/\mathfrak{Q}$ (under the canonical injection $R/\mathfrak{p} \to S^d/\mathfrak{Q}$).*

Proof. If σ is an element of G, not in $G_{\mathfrak{P}}$, then $\sigma\mathfrak{P} \neq \mathfrak{P}$ and $\sigma^{-1}\mathfrak{P} \neq \mathfrak{P}$. Let

$$\mathfrak{Q}_\sigma = \sigma^{-1}\mathfrak{P} \cap S^d.$$

Then $\mathfrak{Q}_\sigma \neq \mathfrak{Q}$. Let x be an element of S^d. There exists an element y of S^d such that

$$y \equiv x \pmod{\mathfrak{Q}}$$
$$y \equiv 1 \pmod{\mathfrak{Q}_\sigma}$$

for each σ in G, but not in $G_{\mathfrak{P}}$. Hence in particular,

$$y \equiv x \pmod{\mathfrak{P}}$$
$$y \equiv 1 \pmod{\sigma^{-1}\mathfrak{P}}$$

for each σ not in $G_{\mathfrak{P}}$. This second congruence yields

$$\sigma y \equiv 1 \pmod{\mathfrak{P}}$$

for all $\sigma \notin G_{\mathfrak{P}}$. The norm of y from L^d to K is a product of y and other factors σy with $\sigma \notin G_{\mathfrak{P}}$. Thus we obtain

$$N_K^{L^d}(y) \equiv x \pmod{\mathfrak{P}}.$$

But the norm lies in K, and even in R, since it is a product of elements integral over R. This last congruence holds mod $\mathfrak{Q}$, since both x and the norm lie in S^d. This is precisely the meaning of the assertion in our proposition.

If x is an element of S, we shall denote by $\bar{x}$ its image under the homomorphism $S \to S/\mathfrak{P}$. Then $\bar{\sigma}$ is the automorphism of $S/\mathfrak{P}$ satisfying the relation

$$\bar{\sigma}\bar{x} = \overline{\sigma x}.$$

If $f(X)$ is a polynomial with coefficients in S, we denote by $\bar{f}(X)$ its natural image under the above homomorphism. Thus, if

$$f(X) = b_n X^n + \cdots + b_0,$$

then

$$\bar{f}(X) = \bar{b}_n X^n + \cdots + \bar{b}_0.$$

Proposition 4. *Let R be integrally closed in its quotient field K, and let S be its integral closure in a finite Galois extension L of K, with group G. Let $\mathfrak{p}$ be a maximal ideal of R, and $\mathfrak{P}$ a maximal ideal of S lying above $\mathfrak{p}$. Then $S/\mathfrak{P}$ is a normal extension of $R/\mathfrak{p}$, and the map $\sigma \mapsto \bar{\sigma}$ induces a homomorphism of $G_{\mathfrak{P}}$ onto the Galois group of $S/\mathfrak{P}$ over $R/\mathfrak{p}$.*

Proof. Let $\bar{S} = S/\mathfrak{P}$ and $\bar{R} = R/\mathfrak{p}$. Any element of $\bar{S}$ can be written as $\bar{x}$ for some $x \in S$. Let $\bar{x}$ generate a separable subextension of $\bar{S}$ over $\bar{R}$, and let f be the irreducible polynomial for x over K. The coefficients of f lie in R because x is integral over R, and all the roots of f are integral over R. Thus

$$f(X) = \prod_{i=1}^{m} (X - x_i)$$

splits into linear factors in S. Since

$$\bar{f}(X) = \prod (X - \bar{x}_i)$$

and all the $\bar{x}_i$ lie in $\bar{S}$, it follows that $\bar{f}$ splits into linear factors in $\bar{S}$. We observe that $f(x) = 0$ implies $\bar{f}(\bar{x}) = 0$. Hence $\bar{S}$ is normal over $\bar{R}$, and

$$[\bar{R}(\bar{x}) : \bar{R}] \leqq [K(x) : K] \leqq [L : K].$$

This implies that the maximal separable subextension of $\bar{R}$ in $\bar{S}$ is of finite degree over $\bar{R}$ (using the primitive element theorem of elementary field theory). This degree is in fact bounded by $[L : K]$.

There remains to prove that the map $\sigma \mapsto \bar{\sigma}$ gives a surjective homomorphism of $G_{\mathfrak{P}}$ onto the Galois group of $\bar{S}$ over $\bar{R}$. To do this, we shall give an argument which reduces our problem to the case when $\mathfrak{P}$ is the only prime ideal of S lying above $\mathfrak{p}$. Indeed, by Proposition 3, the residue class fields of the ground ring and the ring S^d in the decomposition field are the same. This means that to prove our surjectivity, we may take L^d as ground field. This is the desired reduction, and we can assume $K = L^d$, $G = G_{\mathfrak{P}}$.

This being the case, take a generator of the maximal separable subextension of $\bar{S}$ over $\bar{R}$, and let it be $\bar{x}$, for some element x in S. Let f be the irreducible polynomial of x over K. Any automorphism of $\bar{S}$ is determined by its effect on $\bar{x}$, and maps $\bar{x}$ on some root of $\bar{f}$. Suppose that $x = x_1$. Given any root x_i of f, there exists an element σ of $G = G_{\mathfrak{P}}$ such that $\sigma x = x_i$. Hence $\bar{\sigma}\bar{x} = \bar{x}_i$. Hence the automorphism of $\bar{S}$ over $\bar{R}$ induced by elements of G operate transitively on the roots of $\bar{f}$. Hence they give us all automorphisms of the residue class field, as was to be shown.

Corollary 1. *Let R be a ring integrally closed in its quotient field K. Let L be a finite Galois extension of K, and S the integral closure of R in L. Let $\mathfrak{p}$ be a maximal ideal of R. Let $\varphi : R \to R/\mathfrak{p}$ be the canonical homomorphism, and let ψ_1, ψ_2 be two homomorphisms of S extending φ in a given algebraic closure of $R/\mathfrak{p}$. Then there exists an automorphism σ of L over K such that*

$$\psi_1 = \psi_2 \circ \sigma.$$

Proof. The kernels of ψ_1, ψ_2 are prime ideals of S which are conjugate by Proposition 1. Hence there exists an element τ of the Galois group G such that $\psi_1, \psi_2 \circ \tau$ have the same kernel. Without loss of generality, we may therefore assume that ψ_1, ψ_2 have the same kernel $\mathfrak{P}$. Hence there exists an automorphism ω of $\psi_1(S)$ onto $\psi_2(S)$ such that $\omega \circ \psi_1 = \psi_2$. There exists an element σ of $G_{\mathfrak{P}}$ such that $\omega \circ \psi_1 = \psi_1 \circ \sigma$, by the preceding proposition. This proves what we wanted.

Remark. In all the above propositions, we could assume $\mathfrak{p}$ prime instead of maximal. In that case, one has to localize at $\mathfrak{p}$ to be able to apply our proofs. In the application to number fields, this is unnecessary, since every prime is maximal.

In the above discussions, the kernel of the map

$$G_{\mathfrak{P}} \to \bar{G}_{\mathfrak{P}}$$

is called the **inertia group** $T_{\mathfrak{P}}$ of $\mathfrak{P}$. It consists of those automorphisms of $G_{\mathfrak{P}}$ which induce the trivial automorphism on the residue class field. Its fixed field is called the **inertia field,** and is denoted by L^t.

If the inertia group of $\mathfrak{P}$ is trivial, i.e. 1, then we say that $\mathfrak{P}$ is **unramified** over $\mathfrak{p}$. If every prime $\mathfrak{P}$ over $\mathfrak{p}$ is unramified, then we say that $\mathfrak{p}$ is **unramified** in L.

Let again $\mathfrak{p}$ be a maximal ideal of R, and let L be a finite Galois extension of K, of degree N, with S the integral closure of R in L. We shall say that $\mathfrak{p}$ **splits completely** in L if there exist exactly N different primes of L lying above $\mathfrak{p}$. Then $\mathfrak{p}$ splits completely in L if and only if $G_{\mathfrak{P}} = 1$ because G permutes the primes $\mathfrak{P}/\mathfrak{p}$ transitively.

When L/K is abelian, then we have the following characterization of the fixed field of the decomposition group.

Corollary 2. *Let L/K be abelian with group G. Let $\mathfrak{p}$ be a prime of K, let $\mathfrak{P}$ be a prime of L lying above $\mathfrak{p}$, and let $G_{\mathfrak{P}}$ be its decomposition group. Let E be the fixed field of $G_{\mathfrak{P}}$. Then E is the maximal subfield of L containing k in which $\mathfrak{p}$ splits completely.*

Proof. Let

$$G = \bigcup_{i=1}^{r} \sigma_i G_{\mathfrak{P}}$$

be a coset decomposition. Let $\mathfrak{q} = \mathfrak{P} \cap E$. Since a Galois group permutes the primes lying above a given prime transitively, we know that $\mathfrak{P}$ is the only prime of L lying above $\mathfrak{q}$. For each i, the prime $\sigma_i\mathfrak{P}$ is the only prime lying above $\sigma_i\mathfrak{q}$, and since $\sigma_1\mathfrak{P}, \ldots, \sigma_r\mathfrak{P}$ are distinct, it follows that the primes $\sigma_1\mathfrak{q}, \ldots, \sigma_r\mathfrak{q}$ are distinct. Since G is abelian, the primes $\sigma_i\mathfrak{q}$ are primes of E, and $[E : K] = r$, so that $\mathfrak{p}$ splits completely in E. Conversely, let F be an intermediate field between K and L in which $\mathfrak{p}$ splits completely, and let H be the Galois group of L/F. If $\sigma \in G_{\mathfrak{P}}$ and $\mathfrak{P} \cap F = \mathfrak{P}_F$, then σ leaves $\mathfrak{P}_F$ fixed. However, the decomposition group of $\mathfrak{P}_F$ over $\mathfrak{p}$ must be trivial since $\mathfrak{p}$ splits completely in F. Hence the restriction of σ to F is the identity, and therefore $G_{\mathfrak{P}} \subset H$. This proves that $F \subset E$, and concludes the proof of our corollary.

Let L/K be an arbitrary Galois extension again.

Assume now that the residue class field $R/\mathfrak{p}$ is finite, with q elements. We also write $q = \mathrm{N}\mathfrak{p}$. It is a power of the prime number p lying in $\mathfrak{p}$. By the theory of finite fields, there exists a unique automorphism of $S/\mathfrak{P}$ over $R/\mathfrak{p}$ which generates the Galois group of the residue class field extension, and has the effect

$$x \mapsto x^{\mathrm{N}\mathfrak{p}}.$$

In terms of congruences, we can write this automorphism $\bar{\sigma}$ as

$$\sigma x \equiv x^{\mathrm{N}\mathfrak{p}}, \qquad x \in S.$$

By what we have just seen, there exists a coset $\sigma T_{\mathfrak{P}}$ of $T_{\mathfrak{P}}$ in $G_{\mathfrak{P}}$ which induces $\bar{\sigma}$ on the residue class field extension. Any element of this coset will be called a **Frobenius automorphism** of $\mathfrak{P}$, and will be denoted by $(\mathfrak{P}, L/K)$. If the inertia group $T_{\mathfrak{P}}$ is trivial, then $(\mathfrak{P}, L/K)$ is uniquely determined as an element of the decomposition group $G_{\mathfrak{P}}$.

If $\mathfrak{Q}$ is another prime lying above $\mathfrak{p}$, and $\eta \in G$ is such that $\eta\mathfrak{P} = \mathfrak{Q}$, then the decomposition group of $\mathfrak{Q}$ is given by

$$G_{\mathfrak{Q}} = G_{\eta\mathfrak{P}} = \eta G_{\mathfrak{P}} \eta^{-1}.$$

Similarly for the inertia group, and for a Frobenius automorphism,

$$(\eta\mathfrak{P}, L/K) = \eta(\mathfrak{P}, L/K)\eta^{-1}.$$

This is immediately verified from the definitions. Furthermore, if $T_{\mathfrak{P}}$ is trivial, we see that $(\mathfrak{P}, L/K) = 1$ if and only if $\mathfrak{p}$ splits completely, meaning that $G_{\mathfrak{P}} = 1$.

If L/K is abelian, and if the inertia group $T_{\mathfrak{P}}$ is trivial for one of the $\mathfrak{P}|\mathfrak{p}$ (and hence for all $\mathfrak{P}|\mathfrak{p}$), it follows that to each $\mathfrak{p}$ in K we are able to associate a uniquely determined element of G, lying in $G_{\mathfrak{P}}$ (the same for all $\mathfrak{P}|\mathfrak{p}$), which we denote by

$$\sigma = (\mathfrak{p}, L/K),$$

and call the **Artin automorphism** of $\mathfrak{p}$ in G. It is characterized by the congruence

$$\sigma x \equiv x^{\mathrm{N}\mathfrak{p}} \pmod{\mathfrak{P}}, \qquad x \in S.$$

By using Zorn's lemma, one can easily extend the above results to infinite Galois algebraic extensions L/K. Propositions 1 through 4 are valid in this case, and we therefore also get a Frobenius automorphism $(\mathfrak{P}, L/K)$, well-defined modulo the inertia group $T_{\mathfrak{P}}$.

Consider the finite Galois case, not necessarily abelian, and let

$$\sigma_{\mathfrak{P}} = (\mathfrak{P}, L/K)$$

be the Frobenius automorphism of $\mathfrak{P}$. We assume that $\mathfrak{P}$ is unramified, so $\sigma_{\mathfrak{P}}$ is well defined as an element of $G_{\mathfrak{P}}$. Suppose that S is given by generators over R,

$$S = R[x_1, \ldots, x_n].$$

Let τ be an element of the Galois group G such that

$$\tau x_i \equiv x_i^{N\mathfrak{p}} \pmod{\mathfrak{P}}$$

for all $i = 1, \ldots, n$. Suppose also that $\mathfrak{P}$ does not divide the discriminant of any x_i, i.e. does not divide the non-zero differences

$$\lambda_\nu x_i - \lambda_\mu x_i, \qquad \lambda_\nu, \lambda_\mu \in G.$$

Then $\tau = \sigma_{\mathfrak{P}}$ because

$$\tau x_i \equiv \sigma_{\mathfrak{P}} x_i \pmod{\mathfrak{P}}$$

whence $\tau x_i = \sigma_{\mathfrak{P}} x_i$ for all i, whence $\tau = \sigma_{\mathfrak{P}}$ because the x_i generate L over K.

§4. SUMMARY OF CLASS FIELD THEORY

The treatment of class field theory given in my *Algebraic Number Theory* is classical, and is the most suitable for the applications to this book. We summarize briefly the main theorems.

Let k be a number field which we assume for simplicity has no real conjugate, and let K be an abelian extension, say finite to begin with. If $\mathfrak{p}$ is a prime of k, and is unramified in K, then we can associate with $\mathfrak{p}$ the Frobenius automorphism

$$\sigma_{\mathfrak{p}} = (\mathfrak{p}, K/k)$$

in $\mathrm{Gal}(K/k)$. Let $\mathfrak{c}$ be an ideal of $\mathfrak{o}_k$, sufficiently highly divisible by all the primes of k which ramify in K, and let $I_k(\mathfrak{c})$ be the group of fractional ideals prime to $\mathfrak{c}$. We can extend the map $\mathfrak{p} \mapsto (\mathfrak{p}, K/k)$ to $I_k(\mathfrak{c})$ by multiplicativity, and then get a homomorphism called the Artin map,

$$I_k(\mathfrak{c}) \to \mathrm{Gal}(K/k)$$

which can be proved to be surjective.

Let $P_1(\mathfrak{c})$ be the subgroup of $I_k(\mathfrak{c})$ consisting of those principal ideals (α), where

$$\alpha \equiv 1 \pmod{^* \mathfrak{c}}.$$

This means that $\alpha \equiv 1 \pmod{\mathfrak{m}_{\mathfrak{p}}^{r(\mathfrak{p})}}$, where $\mathfrak{m}_{\mathfrak{p}}$ is the maximal ideal of the local ring $\mathfrak{o}_{\mathfrak{p}}$ at $\mathfrak{p}$, for $\mathfrak{p}|\mathfrak{c}$, and $r(\mathfrak{p})$ is the order of $\mathfrak{c}$ at $\mathfrak{p}$. Let $\mathfrak{N}(\mathfrak{c})$ denote the group generated by the norms of all prime ideals of K, relatively prime to $\mathfrak{c}$. Then the kernel of the Artin map is precisely

$$P_1(\mathfrak{c})\mathfrak{N}(\mathfrak{c}),$$

and this is Artin's reciprocity law.

This can also be formulated in terms of ideles. An idele s is an element of the restricted product

$$\prod{}' k_v^*$$

of the multiplicative groups of the completions k_v, at all absolute values of k, extending the ordinary absolute value on $\mathbf{Q}$, or the p-adic absolute value on $\mathbf{Q}$, such that $|p|_p = 1/p$. The non-archimedean absolute values of k correspond then to the prime ideals of $\mathfrak{o}_k$. The restriction in the product means that we take elements

$$s = (\ldots, s_{\mathfrak{p}}, \ldots)$$

such that $s_{\mathfrak{p}}$ is a $\mathfrak{p}$-unit for almost all (all but a finite number of) primes $\mathfrak{p}$. We define the Artin symbol for ideles $(s, K/k)$ as follows. We select $\alpha \in k$ such that the idele αs having $\mathfrak{p}$-component $\alpha s_{\mathfrak{p}}$, is such that $\alpha s_{\mathfrak{p}}$ is very close to 1 at all $\mathfrak{p}$ ramified in K. (Close to 1 is determined by the same type of congruence that defines $P_1(\mathfrak{c})$.) We then define the ideal

$$(\alpha s) = \prod_{\mathfrak{p}} \mathfrak{p}^{m(\mathfrak{p})}$$

where $m(\mathfrak{p})$ is the order of $\alpha s_{\mathfrak{p}}$ at $\mathfrak{p}$. The symbol $(s, K/k)$ is defined to be

$$(s, K/k) = ((\alpha s), K/k).$$

This symbol is well defined, and gives a homomorphism of the idele group J_k onto $\mathrm{Gal}(K/k)$, whose kernel is generated by k^*, embedded on the diagonal in the ideles, and the group of norms of ideles from K, i.e. the kernel is

$$k^* N_{K/k} \mathbf{A}_K^*.$$

The norm is defined in a natural way, which is irrelevant for us here.

If $K' \supset K \supset k$ are abelian extensions, then the restriction of $(s, K'/k)$ to K is exactly $(s, K/k)$. This consistency allows us to define (s, k) in the Galois group of k_{ab} over k, i.e. the Galois group of the maximal abelian extension of k.

Given a prime $\mathfrak{p}$, we can consider the values of the Artin map at ideles

$$(\ldots, 1, 1, s_{\mathfrak{p}}, 1, 1, \ldots)$$

with component 1 except at $\mathfrak{p}$. These values lie in the decomposition group of a prime $\mathfrak{P}$ lying above $\mathfrak{p}$, and give an injective homomorphism of $k_{\mathfrak{p}}^*$ onto a dense subgroup of this decomposition group. The mapping is surjective for every finite abelian extension. This local fact will not be needed, except for one application, and the reader may disregard it until he needs it.

Over the rational numbers, it is easy to describe what's going on in elementary terms. Consider a cyclotomic extension $\mathbf{Q}_n = \mathbf{Q}(\zeta_n)$ where ζ_n is a primitive n-th root of unity. The ideals of $\mathbf{Z}$ are all principal. We get the Artin map as follows. If $a \in \mathbf{Z}$ and $a > 0$, and a is prime to n, then $((a), \mathbf{Q}_n/\mathbf{Q})$ is that automorphism σ such that

$$\zeta_n^\sigma = \zeta_n^a.$$

In particular, for a prime $p \nmid n$ we have

$$\zeta_n^\sigma = \zeta_n^p.$$

We see that the decomposition law of p in $\mathbf{Q}_n$ takes place according to an arithmetic progression. The congruence relations defining the generalized ideal class groups extend this notion to arbitrary number fields.

Finally we recall the characterization of Galois extensions by the nature of the primes which split completely in them. Let M be a set of primes. One defines the limit

$$\lim_{s \to 1+} \frac{\displaystyle\sum_{\mathfrak{p} \in M} \frac{1}{\mathbf{N}\mathfrak{p}^s}}{\log \dfrac{1}{s-1}}$$

to be the **Dirichlet density** of M (if it exists). It is provable (e.g. from class field theory, cf. my *Algebraic Number Theory*, Chapter VIII, §4) that an ideal class of $I(\mathfrak{c})/P_1(\mathfrak{c})$ always has such a density, and that this density is equal to $1/h_{\mathfrak{c}}$ (where $h_{\mathfrak{c}}$ is the order of $I(\mathfrak{c})/P_1(\mathfrak{c})$). Let S, T be sets of primes in k. Let us write $S \prec T$ if there exists a set Z of primes of Dirichlet density 0, contained in S, such that $S - Z \subset T$. Thus S is contained in T except for a set of primes of density 0.

Let K/k be a Galois extension and let $S_{K/k}$ be the set of primes of k which split completely in K. If $L \supset K$ is another Galois extension of k, then trivially, $S_{L/k} \subset S_{K/k}$. If

$$S_{K/k} \prec S_{L/k},$$

then $L = K$. Indeed, $S_{L/k}$ has density $1/[L:k]$, and hence

$$[L:k] \leqq [K:k],$$

so $L = K$. One can then prove (see e.g. [B7], Theorem 9, Chapter 8, §4):

Let K/k be a Galois extension, and E a finite extension of k. Then $S_{K/k} \prec S_{E/k}$ if and only if $E \subset K$.

One can then characterize the **ray class field** *belonging to an ideal* $\mathfrak{c}$ (or as one says, with conductor $\mathfrak{c}$) *as the abelian extension K of k such that $S_{K/k}$ consists precisely of those primes lying in the unit class of $I(\mathfrak{c})/P_1(\mathfrak{c})$, i.e. those primes which are principal, generated by an element $\alpha \equiv 1 \pmod{\mathfrak{c}}$.*

For some purposes (i.e. for our construction of abelian extensions in Chapter 10, §1) this characterization suffices. Later, when we analyze the nature of the Artin automorphism in terms of its effect on the values of certain analytic functions, such a characterization is of course insufficient, and one must know some of the other statements of class field theory as well to understand fully what's going on.

9 Reduction of Elliptic Curves

§1. NON-DEGENERATE REDUCTION, GENERAL CASE

The properties of reduction in this chapter, except for §3, are due to Deuring, who used them to give his algebraic proofs for complex multiplication. We shall not give any proofs. These can be given ad hoc, as Deuring did, for the elliptic curves, or one can develop a general reduction theory, as in Shimura [39]. No matter what, it is a pain to lay these foundations, but the results can be stated simply. Although classically one reduces over a discrete valuation ring, it is useful to deal with an arbitrary local ring.

Let $\mathfrak{o}$ be a local ring (always without divisors of zero), with maximal ideal $\mathfrak{m}$. An elliptic curve A defined by an irreducible non-singular equation

$$f(X_0, X_1, X_2) = 0$$

in projective space, with coefficients in $\mathfrak{o}$, is said to have **non-degenerate reduction** mod $\mathfrak{m}$ if when we reduce f mod $\mathfrak{m}$ we obtain again an absolutely irreducible equation, defining again a curve without singularities, denoted by $\bar{A}$.

If the curve is defined by a Weierstrass equation

$$y^2 = 4x^3 - g_2 x - g_3,$$

with $g_2, g_3 \in \mathfrak{o}$, and the characteristic of $\mathfrak{o}/\mathfrak{m}$ is not 2 or 3, then non-degenerate reduction means that the discriminant Δ is a unit in $\mathfrak{o}$. For our purposes, the reader can always restrict himself to this case.

If K is a field containing $\mathfrak{o}$ and

$$w \mapsto \bar{w}$$

denotes a place of K extending the canonical homomorphism $\mathfrak{o} \to \mathfrak{o}/\mathfrak{m}$, then this place induces a homomorphism

$$A_K \to \bar{A}_{\bar{K}}$$

of the K-rational points of A into the $\bar{K}$ rational points of $\bar{A}$, by applying the bar to the coordinates of points. If the curve is given in Weierstrass form as above, the map on points is given by

$$(x, y) \mapsto (\bar{x}, \bar{y}),$$

If $\bar{x}$ or $\bar{y} = \infty$, then the point with coordinates (x, y) lies in the kernel of our homomorphism. Suppose that the points of period N on A are rational over K. Let p be the characteristic of $\mathfrak{o}/\mathfrak{m}$. If N is a positive integer prime to p, then the map

$$A_N \to \bar{A}_N$$

is an isomorphism. Essentially this is due to the fact that we can obtain the points of A_N as an inverse image

$$(N\delta)^{-1}(0) = pr_1[\Gamma_{N\delta} \,.\, (A \times 0)]$$

and that reduction mod $\mathfrak{m}$ commutes with the operations of algebraic geometry, especially inverse images. This shows that the points of A_N map onto the points of $\bar{A}_N$, and since these two abelian groups have the same number of elements, we must get an isomorphism between them.

§2. REDUCTION OF HOMOMORPHISMS

Let A, B be elliptic curves with non-degenerate reductions $\bar{A}$ and $\bar{B}$ over a local ring $\mathfrak{o}$ as before. We know that $\mathrm{Hom}(A, B)$ is finitely generated. In fact, in characteristic 0, it has at most rank 2 over $\mathbf{Z}$, and this will be the main case of interest to us. If $\lambda\colon A \to B$ is a homomorphism, then λ is defined over an algebraic extension L of the quotient field K of $\mathfrak{o}$. However, it can be shown that for any place extending the canonical homomorphism $\mathfrak{o} \to \mathfrak{o}/\mathfrak{m}$ to L, λ has a non-degenerate reduction $\bar{\lambda}\colon \bar{A} \to \bar{B}$, and that the association

$$\lambda \mapsto \bar{\lambda}$$

is an injective homomorphism

$$\mathrm{Hom}(A, B) \to \mathrm{Hom}(\bar{A}, \bar{B}).$$

Warning. This last map is not necessarily a surjection. Two significant cases arise: when A, B have transcendental j-invariant, but reduce to special elliptic curves over the complex numbers, having invariant $j(\tau)$ with imaginary quadratic τ; and when A, B are already special, but reduce to elliptic curves in characteristic p, and then pick up new endomorphisms besides those arising from complex multiplications. We shall study both cases. The first is the theory of complex multiplication proper. The second has its genesis in the Deuring theory as in Chapter 13.

We can give a heuristic motivation for the fact that the reduction of a homomorphism is also a homomorphism. Let Γ be the graph of λ. Then $\Gamma \subset A \times B$ and $pr_1\Gamma = A$. If reduction is to preserve the operations of algebraic geometry, we must have $pr_1\bar{\Gamma} = \bar{A}$. Also $\bar{\Gamma}$ has to be connected (being a deformation of Γ), whence $\bar{\Gamma}$ is also the graph of a mapping from A into B.

Considering the intersection

$$\Gamma \cdot (A \times Q)$$

with a general point Q of B, and the fact that the degree of this cycle (i.e. the number of points in it, counting multiplicities) is the degree of λ, we see that reduction being compatible with intersection implies that $\bar{\lambda}$ and λ have the same degree.

Suppose that the characteristic of the residue class field is not 2 or 3, and that $j \in \mathfrak{o}$ but $j \not\equiv 0$ or 1728 mod $\mathfrak{m}$. We can find an elliptic curve defined by the equation

$$y^2 = 4x^3 - cx - c,$$

having the given invariant j, and non-degenerate reduction mod $\mathfrak{m}$, by solving linearly

$$c = \frac{27j}{j - 1728},$$

and we see that this gives a "universal" parametrization for such curves. For the other two cases, we can always take

$$y^2 = 4x^3 - x \qquad \text{and} \qquad y^2 = 4x^3 - 1.$$

Let A be an elliptic curve in characteristic 0, defined over the local ring $\mathfrak{o}$, and with non-degenerate reduction. Let $\mathfrak{g}$ be a finite subgroup of A. Then $A/\mathfrak{g}$ has many models. Its invariant is integral over $\mathbf{Z}[j_A]$, and therefore integral over $\mathfrak{o}$, because $j_A \in \mathfrak{o}$. We can therefore find a model B for $A/\mathfrak{g}$ defined over an integral extension S of $\mathfrak{o}$, and having non-degenerate reduction at every maximal ideal of S lying above $\mathfrak{m}$, by writing down the usual simple equations as in Chapter 1, §4 (and assuming for our purposes that the characteristic of $\mathfrak{o}/\mathfrak{m}$ is $\neq 2, 3$, although one can also give normalized equations valid in these cases.)

§3. COVERINGS OF LEVEL N

Theorem 1. *Let A be an elliptic curve defined over an integrally closed local ring $\mathfrak{o}$, with non-degenerate reduction modulo the maximal ideal $\mathfrak{m}$. Let p be the characteristic of $\mathfrak{o}/\mathfrak{m}$, and let N be prime to p. Let K be the quotient field of $\mathfrak{o}$. Let $G = \mathrm{Gal}(K(A_N)/K)$. Let $\mathfrak{M}$ be a maximal ideal of the* *integral closure* S *of* $\mathfrak{o}$ *in* $K(A_N)$, *and let the bar*, $w \mapsto \bar{w}$, *denote reduction* mod $\mathfrak{M}$, *for* $w \in S$. *Then:*

i) *The ideal* $\mathfrak{m}$ *is unramified in* $K(A_N)$.

ii) *For any* $\sigma \in G_{\mathfrak{M}}$ *and* $P \in A_N$ *we have*

$$\overline{\sigma P} = \bar{\sigma}\bar{P}.$$

iii) *If* $\sigma \in G$ *and* $\overline{\sigma P} = \bar{P}$ *for all* $P \in A_N$, *then* $\sigma = 1$.

Proof. The formula of (ii) holds by the definition of the effect $\bar{\sigma}$ on the residue class field extension. Since the map $P \mapsto \bar{P}$ is an injection on A_N, we conclude from the hypotheses of (iii) that $\sigma P = P$, whence $\sigma = 1$ because the coordinates of points in A_N generate $K(A_N)$. *Note:* In (iii), we do not assume that σ is necessarily in $G_{\mathfrak{M}}$. This is useful in applications. The fact that $\mathfrak{m}$ is unramified in $K(A_N)$ follows from (iii).

In applications, we are sometimes given elements $\sigma_1, \sigma_2 \in G$ such that

$$\overline{\sigma_1 P} = \overline{\sigma_2 P}$$

for all $P \in A_N$. Considering $\sigma_2^{-1}\sigma_1$ shows that $\sigma_1 = \sigma_2$.

Corollary. *Let* A *have invariant* $j \in \mathfrak{o}$, *and such that* $\bar{j} \neq 0$, 12^3, *and the characteristic of* $\mathfrak{o}/\mathfrak{m}$ *is* $\neq 2, 3$. *Let* h *be the first Weber function*, i.e. $g_2 g_3 x/\Delta$. *If* $\sigma \in G$ *is such that*

$$\overline{\sigma h(P)} = \overline{h(P)}$$

for all $P \in A_N$, *then* σ *is the identity on* $K(h(A_N))$.

Proof. We have $\sigma h(P) = h(Q)$ for some point $Q \in A_N$. By hypothesis, we get

$$\overline{h(P)} = \overline{h(Q)}, \qquad \text{i.e.} \quad \bar{h}(\bar{P}) = \bar{h}(\bar{Q}),$$

where $\bar{h}$ is the Weber function of the reduced curve $\bar{A}$. This means that $\bar{Q} = \pm\bar{P}$ (because the x-coordinates of $\bar{Q}$ and $\bar{P}$ are the same), whence $Q = \pm P$. Hence $h(Q) = h(P)$, so that

$$\sigma h(P) = h(P).$$

This being true for all $P \in A_N$, we conclude that $\sigma = 1$ on $K(h(A_N))$.

Next we deal with the two exceptional cases, and we shall take values in characteristic zero.

We shall see later that $J^{\frac{1}{3}}$ and $\sqrt{J-1}$ are modular functions of level 6, essentially from the product expansion for Δ, which shows that $\Delta^{\frac{1}{3}}$ and $\Delta^{\frac{1}{2}}$ are holomorphic on the upper half plane. This means in terms of points of finite order that the field F_N is ramified of order 3 over $J = 0$, and ramified of order 2 over $J = 1$ if $6|N$. See Chapter 18, §5. From this, we shall prove it is true for all N, and we shall determine the decomposition group.

No other ramification than the above can occur (for finite values of j), because we can define an elliptic curve by the equation

$$y^2 = 4x^3 - 3\sqrt[3]{J}x - \sqrt{J-1},$$

which has obviously non-degenerate reduction over $J \mapsto 1$ and $J \mapsto 0$. Theorem 1 shows that the extension by the coordinates of the points of order N over $\mathbf{Q}(\sqrt[3]{J}, \sqrt{J-1})$ is then unramified, except at infinity.

The ramification at infinity will become clear in Chapter 15, using a different parametrization for the points of the elliptic curve.

As before, we call

$$\frac{g_2^3}{\Delta}x^2 \quad \text{and} \quad \frac{g_3}{\Delta}x^3$$

the **second and third Weber functions** respectively, defined for an elliptic curve in Weierstrass form by the above formulas. We let F_N be the field of modular functions of level N, identified with the field $\mathbf{Q}(j, h(A_N))$, where A is an elliptic curve with invariant j, and h is the first Weber function.

Theorem 2. The field F_N (for $N > 1$) is ramified over $\mathbf{Q}(j)$ at $j = 12^3$, with ramification index 2. Let h be the second Weber function. Let $\mathfrak{M}$ be a maximal ideal of the integral closure of $\mathbf{Q}[j]$ in F_N lying above the ideal $(j - 12^3)$, and let the bar denote reduction mod $\mathfrak{M}$. Let $T_{\mathfrak{M}}$ be the inertia group. An element $\sigma \in \mathrm{Gal}(F_N/\mathbf{Q}(j))$ is such that

$$\overline{\sigma h(P)} = \overline{h(P)}$$

for all $P \in A_N$, if and only if σ lies in $T_{\mathfrak{M}}$.

Proof. Take N sufficiently large first (with respect to divisibility) to insure that F_N is ramified at $j = 12^3$. We can represent σ by a matrix operating on A_N. If $\overline{\sigma h(P)} = \overline{h(P)}$ for all $P \in A_N$, then we must have in the analytic representation

$$\wp^2\left(\left(\frac{r}{N}, \frac{s}{N}\right)\begin{pmatrix} i \\ 1 \end{pmatrix}\right) = \wp^2\left(\left(\frac{r}{N}, \frac{s}{N}\right)\begin{pmatrix} a & b \\ c & d \end{pmatrix}\begin{pmatrix} i \\ 1 \end{pmatrix}\right)$$

for all integers r, s not both 0 (mod N), and a, b, c, d are the components of the matrix representing σ. Putting r, s equal to 0, 1 respectively, one sees that this can be so if and only if the matrix represents multiplication by ± 1, $\pm i$. In the case of i, the matrix is

$$\begin{pmatrix} 0 & -1 \\ 1 & 0 \end{pmatrix}.$$

Since ± 1 operates trivially on F_N, we see that only $\pm i$ yields a possible non-trivial automorphism of F_N. We know that there is a non-trivial inertia group $T_{\mathfrak{M}}$, whence its generator is necessarily represented by such a matrix. Now for any $n|N$, the same matrix operating on A_n represents the restriction of σ to F_n, and operates non-trivially, so that we must also have ramification of order 2 in

F_n. Given n, we can always find N divisible by n so that we can argue as above. This proves our theorem.

Theorem 3. *The field F_N (for $N > 1$) is ramified over $\mathbf{Q}(j)$ at $j = 0$ with ramification index 3. Let h be the third Weber function. Let $\mathfrak{M}$ be a maximal ideal of the integral closure of $\mathbf{Q}[j]$ in F_N lying above the ideal (j), and let the bar denote reduction* mod $\mathfrak{M}$. *Let $T_{\mathfrak{M}}$ be the inertia group of $\mathfrak{M}$. An element $\sigma \in \mathrm{Gal}(F_N/\mathbf{Q}(j))$ is such that*

$$\overline{\sigma h(P)} = \overline{h(P)}$$

for all $P \in A_N$ if and only if σ lies in $T_{\mathfrak{M}}$.

Proof. The proof is completely analogous, except that this time σ is represented by the matrix corresponding to multiplication by ρ or ρ^2, where $\rho = e^{2\pi i/3}$, e.g.

$$\begin{pmatrix} -1 & -1 \\ 1 & 0 \end{pmatrix}.$$

We stated Theorem 2 and Theorem 3 in terms of points of finite order. We can also state them in terms of modular functions.

Theorem 2′. *Let F_N be the field of modular functions of level $N > 1$. Let z be equivalent to i under the modular group in $\mathfrak{H}$. Let*

$$f_a(\tau) = \frac{g_2^2(\tau)}{\Delta(\tau)} \wp^2\left(a\begin{pmatrix} \tau \\ 1 \end{pmatrix}; \tau, 1\right)$$

be the second Fricke functions, with $a \in (\mathbf{Q}^2/\mathbf{Z}^2)_N$, $a \neq 0$. If $\sigma \in \mathrm{Gal}(F_N/F_1)$ is such that

$$(\sigma f_a)(z) = f_a(z)$$

for all $a \in (\mathbf{Q}^2/\mathbf{Z}^2)_N$, $a \neq 0$, then

$$(\sigma f)(z) = f(z)$$

for all functions $f \in F_N$ which are defined at z. The group of such σ is cyclic of order 2, and consists of those elements represented by matrices $\gamma \in SL_2(\mathbf{Z})$ such that $\gamma z = z$.

Theorem 3′. *Let F_N be as above, and let z be equivalent to ρ under the modular group in $\mathfrak{H}$. Let*

$$f_a(\tau) = \frac{g_3(\tau)}{\Delta(\tau)} \wp^3\left(a\begin{pmatrix} \tau \\ 1 \end{pmatrix}; \tau, 1\right)$$

be the third Fricke functions with $a \in (\mathbf{Q}^2/\mathbf{Z}^2)_N$, $a \neq 0$. If $\sigma \in \mathrm{Gal}(F_N/F_1)$ is such that

$$(\sigma f_a)(z) = f_a(z)$$

for all $a \in (\mathbf{Q}^2/\mathbf{Z}^2)_N$, $a \neq 0$, then

$$(\sigma f)(z) = f(z)$$

for all functions $f \in F_N$ which are defined at z. The group of such σ is cyclic of order 3, and consists of those elements represented by matrices $\gamma \in SL_2(\mathbf{Z})$ such that $\gamma z = z$.

§4. REDUCTION OF DIFFERENTIAL FORMS

Let V be a curve (always projective non-singular) over a field k_0. One can define differential forms in the function field $k_0(V)$ as the dual space of the derivations of $k_0(V)$ which are trivial on k_0. Suppose that $k_0(V) = k_0(x, y)$ where x is transcendental over k and y is separable algebraic over $k_0(x)$. Then the differential forms are a 1-dimensional space over $k_0(V)$ and dx is a $k_0(V)$-basis for this space, where dx has the effect Dx on the derivation D. Any differential form of $k_0(V)$ is of type zdx for some $z \in k_0(V)$.

One can define in the usual manner the zeros and poles of a differential form, expanding in a power series with a local parameter at a given point.

If A is an elliptic curve in Weierstrass form

$$y^2 = 4x^3 - g_2x - g_3,$$

then dx/y is a differential form of the first kind, in other words, it has no pole. Over the complex numbers, under the Weierstrass parametrization

$$u \mapsto (1, \wp(u), \wp'(u)),$$

the differential form dx/y corresponds to the differential form du on $\mathbf{C}/L$, as one sees immediately from $x = \wp(u)$ and $y = \wp'(u)$.

Back to a general curve over a field k_0. Let

$$f\colon V \to W$$

be a rational map of V onto another curve, and suppose that f is not constant. Then $k_0(W)$ is contained in $k_0(V)$, and a differential form on W pulls back to a differential form on V. If

$$\omega = z\,dx$$

with $z, x \in k_0(W)$, then we may view z, x as functions on V (i.e. as $z \circ f, x \circ f$) and we then get a differential form

$$\omega \circ f = f^*\omega = (z \circ f)\, d(x \circ f)$$

which we also write as $z\,dx$ by abuse of notation. If the map f is separable, i.e. $k_0(V)$ is a separable extension of $k_0(W)$, and $z\,dx \neq 0$, then $f^*\omega \neq 0$. On the other hand, if $z\,dx \neq 0$ but f is not separable, i.e. $k_0(V)$ over $k_0(W)$ has inseparable degree > 1, then $f^*\omega = 0$.

A typical example of an inseparable extension is obtained as follows. Suppose that k_0 has characteristic p, and let $k_0(V) = k_0(x, y)$ where y is separable over $k_0(x)$. Then $k_0(x)$ is purely inseparable over $k_0(x^p)$, of degree p. Furthermore, since y is separable over $k_0(x)$, we have $k_0(x, y) = k_0(x, y^p)$, and y^p is separable over $k_0(x^p)$. From the diagram

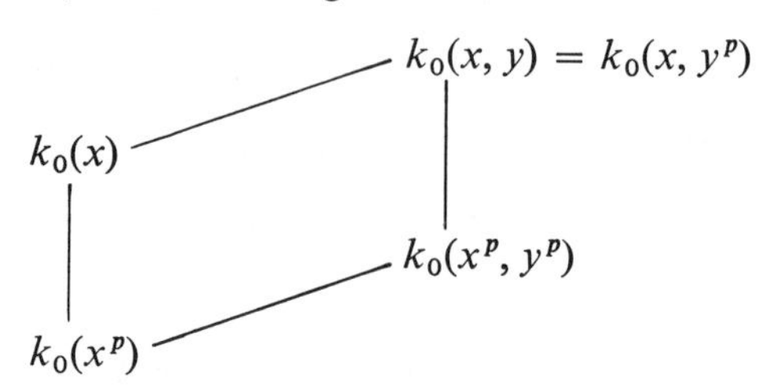

we conclude that $[k_0(x^p, y^p) : k_0(x^p)] = [k_0(x, y) : k_0(x)]$, and that

$$[k_0(x, y) : k_0(x^p, y^p)] = p.$$

The subfield $k_0(x^p, y^p)$ is the function field of a curve denoted by $V^{(p)}$ and we have a purely inseparable rational map

$$\pi_p \colon V \to V^{(p)},$$

called the **Frobenius** map. Similarly, if $q = p^r$ is a power of p, we get a rational map π_q of degree q, purely inseparable, sending

$$(x, y) \mapsto (x^q, y^q).$$

Suppose that k_0 is perfect, so that $k_0^p = k_0$. Then raising to the q-th power gives an isomorphism of $k_0(x, y)$ onto $k_0(x^q, y^q)$. It follows that there is precisely one subfield of $k_0(x, y)$ over which $k_0(x, y)$ is purely inseparable of degree q, and that is $k_0(x^q, y^q)$.

If $V = A$ is an elliptic curve, the map π_q is a homomorphism of elliptic curves.

Let

$$\lambda \colon A \to B$$

be an isogeny, defined over a field k_0. It can be shown that the space of differential forms on A (or on B) defined over k_0, and of the **first kind** (i.e., without pole), is 1-dimensional *over* k_0. Consequently if ω_B is a non-zero differential form of the first kind on B, we conclude that

$$\lambda^*\omega_B = c\omega_A,$$

where $c \in k_0$. Furthermore λ is separable if and only if $c \neq 0$.

Actually we want the dependence of c on λ, so let us write

$$\lambda^*\omega_B = c_\lambda\omega_A.$$

Then $\lambda \mapsto c_\lambda$ is a homomorphism of the subgroup of $\mathrm{Hom}(A, B)$ consisting of those homomorphisms defined over k_0, into the constant field. Observe that $c = c_\lambda$ is independent of the choice of differential form $\omega_B \neq 0$, because any other form of the first kind on B is a constant multiple of ω_B. A similar remark applies to ω_A. Thus $\lambda \mapsto c_\lambda$ is a well-defined representation.

Let $\varphi: \mathbf{C}/L \to A_\mathbf{C}$ be an analytic representation. Let α be a complex number such that $\alpha L \subset L$, and $\lambda: A \to A$ an endomorphism of A making the following diagram commutative.

$$\begin{array}{ccc} \mathbf{C} & \to & A_\mathbf{C} \\ {\scriptstyle\alpha}\downarrow & & \downarrow{\scriptstyle\lambda} \\ \mathbf{C} & \to & A_\mathbf{C} \end{array}$$

Then for any differential form of the first kind ω on A we have

$$\omega \circ \lambda = \alpha\omega.$$

Thus the number α is the number c_λ mentioned above.

Suppose that $\mathrm{End}(A)_\mathbf{Q}$ is isomorphic to an imaginary quadratic field k (a subfield of the complex numbers). We can therefore define an isomorphism

$$\theta: k \to \mathrm{End}(A)_\mathbf{Q}$$

such that

$$\omega \circ \theta(\alpha) = \alpha\omega$$

for all differential forms of the first kind ω, and all $\alpha \in k$. If this condition is satisfied, we call the pair (A, θ) a **normalized pair.** We have some easy functorial properties.

DIFF 1. *If (A, θ) and (A', θ') are normalized pairs, and*

$$\lambda: A \to A'$$

is a homomorphism, then

$$\lambda \circ \theta(\alpha) = \theta'(\alpha) \circ \lambda$$

for all $\alpha \in k$.

This is obvious, because $\omega \circ \lambda \circ \theta(\alpha) = \alpha\omega \circ \lambda$ and $\omega \circ \theta'(\alpha) \circ \lambda = \alpha\omega \circ \lambda$, by the definitions of normalization.

DIFF 2. *If (A, θ) is normalized and if σ is an isomorphism of the field over which A and all elements of* End(A) *are defined, then $(A^\sigma, \theta^\sigma)$ is also normalized.*

The proof is immediate.

DIFF 3. *Let (A, θ) be normalized. If A is defined over $k_0 \subset \mathbf{C}$, then every element of* End(A) *is defined over k_0k.*

Proof. We can find a differential form of the first kind defined over k_0. Let σ be an automorphism of $\mathbf{C}$ over k_0k. Then $\omega^\sigma = \omega$, whence

$$\omega \circ \theta(\alpha)^\sigma = \omega^\sigma \circ \theta(\alpha)^\sigma = (\omega \circ \theta(\alpha))^\sigma = (\alpha\omega)^\sigma = \alpha\omega = \omega \circ \theta(\alpha).$$

Hence $\theta(\alpha) = \theta(\alpha)^\sigma$ for all σ, whence $\theta(\alpha)$ is defined over k_0k.

To be able to reduce differential forms when we are given a non-degenerate reduction of $A \bmod \mathfrak{m}$, we must give A an integral structure over the local ring $\mathfrak{o}$. One can lay foundations working with rings instead of fields for differential forms, i.e. within the framework of schemes, or one can select generators (x, y) for the function field of A, as for instance in the Weierstrass form when the characteristic of $\mathfrak{o}/\mathfrak{m}$ is $\neq 2$ or 3, and then work ad hoc, as Deuring did, using such simple equations. It is then "clear" that the differential form $\omega = dx/y$ reduces properly, to the differential form $\bar{\omega} = d\bar{x}/\bar{y}$. For any element $c \in \mathfrak{o}$ we have

$$\overline{c\omega} = \bar{c}\bar{\omega}.$$

If A, B have non-degenerate reduction over $\mathfrak{o}$, with quotient field K, and

$$\lambda \colon A \to B$$

is a homomorphism defined over K (whence over $\mathfrak{o}$), and if ω is a differential form on B such that $\bar{\omega} \neq 0$, $\omega \circ \lambda = c\omega$, with $c \in \mathfrak{o}$, then

$$\overline{\omega \circ \lambda} = \bar{\omega} \circ \bar{\lambda} = \bar{c}\bar{\omega}.$$

In particular,

$$\overline{\omega \circ \lambda} = 0$$

if and only if $\bar{\lambda}$ is not separable. This is the case when, for instance, $\bar{\lambda}$ is the Frobenius endomorphism π_q for some $q = p^r$.

If an elliptic curve A is defined over K, and if we have a family of discrete valuations of K such that an element of K has only a finite number of zeros and poles in this family, then given a non-zero differential form on A, for all but a finite number of the discrete valuation rings in the family, the reduction of A is non-degenerate, and the differential form reduces to a non-zero differential form on $\bar{A}$. In all the sequel, we shall use reduction mostly in this case, omitting a finite set of bad primes. The family of all valuation rings in a number field K gives an example of such a family.

We already mentioned that $\bar{A}$ may have more endomorphisms than A. It is important in certain cases to know when an endomorphism of $\bar{A}$ is the reduction of an element in End(A). If (A, θ) is normalized, then we define

$$\bar{\theta} \colon k \to \mathrm{End}(\bar{A})$$

by

$$\bar{\theta}(\alpha) = \overline{\theta(\alpha)}.$$

DIFF 4. *Assume that* $\text{End}(A)_{\mathbf{Q}}$ *is an imaginary quadratic field. If an element of* $\text{End}(\bar{A})_{\mathbf{Q}}$ *commutes with all elements of* $\overline{\text{End}(A)}$, i.e. *with all the reduced endomorphisms of* A, *then it lies in* $\overline{\text{End}(A)}_{\mathbf{Q}}$.

To prove this, one has to know that $\text{End}(\bar{A})_{\mathbf{Q}}$ is either a quadratic field, or a division algebra of dimension 4 over $\mathbf{Q}$. This will be proved later, with ℓ-adic representations, assuming only that $\nu(N\delta) = N^2$. If one knows this result, we see that $\overline{\text{End}(A)}_{\mathbf{Q}}$ already provides a quadratic subfield of $\text{End}(\bar{A})_{\mathbf{Q}}$, whence **DIFF 4** follows.

10 Complex Multiplication

§1. GENERATION OF CLASS FIELDS, DEURING'S APPROACH

We first consider values of the j-function at quadratic imaginary numbers. We shall see that these values generate abelian extensions of quadratic fields.

Let k be an imaginary quadratic field and $\mathfrak{o}_k$ its ring of algebraic integers. We view j as the isomorphism invariant of elliptic curves. We don't need analysis, and if $A_{\mathbf{C}} \approx \mathbf{C}/L$ where $L = [z_1, z_2]$, $z = z_1/z_2 \in k \cap \mathfrak{H}$, then we write

$$j_A = j(L) = j(z) = j(\lambda L), \qquad \text{all } \lambda \in k^*.$$

Theorem 1. *Let $\mathfrak{a}$ be an ideal of $\mathfrak{o}_k$. Then $j(\mathfrak{a})$ generates an abelian extension of k, and in fact generates the maximal unramified abelian extension of k. If $\mathfrak{a}_i$ $(i = 1, \ldots, h)$ are representative ideals for the ideal classes in k, then the numbers $j(\mathfrak{a}_i)$ are all conjugate over k, and for all but a finite number of primes $\mathfrak{p}$ of k such that $(p) = \mathfrak{p}\mathfrak{p}'$ in k, $\mathfrak{p} \neq \mathfrak{p}'$ and $\mathbf{N}\mathfrak{p} = p$, we have the Kronecker congruence relation,*

$$j(\mathfrak{p}^{-1}\mathfrak{a}) \equiv j(\mathfrak{a})^p \pmod{\mathfrak{P}}$$

for any prime $\mathfrak{P}$ in $k(j(\mathfrak{a}))$ over $\mathfrak{p}$. Therefore, if $\sigma_{\mathfrak{p}}$ is the Artin automorphism of $\mathfrak{p}$ in $k(j(\mathfrak{a}))$, then

$$j(\mathfrak{p}^{-1}\mathfrak{a}) = \sigma_{\mathfrak{p}} j(\mathfrak{a}).$$

Proof. Let K be the smallest Galois extension of k containing all the numbers $j(\mathfrak{a}_i)$. For each $j(\mathfrak{a}_i)$ select an elliptic curve defined by a Weierstrass equation over K and having invariant $j(\mathfrak{a}_i)$. For any $\mathfrak{a}$ among the $\mathfrak{a}_i$, the corresponding elliptic curve is analytically isomorphic to $\mathbf{C}/\mathfrak{a}$ and we suppose given an analytic representation

$$\mathbf{C}/\mathfrak{a} \to A_{\mathbf{C}}.$$

Select a prime $p \neq 2, 3$, such that $(p) = \mathfrak{p}\mathfrak{p}'$ in k, $\mathfrak{p} \neq \mathfrak{p}'$, $\mathbf{N}\mathfrak{p} = p$, such that all the above elliptic curves have non-degenerate reduction at a prime $\mathfrak{P}$ lying above $\mathfrak{p}$ in K, and such that $\mathfrak{p}$ is relatively prime to the discriminants of the numbers $j(\mathfrak{a}_i)$, all i. If B is the elliptic curve chosen above whose invariant is

$j(\mathfrak{p}^{-1}\mathfrak{a})$, then there exists an isogeny $\lambda\colon A \to B$ such that the following diagram is commutative,

$$\begin{array}{ccc} \mathbf{C}/\mathfrak{a} & \longrightarrow & A_{\mathbf{C}} \\ \downarrow & & \downarrow \lambda \\ \mathbf{C}/\mathfrak{p}^{-1}\mathfrak{a} & \rightarrow & B_{\mathbf{C}} \end{array}$$

and the left vertical map is the canonical map arising from the inclusion $\mathfrak{a} \subset \mathfrak{p}^{-1}\mathfrak{a}$. Let $\mathfrak{b}$ be an ideal prime to $\mathfrak{p}$ and such that $\mathfrak{p}\mathfrak{b} = (\alpha)$ is principal. Then our diagram becomes

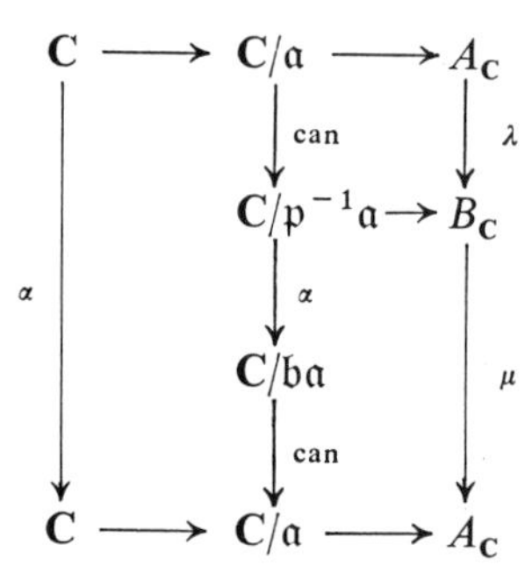

with some isogeny $\mu\colon B \to A$ which makes the diagram commutative. On the left we have multiplication by α. We compute degrees:

$$\nu(\lambda) = (\mathfrak{p}^{-1}\mathfrak{a} : \mathfrak{a}) = (\mathfrak{o}_k : \mathfrak{p}) = \mathbf{N}\mathfrak{p} = p$$
$$\nu(\mu) = (\mathfrak{a} : \mathfrak{b}\mathfrak{a}) = \mathbf{N}\mathfrak{b} \quad \text{prime to } p.$$

We contend that $\bar{\lambda}$ (reduction mod $\mathfrak{P}$) is purely inseparable. Let ω be a differential form of the first kind on B, say $\omega = dx/y$. Then

$$\omega \circ \mu \circ \lambda = \alpha\omega,$$

whence

$$\bar{\omega} \circ \bar{\mu} \circ \bar{\lambda} = \bar{\alpha}\bar{\omega} = 0,$$

because $\alpha \in \mathfrak{p}$. Hence $\bar{\mu} \circ \bar{\lambda}$ is not separable. But the degree of $\bar{\mu}$ (which is the degree of μ) is prime to p, and hence λ is not separable. Since λ has degree p, it follows that λ is purely inseparable.

It follows that $\bar{B}$ is isomorphic to $\bar{A}^{(p)}$. But the invariant of $A^{(p)}$ is $j_{\bar{A}}^{p}$ (by first principles, applying the isomorphism "raising to the p-th power"). Hence

$$\overline{j(\mathfrak{p}^{-1}\mathfrak{a})} = \overline{j(\mathfrak{a})}^{p}.$$

This means precisely the congruence relation of Kronecker. The Frobenius automorphism $\sigma_{\mathfrak{P}}$ has the same effect on $j(\mathfrak{a})$, and hence we must have the final equality

$$j(\mathfrak{p}^{-1}\mathfrak{a}) = \sigma_{\mathfrak{P}} j(\mathfrak{a}),$$

having chosen p prime to the discriminant of the numbers $j(\mathfrak{a})$.

We now see that the numbers $j(\mathfrak{a}_i)$ are all conjugate, because there is a prime of k of degree 1 over $\mathbf{Q}$ in every ideal class of k. Furthermore, a prime $\mathfrak{p}$ as above splits completely in K if and only if $\mathfrak{p}$ is principal, because j takes on the same value on two lattices if and only if they are linearly equivalent. Hence by class field theory, we conclude that K is the maximal unramified abelian extension of k. This proves Theorem 1.

The next thing to do is to prove a theorem analogous to that of Theorem 1 for points of finite order on an elliptic curve A having invariant $j_A = j(\mathfrak{a})$, where $\mathfrak{a}$ is an ideal of $\mathfrak{o}_k$. Recall that two analytic representations $\mathbf{C}/\mathfrak{a} \to A_{\mathbf{C}}$ differ by an automorphism of A. The points of finite order in $\mathbf{C}/\mathfrak{a}$ are obviously the points $k/\mathfrak{a}$. Those of order N are denoted as usual by $(k/\mathfrak{a})_N$.

We have the commutative diagram as in Theorem 1,

$$\begin{array}{ccc} \mathbf{C}/\mathfrak{a} & \longrightarrow & A_{\mathbf{C}} \\ {\scriptstyle\text{can}}\downarrow & & \downarrow{\scriptstyle\lambda} \\ \mathbf{C}/\mathfrak{p}^{-1}\mathfrak{a} & \rightarrow & A^{\sigma}_{\mathbf{C}} \end{array}$$

where the map on the left is the canonical one and λ is an isogeny such that $\bar{\lambda}$ is purely inseparable of degree p. We also have, by the definition of σ, that

$$\overline{A^{\sigma}} = \pi_p(\bar{A}) = \bar{A}^{(p)}.$$

Indeed, if A is defined by

$$y^2 = 4x^3 - g_2 x - g_3,$$

then A^{σ} is defined by

$$y^2 = 4x^3 - g_2^{\sigma} x - g_3^{\sigma},$$

and reducing mod $\mathfrak{P}$ yields the equation for $\overline{A^{\sigma}}$, namely

$$y^2 = 4x^3 - \bar{g}_2^{p} x - \bar{g}_3^{p}.$$

Hence there exists an automorphism $\bar{\varepsilon}$ of $\bar{A}^{(p)}$ such that

$$\bar{\lambda} = \bar{\varepsilon} \circ \pi.$$

We contend that $\bar{\varepsilon}$ is the reduction of some element in $\operatorname{Aut}(A^{\sigma})$. Since $\operatorname{End}(A^{\sigma}) \approx \mathfrak{o}_k$ is integrally closed, it will suffice to prove that $\bar{\varepsilon}$ lies in $\overline{\operatorname{End}(A^{\sigma})}_{\mathbf{Q}}$, and for that it will suffice to prove by **DIFF 4** that $\bar{\varepsilon}$ commutes with all the endomorphism of $\bar{A}^{(p)}$ obtained by reducing the endomorphism of A^{σ}. We may assume that (A, θ) is normalized, so that $(A^{\sigma}, \theta^{\sigma})$ is also normalized. Then by **DIFF 1** and **DIFF 2**, we get for any $\gamma \in k$,

$$\lambda \circ \theta(\gamma) = \theta(\gamma)^{\sigma} \circ \lambda,$$

whence

$$\bar{\lambda} \circ \overline{\theta(\gamma)} = \overline{\theta(\gamma)^{\sigma}} \circ \bar{\lambda}$$

and

$$\bar{\varepsilon} \circ \pi \circ \overline{\theta(\gamma)} = \overline{\theta(\gamma)^{\sigma}} \circ \bar{\varepsilon} \circ \pi. \tag{1}$$

But from the definition of the Frobenius mapping, we have

$$\overline{\theta(\gamma)^{\sigma}} = \overline{\theta(\gamma)}^{(p)},$$

whence

$$\bar{\varepsilon} \circ \pi \circ \overline{\theta(\gamma)} = \bar{\varepsilon} \circ \overline{\theta(\gamma)^{\sigma}} \circ \pi. \tag{2}$$

Comparing the right-hand sides of (1) and (2) proves what we want.

We can then change λ by ε^{-1} in order to achieve the more precise relation

$$\bar{\lambda} = \pi,$$

at the cost of changing the bottom arrow, giving the analytic representation of $A^{\sigma}_{\mathbf{C}}$, by an automorphism of A^{σ}. We have therefore proved the following result.

Lemma 1. *Let A be an elliptic curve with $j_A = j(\mathfrak{a})$, where $\mathfrak{a}$ is an ideal of $\mathfrak{o}_k$ in k. Let*

$$\varphi: \mathbf{C}/\mathfrak{a} \to A_{\mathbf{C}}$$

be an analytic representation. Assume that A is defined over $k(j_A)$. Then for all but a finite number of primes $\mathfrak{p}$ of degree 1 in k, if $\sigma = \sigma_{\mathfrak{p}}$ is the Frobenius automorphism of $\mathfrak{p}$ in $k(j_A)$, we can find an analytic representation

$$\psi: \mathbf{C}/\mathfrak{p}^{-1}\mathfrak{a} \to A^{\sigma}_{\mathbf{C}}$$

and an isogeny λ such that the following diagram commutes,

$$\begin{array}{ccc} \mathbf{C}/\mathfrak{a} & \xrightarrow{\varphi} & A_{\mathbf{C}} \\ {\scriptstyle \mathrm{can}}\downarrow & & \downarrow{\scriptstyle \lambda} \\ \mathbf{C}/\mathfrak{p}^{-1}\mathfrak{a} & \xrightarrow[\psi]{} & A^{\sigma}_{\mathbf{C}} \end{array}$$

and such that if the bar denotes reduction with respect to some prime $\mathfrak{P}$ extending $\mathfrak{p}$ in $k(j_A)$, then $\bar{\lambda} = \pi_p$.

Theorem 2. *Let A be an elliptic curve whose ring of endomorphisms is the ring of algebraic integers $\mathfrak{o}_k$ in an imaginary quadratic field k, and A is defined over $k(j_A)$. Let h be the Weber function on A, giving the quotient of A by its group of automorphisms. Then $k(j_A, h(A_N))$ is the ray class field of k with conductor N.*

Proof. Let K be the smallest Galois extension of k containing $j_A = j(\mathfrak{a})$ and all coordinates $h(A_N)$. We take a prime $\mathfrak{p}$ of k of degree 1 as before, omitting only a finite number of them, e.g. those which ramify in K, all $\mathfrak{p}|N$ and $\mathfrak{p}|\mathfrak{a}$, and all $\mathfrak{p}$ where we might have bad reduction of A and its conjugates. We can now take the elliptic curve $B = A^{\sigma}$ to have invariant j_A^{σ} where $\sigma = \sigma_{\mathfrak{P}}$ is the Frobenius automorphism of some prime $\mathfrak{P}$ in K lying above $\mathfrak{p}$. The bar reduction will again be with respect to $\mathfrak{P}$.

If $t \in A_N$, then

$$\overline{\lambda t} = \bar{\lambda}\bar{t} = \pi(\bar{t}) = \overline{\sigma t},$$

by the definition of the Frobenius automorphism applied to the coordinates of t. Since reduction induces an injection on A_N, we conclude that $\lambda = \sigma$ on A_N. Therefore the commutative diagram of Lemma 1 now reads

$$\begin{array}{ccc} (k/\mathfrak{a})_N & \xrightarrow{\varphi} & A_N \\ {\scriptstyle \text{can}}\downarrow & & \downarrow{\scriptstyle \sigma_{\mathfrak{P}}} \\ (k/\mathfrak{p}^{-1}\mathfrak{a})_N & \xrightarrow[\psi]{} & A_N \end{array} \tag{3}$$

We shall prove that $\mathfrak{p}$ splits completely in K if and only if $\mathfrak{p} = (\alpha)$ with some $\alpha \in \mathfrak{o}_k$ such that $\alpha \equiv 1 \pmod{^*\ N\mathfrak{o}_k}$.

Suppose first that $\mathfrak{p} = (\alpha)$ with $\alpha \in k$. Then $\mathfrak{p}$ splits completely in $k(j_A)$ and hence $A^\sigma = A$. Following the left vertical arrow in Lemma 1 by multiplication with α, we get a commutative diagram with an analytic representation ψ' of $A_{\mathbf{C}}$,

$$\begin{array}{ccc} (k/\mathfrak{a})_N & \xrightarrow{\varphi} & A_N \\ {\scriptstyle \text{can}}\downarrow & & \downarrow{\scriptstyle \sigma_{\mathfrak{P}} = \sigma} \\ (k/\mathfrak{p}^{-1}\mathfrak{a})_N & \xrightarrow[\psi]{} & A_N \\ {\scriptstyle \alpha}\downarrow & & \downarrow{\scriptstyle \text{id}} \\ (k/\mathfrak{a})_N & \xrightarrow[\psi']{} & A_N \end{array} \tag{4}$$

If $\alpha \equiv 1 \pmod{^*\ N\mathfrak{o}_k}$, then the composite vertical map on the left is the identity. Furthermore, ψ' differs from φ by an automorphism of A. It follows that $\sigma_{\mathfrak{P}}$ acts as the identity on the Weber coordinates $h(t)$ for all $t \in A_N$, and hence $\sigma_{\mathfrak{P}} = 1$ on $K = k(j_A, h(A_N))$.

Conversely, suppose that $\mathfrak{p}$ splits completely in K, and in particular splits completely in $k(j_A)$. Then by class field theory, $\mathfrak{p} = (\alpha)$ is principal, and $A^\sigma = A$ in Lemma 1. We obtain the same two-storied diagram (4) above, using multiplication by α. For the Weber function h we have $h^\sigma = h$ because h can be defined over $k(j_A)$. For any element $u \in (k/\mathfrak{a})_N$ we get:

$$\begin{aligned} h(\varphi(u)) = h(\varphi(u))^\sigma &= h^\sigma(\varphi(u)^\sigma) \\ &= h(\varphi(u)^\sigma) \\ &= h(\psi'(\alpha u)) \qquad \text{(by the commutative diagram)} \\ &= h(\varphi(\alpha u)). \end{aligned}$$

Observe that $k/\mathfrak{a}$ is an $\mathfrak{o}_k$-module, and by localizing one sees that $(k/\mathfrak{a})_N$ is principal, generated by an element u_0, say. Our final equality above implies that there exists a root of unity ζ such that

$$\alpha u_0 = \zeta u_0$$

(the h-coordinates of the two points $\varphi(u_0)$ and $\varphi(\alpha u_0)$ being equal, the two points differ by an automorphism of A). We change the generator α of $\mathfrak{p}$ by the inverse of this root of unity. We then get

$$\alpha u_0 = u_0,$$

and hence $\alpha u = u$ for all $u \in (k/\mathfrak{a})_N$ because any u can be written as λu_0 for some $\lambda \in \mathfrak{o}_k$, and $\alpha\lambda$ commutes. It follows that $\alpha \equiv 1 \pmod{N\mathfrak{o}_k}$.

By class field theory, we now conclude that K is the ray class field of k with conductor N. (Cf. Theorem 9 of Chapter VIII, §4 in my *Algebraic Number Theory*.) This proves Theorem 2.

Corollary. *Let F_N be the field of modular functions of level N, and let k be an imaginary quadratic field. Let kF_N be the composite field. Let $\mathfrak{a}$ be an $\mathfrak{o}_k$-ideal, $\mathfrak{a} = [z_1, z_2]$ and $z = z_1/z_2 \in \mathfrak{H}$. Then the field $kF_N(z)$ generated over k by all values $f(z)$, with $f \in F_N$, and f defined at z, is the ray class field over k with conductor N.*

Proof. Let $\mathfrak{M}$ be the kernel of the place $f \mapsto f(z)$ for $f \in kF_N$. Let

$$G = \mathrm{Gal}(kF_N/k(j)),$$

and let $G_{\mathfrak{M}}$ be the decomposition group. From general decomposition group theory we know that the induced group $\bar{G}_{\mathfrak{M}}$ is the Galois group of the residue class field extension. We also know by Theorem 2 that the residue class field contains the above mentioned ray class field. Let $\sigma \in G_{\mathfrak{M}}$ be such that $\bar{\sigma}$ is the identity on this ray class field. In particular, $\bar{\sigma}$ is the identity on all elements $f_a(z)$ and $j(z)$,w here f_a are the Fricke functions ($a \in (\mathbf{Q}^2/\mathbf{Z}^2)_N$, $a \neq 0$). By Theorems 2′ and 3′ of Chapter 9, §3 we conclude that σ lies in the inertia group, whence $\bar{\sigma}$ is the identity on the residue class field. This means that the residue class field is precisely the stated ray class field, and concludes the proof.

As is well known, Kronecker started the whole business of complex multiplication, and Weber gave a first systematization of the results known at the time. They were considerably incomplete, for instance the so-called Kronecker congruence relation of Theorem 1 was known only in a weaker form, namely

$$\Phi_p(X) \equiv (X^p - j)(X - j^p) \pmod{p}$$

actually proved by Weber (*Acta Mathematica* 6, 1885, p. 390). Hasse proved it in the form we stated it [19], and also obtained all the abelian extensions of k from values of the Weber function. Weber himself needed some quadratic extensions in addition. Fricke [B2] and Fueter [B5] gave treatments before that which are still of some interest, for the special cases which they discuss, and for the analytic methods.

The Institute Seminar [B17] is also a convenient reference for a quick introduction to some basic results, using the analytic approach, and some useful chapters on computational aspects.

Deuring [B1] simplified considerably some of Hasse's proofs in his monograph, which is an exceedingly good reference for the analytic development of the complex multiplication. For the convenience of the reader we shall reproduce the analytic proof of the congruence relation in Chapter 12, §3. It will not require any knowledge beyond Chapter 12, §2 which is self-contained. Hence the reader can read these sections as an alternative approach to part of the results of complex multiplication.

Deuring's major contribution, however, was to have found the algebraic development which we have followed, using reduction mod p, cf. all his papers in the bibliography. This was extended to abelian varieties by Shimura and Taniyama [B13], see also Shimura's book [B12].

§2. IDELIC FORMULATION FOR ARBITRARY LATTICES

In the first section we derived the basic theorem of complex multiplication using ordinary ideals of $\mathfrak{o}_k$. For a number of technical reasons, and also in order to tie up the situation over the quadratic field k with the generic situation, it is necessary to have a formulation describing the values of $j(L)$ for arbitrary lattices L, and also to know the relation with class field theory through the ideles. For this, we shall give a theorem as in Shimura [B12], who did it for a finite number of points, but whose final formulation is due to A. Robert. We now assume that the reader knows Chapter 8, §1 and §2, especially how the ideles operate on $k/\mathfrak{a}$ where $\mathfrak{a}$ is an arbitrary lattice in k. If s is an idele, (s, k) is the Artin symbol on the maximal abelian extension k_{ab}.

Theorem 3. *Let φ: $\mathbf{C}/\mathfrak{a} \to A_{\mathbf{C}}$ be an analytic representation of the elliptic curve A, where $\mathfrak{a}$ is a lattice in k. Let s be an idele of k and let σ be an automorphism of the complex numbers whose restriction to k_{ab} is (s, k). Then there exists an analytic representation*

$$\psi: \mathbf{C}/s^{-1}\mathfrak{a} \to A^{\sigma}_{\mathbf{C}}$$

such that the following diagram is commutative.

$$\begin{array}{ccc} k/\mathfrak{a} & \xrightarrow{\varphi} & A_{\mathbf{C}} \\ {\scriptstyle s^{-1}}\downarrow & & \downarrow{\scriptstyle \sigma} \\ k/s^{-1}\mathfrak{a} & \xrightarrow[\psi]{} & A^{\sigma}_{\mathbf{C}} \end{array}$$

Proof. Our first task is to reduce the theorem to the case when $\mathfrak{a}$ is an ideal of $\mathfrak{o}_k$. Let $\mathfrak{b}$ be an ideal of $\mathfrak{o}_k$ contained in $\mathfrak{a}$. Let ξ: $\mathbf{C}/\mathfrak{b} \to B_{\mathbf{C}}$ be an analytic

representation of an elliptic curve with lattice $\mathfrak{b}$, and let $\lambda: B \to A$ be the isogeny which makes the top of the following diagram commutative.
The back side is trivially commutative. Putting λ^σ on the lower right makes the right side commutative. Assuming our problem solved for $\mathfrak{b}$ we can find ξ' making the front square commutative. We define ψ such that the bottom is

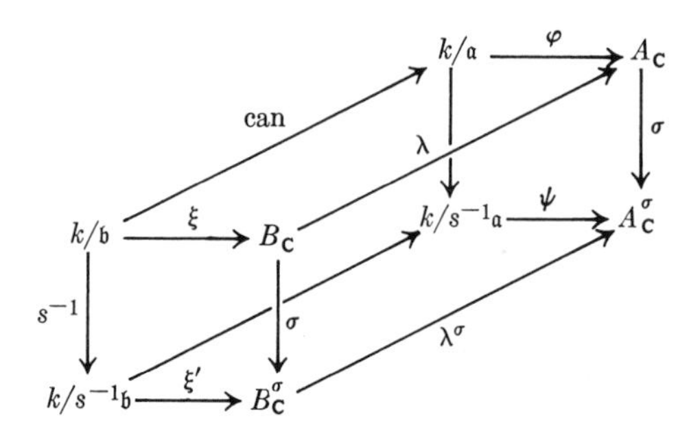

Fig. 10-1

commutative. This can be done by checking that the kernels of the two bottom maps are the same. It then follows that ψ makes the back face commutative, and this solves our problem for $\mathfrak{a}$.

Observe that the above reduction shows that if we can solve our problem for one elliptic curve $A \approx \mathbf{C}/\mathfrak{a}$, then we can solve it for any other elliptic curve A' isomorphic to A. Of course, a simpler direct argument can also be given in this case.

We now assume that $\mathfrak{a}$ is an ideal of $\mathfrak{o}_k$. A positive integer m will be said to be **freezing** for A if any automorphism of A which leaves A_m fixed (pointwise) must be the identity. Since A has only a finite number of automorphisms, there always exists such an integer. Let N be a positive integer such that $m|N$. We shall prove that there exists $\psi: \mathbf{C}/s^{-1}\mathfrak{a} \to A^\sigma_{\mathbf{C}}$ such that the desired diagram commutes on $(k/\mathfrak{a})_N$, i.e. such that the following diagram commutes.

$$\begin{array}{ccc} (k/\mathfrak{a})_N & \xrightarrow{\varphi} & A_N \\ {\scriptstyle s^{-1}}\downarrow & & \downarrow{\scriptstyle \sigma} \\ (k/s^{-1}\mathfrak{a})_N & \xrightarrow[\psi]{} & A_N^\sigma \end{array}$$

Since we could prove the theorem for any A in an isomorphism class, we can select A defined over $k(j_A)$ and we can then proceed as in Theorem 1 and the first part of Theorem 2. We select K Galois over k containing $k(j(\mathfrak{a}))$, containing the ray class field with conductor N and such that $k(A_N) \subset K$. Actually, using Theorem 2 shows that this last condition implies the one preceding it. There exists a prime p which splits completely in k, that is $(p) = \mathfrak{p}\mathfrak{p}'$, $\mathfrak{p} \neq \mathfrak{p}'$, such that $\mathfrak{p}$ is unramified in K, and for some $\mathfrak{P}|\mathfrak{p}$ in K, the finite number of elliptic curves under consideration have good reduction mod $\mathfrak{P}$. We also take $\mathfrak{p}$ prime to N and to $\mathfrak{a}$. Finally we require that $\sigma = \sigma_{\mathfrak{p}}$ on K.

By the Kronecker congruence relation we have $j(\mathfrak{p}^{-1}\mathfrak{a}) = j(\mathfrak{a})^{\sigma}$, and the first part of the proof of Theorem 2 shows that we have a commutative diagram (3) which is almost the one we want, except that we have to relate $\mathfrak{p}^{-1}\mathfrak{a}$ and $s^{-1}\mathfrak{a}$.

Let $c = (\ldots, 1, 1, c_{\mathfrak{p}}, 1, 1, \ldots)$ be an idele with component 1 except at $\mathfrak{p}$, and with component $c_{\mathfrak{p}}$ at $\mathfrak{p}$ having order 1 at $\mathfrak{p}$, so that we have

$$(s, k) = (c, k) = (\mathfrak{p}, k)$$

on the maximal abelian subfield of K over k, which we denote by $(K/k)_{ab}$. Then we have

$$c = s\beta b,$$

with some element $\beta \in k$ and some idele $b \equiv 1 \pmod{^* N}$, i.e. b has unit component outside N and its components at primes dividing N satisfy the desired congruence. Since $s^{-1}\mathfrak{a} = \beta\mathfrak{p}^{-1}\mathfrak{a}$ we get a commutative diagram

$$\begin{array}{ccc} \mathbf{C}/\mathfrak{a} & \xrightarrow{\varphi} & A_{\mathbf{C}} \\ {\scriptstyle \text{can}}\downarrow & & \downarrow{\scriptstyle \lambda} \\ \mathbf{C}/\mathfrak{p}^{-1}\mathfrak{a} & \xrightarrow[\psi]{} & A_{\mathbf{C}}^{\sigma} \\ {\scriptstyle \beta}\downarrow & & \downarrow{\scriptstyle \text{id}} \\ \mathbf{C}/s^{-1}\mathfrak{a} & \xrightarrow[\psi_1]{} & A_{\mathbf{C}}^{\sigma} \end{array}$$

with some analytic representation ψ_1, and the lower left vertical map being multiplication by β.

As we saw in the proof of Theorem 2, we can choose the map λ so that $\bar{\lambda} = \pi$ and hence on the points of order N, λ has the same effect as σ. So we get a commutative diagram

$$\begin{array}{ccc} (k/\mathfrak{a})_N & \xrightarrow{\varphi} & A_N \\ {\scriptstyle \beta}\downarrow & & \downarrow{\scriptstyle \sigma} \\ (k/s^{-1}\mathfrak{a})_N & \xrightarrow[\psi_1]{} & A_N^{\sigma} \end{array}$$

and there remains but to prove that multiplication by β on the points of period N on the left is the same as multiplication by s^{-1}, i.e. that if $u \in (k/\mathfrak{a})_N$, then $\beta u = s^{-1}u$. This has to be checked locally for each prime $\mathfrak{q}$ of k (since we work with an ideal $\mathfrak{a}$ of $\mathfrak{o}_k$, we can use $\mathfrak{q}$-components.) If $\mathfrak{q} \neq \mathfrak{p}$ then $c_{\mathfrak{q}} = 1$ and $s_{\mathfrak{q}}^{-1} = \beta_{\mathfrak{q}} b_{\mathfrak{q}}$. Since $b_{\mathfrak{q}} \equiv 1 \pmod{N\mathfrak{o}_{\mathfrak{q}}}$, we see that $b_{\mathfrak{q}} u_{\mathfrak{q}} = u_{\mathfrak{q}}$, and what we want is true at $\mathfrak{q}$. If $\mathfrak{q} = \mathfrak{p}$, then $\mathfrak{p} \nmid N$, and $u_{\mathfrak{q}} = 0$. This proves Theorem 3, for $(k/\mathfrak{a})_N$ instead of $k/\mathfrak{a}$.

However, if we have found two analytic representations

$$\psi_1 \quad \text{and} \quad \psi_2 \colon \mathbf{C}/s^{-1}\mathfrak{a} \to A^{\sigma}_{\mathbf{C}}$$

which make the diagram commutative on $(k/\mathfrak{a})_m$, they must be equal because m is freezing for A. This means that the solution of our problem at the m level is the same solution as that on the N level, for any N divisible by m. This concludes the proof of Theorem 3.

§3. GENERATION OF CLASS FIELDS BY SINGULAR VALUES OF MODULAR FUNCTIONS

We shall give direct applications of Theorem 3. The results are classical, and the exposition follows Shimura [B12].

Theorem 4. *For any lattice $\mathfrak{a}$ in k the number $j(\mathfrak{a})$ lies in k_{ab}, and for any idele s we have*

$$j(s^{-1}\mathfrak{a}) = j(\mathfrak{a})^{(s,k)}.$$

Proof. Take first $s = 1$ and let σ be any automorphism of $\mathbf{C}$ which is the identity on k_{ab}. Then Theorem 3 shows that

$$j(\mathfrak{a}) = j(\mathfrak{a})^{\sigma},$$

whence $j(\mathfrak{a})$ lies in k_{ab}. The formula of our theorem then expresses the fact that if $A \approx \mathbf{C}/\mathfrak{a}$, then $A^{\sigma} \approx \mathbf{C}/s^{-1}\mathfrak{a}$, which is also contained in the statement of Theorem 3.

Remark. For any proper $\mathfrak{o}$-lattice $\mathfrak{a}$ and any isomorphism σ of $\mathbf{Q}(j(\mathfrak{a}))$ over $\mathbf{Q}$, $\sigma j(\mathfrak{a}) = j(\mathfrak{b})$ for some proper $\mathfrak{o}$-lattice $\mathfrak{b}$.

Proof. Let $A \approx \mathbf{C}/\mathfrak{a}$. Then $\mathrm{End}(A) \approx \mathfrak{o}$, and consequently $\mathrm{End}(A^{\sigma}) \approx \mathfrak{o}$ also. If $\mathfrak{b}$ is a lattice such that $\mathbf{C}/\mathfrak{b} \approx (A^{\sigma})_{\mathbf{C}}$, then it follows that $\mathfrak{o}$ is the set of complex numbers α such that $\alpha\mathfrak{b} \subset \mathfrak{b}$. Hence $\mathfrak{b}$ is also a proper $\mathfrak{o}$-lattice. We also have

$$\sigma j_A = \sigma j(\mathfrak{a}) = j_{A^{\sigma}} = j(\mathfrak{b}).$$

This proves our remark.

For the next result, we introduce a notation. Let $\mathfrak{b}$ be a proper $\mathfrak{o}$-ideal prime to the conductor of $\mathfrak{o}$. Let $\mathfrak{b}_k = \mathfrak{b}\mathfrak{o}_k$ be the extension of $\mathfrak{b}$ to $\mathfrak{o}_k$. We denote by

$$(\mathfrak{b}, k)$$

the Artin automorphism $(\mathfrak{b}_k, K/k)$ in the maximal abelian extension K of k in which all primes dividing $\mathfrak{b}$ are unramified. This is well defined because of the consistency of the Artin automorphism when restricted to subfields. In particular, $(\mathfrak{b}, k)$ is defined on the ray class field whose conductor is the conductor of $\mathfrak{o}$.

Theorem 5. *Let $\mathfrak{o}$ be an order in k, and let $\{\mathfrak{a}_i\}$ $(i = 1, \ldots, h_{\mathfrak{o}})$ be representatives for the distinct proper $\mathfrak{o}$-lattice classes. Then the numbers $j(\mathfrak{a}_i)$ are all conjugate over k, and over $\mathbf{Q}$. The Galois group of $k(j(\mathfrak{a}))$ for any proper $\mathfrak{o}$-lattice $\mathfrak{a}$ is isomorphic to the group of proper $\mathfrak{o}$-lattice classes, under the map*

$$\mathfrak{b} \mapsto \sigma_{\mathfrak{b}}$$

such that

$$\sigma_{\mathfrak{b}} j(\mathfrak{a}) = j(\mathfrak{b}^{-1}\mathfrak{a}).$$

Furthermore $\sigma_{\mathfrak{b}}$ is the restriction of $(\mathfrak{b}, k)$ to $k(j(\mathfrak{o}))$, so that we have the formula

$$j(\mathfrak{a})^{(\mathfrak{b},k)} = j(\mathfrak{b}^{-1}\mathfrak{a}).$$

Proof. We know from Theorem 4 of Chapter 8, §1 that any proper $\mathfrak{o}$-lattice $\mathfrak{b}$ is locally principal, say $\mathfrak{b}_p = s_p\mathfrak{o}_p$. Let s be the idele whose p-component is s_p. Then $\mathfrak{b}^{-1}\mathfrak{a} = s^{-1}\mathfrak{a}$, and Corollary 1 implies all our assertions, except the last one. To prove it, let s be an idele such that $s_p = 1$ for all primes dividing the conductor of $\mathfrak{o}$, and such that

$$s_p\mathfrak{o}_p = \mathfrak{b}_p$$

for all other primes. Since $\mathfrak{b}_p = (\mathfrak{b}\mathfrak{o}_k)_p$ at all primes dividing b (because such primes do not divide the conductor), it follows that

$$s\mathfrak{o} = \mathfrak{b} \quad \text{and} \quad s\mathfrak{o}_k = \mathfrak{b}\mathfrak{o}_k.$$

Our formula is now a special case of Theorem 4. The fact that the values of j on proper $\mathfrak{o}$-lattice classes are conjugate over $\mathbf{Q}$ was already mentioned in the remark preceding our theorem.

Remark 1. Let L' denote the complex conjugate of a lattice L in k, and w' the complex conjugate of a complex number w. From the original series for g_2 and g_3 we see that for any lattice L in k we have

$$g_2(L') = g_2(L)' \quad \text{and} \quad g_3(L') = g_3(L)',$$

whence

$$j(L') = j(L)'.$$

Since $\mathfrak{o} = \mathfrak{o}'$ for any order $\mathfrak{o}$, we conclude that $j(\mathfrak{o})$ is real. We have seen that all the conjugates of $j(\mathfrak{o})$ over $\mathbf{Q}$ are the same as the conjugates of $j(\mathfrak{o})$ over k. Hence we also find that $\mathbf{Q}(j(\mathfrak{o}))$ *is the real subfield of* $k(j(\mathfrak{o}))$.

Remark 2. Let $K = k(j(\mathfrak{o}))$ where $\mathfrak{o}$ is an order in k. Let ρ be the complex conjugation automorphism of $\mathbf{C}$. Then for any proper $\mathfrak{o}$-lattice $\mathfrak{a}$ we have

$$j(\mathfrak{a})' = j(\mathfrak{a}').$$

Since $\mathfrak{a}\mathfrak{a}' = \lambda\mathfrak{o}$ for some complex number λ, we conclude that

$$\boxed{j(\mathfrak{a})' = j(\mathfrak{a}^{-1}).}$$

We now conclude that the field K is Galois over $\mathbf{Q}$, and we shall prove that for any automorphism $\sigma \in \mathrm{Gal}(K/\mathbf{Q})$ we have the formula

$$\boxed{\rho\sigma\rho^{-1} = \sigma^{-1}.}$$

To see this, we have for any proper $\mathfrak{o}$-ideal $\mathfrak{b}$, and any proper $\mathfrak{o}$-ideal $\mathfrak{a}$ such that $\sigma = \sigma_{\mathfrak{a}}$,

$$\rho\sigma\rho^{-1}j(\mathfrak{b}) = \rho\sigma j(\mathfrak{b}^{-1}) = \rho j(\mathfrak{a}^{-1}\mathfrak{b}^{-1}) = j(\mathfrak{a}\mathfrak{b}) = \sigma_{\mathfrak{a}^{-1}}j(\mathfrak{b}).$$

This proves what we wanted.

Since ρ is not the identity on k, it follows that the Galois group of $k(j(\mathfrak{o}))$ over $\mathbf{Q}$ is a group extension of $\mathrm{Gal}(k(j(\mathfrak{o}))/k)$ by a group of order 2, and that the structure of the Galois group over $\mathbf{Q}$ is completely determined by the Galois group over k and the formula giving the commutation rule between σ and the complex conjugation.

Theorem 6. *Let $\mathfrak{o} \subset \mathfrak{o}'$ be two orders of k. Then*

$$k(j(\mathfrak{o})) \supset k(j(\mathfrak{o}')).$$

Proof. We have to show that an automorphism σ of k_{ab} leaving $j(\mathfrak{o})$ fixed also leaves $j(\mathfrak{o}')$ fixed. Write $\sigma = (s, k)$ for some idele s. Then $s\mathfrak{o} = \alpha\mathfrak{o}$ for some $\alpha \in k$. Changing s by α^{-1} we may assume without loss of generality that $\sigma = (s, k)$ with $s\mathfrak{o} = \mathfrak{o}$. But then for any prime number p, we have $\mathfrak{o}\mathfrak{o}' = \mathfrak{o}'$, $\mathfrak{o}_p\mathfrak{o}'_p = \mathfrak{o}'_p$ and $s_p\mathfrak{o}'_p = \mathfrak{o}'_p$, whence $s\mathfrak{o}' = \mathfrak{o}'$. The commutative diagram of Theorem 3 shows that (s, k) leaves $j(\mathfrak{o}')$ fixed, as desired.

As in Shimura, we can now give a criterion in terms of ideles for the Galois group leaving a point of order N fixed.

Theorem 7. *Let A be an elliptic curve such that $\varphi: \mathbf{C}/\mathfrak{a} \to A_{\mathbf{C}}$ is an analytic representation for some lattice $\mathfrak{a}$ in k. Let h be the Weber function associated with A. Let s be an idele of k. Then (s, k) is the identity on $k(j(\mathfrak{a}))$, $h(\varphi(u))$ for some point $u \in k/\mathfrak{a}$ if and only if $s \in k^* V_{\mathfrak{a},u}$, where $V_{\mathfrak{a},u}$ is the subgroup of ideles b such that*

$$b\mathfrak{a} = \mathfrak{a} \quad \text{and} \quad bu = u.$$

Proof. First consider the case where we deal only with $k(j(\mathfrak{a}))$. Let $V_{\mathfrak{a}}$ be the subgroup of ideles b such that $b\mathfrak{a} = \mathfrak{a}$. If s is an idele such that $s\mathfrak{a} = \mathfrak{a}$, then $s^{-1}\mathfrak{a} = \mathfrak{a}$ and

$$j(\mathfrak{a}) = j(s^{-1}\mathfrak{a}) = j(\mathfrak{a})^{\sigma}, \qquad \sigma = (s, k),$$

whence the image of $k^*V_{\mathfrak{a}}$ in $\mathrm{Gal}(k_{ab}/k)$ leaves $k(j(\mathfrak{a}))$ fixed. Conversely, if $(s, k) = \sigma$ leaves $j(\mathfrak{a}) = j_A$ fixed, then

$$j_{A^{\sigma}} = j(\mathfrak{a})^{\sigma} = j(\mathfrak{a}) = j_A,$$

and $A^{\sigma} \approx A$. Hence $s^{-1}\mathfrak{a}$ and $\mathfrak{a}$ differ by multiplication with an element from k, i.e. $s^{-1}\mathfrak{a} = \alpha\mathfrak{a}$ for some $\alpha \in k$. Hence $s \in k^*V_{\mathfrak{a}}$, thus proving our assertion in the present case.

By Theorem 2, if h is the Weber function, then $h(\varphi(u))$ generates an abelian extension of k. Suppose first that $s\mathfrak{a} = \mathfrak{a}$ and $su = u$. Let $\sigma = (s, k)$. Then

$$\varphi(u)^{\sigma} = \psi(s^{-1}u) = \psi(u) = \varepsilon\varphi(u)$$

for some automorphism ε of A, because φ and ψ differ by an automorphism of A. Since $h^{\sigma} = h$ we conclude that

$$\sigma(h(\varphi(u))) = h(\varphi(u)).$$

Conversely, suppose that $\sigma j_A = j_A$ and $\sigma h(\varphi(u)) = h(\varphi(u))$. From $\sigma j_A = j_A$ we conclude that $s^{-1}\mathfrak{a} = \alpha\mathfrak{a}$ with some $\alpha \in k^*$. Replace s by αs. This reduces our assertion to the case when $s\mathfrak{a} = \mathfrak{a}$ because $(s, k) = (s\alpha, k) = \sigma$. Now

$$\sigma\varphi(u) = \psi(s^{-1}u) = \varepsilon\varphi(s^{-1}u)$$

for some automorphism ε of A. Take h to get

$$h(\varphi(u)) = \sigma h(\varphi(u)) = h(\varphi(s^{-1}u)).$$

It follows that $s^{-1}u$ and u differ by an automorphism of $\mathbf{C}/\mathfrak{a}$, i.e. there exists a root of unity ζ such that $s^{-1}u = \zeta u$, whence $s\zeta u = u$. Hence $s\zeta$ lies in $V_{\mathfrak{a},u}$ and $s \in k^*V_{\mathfrak{a},u}$, thereby proving our theorem.

Corollary. *Let $\mathfrak{c}$ be an ideal of $\mathfrak{o}_k$. The ray class field with conductor $\mathfrak{c}$ over k is obtained as*

$$k(j(\mathfrak{a}), h(\varphi(\mathfrak{c}^{-1}\mathfrak{a}/\mathfrak{a})))$$

for any ideal $\mathfrak{a}$ of $\mathfrak{o}_k$.

Proof. The proof is obvious using Theorem 3 and the remarks at the end of Chapter 8, §2. Note that as a module over $\mathfrak{o}_k$, $\mathfrak{c}^{-1}\mathfrak{a}/\mathfrak{a}$ is principal, and hence the above class field can be obtained by the image of one point in $\mathfrak{c}^{-1}\mathfrak{a}/\mathfrak{a}$, together with $j(\mathfrak{a})$, adjoined to k.

§4. THE FROBENIUS ENDOMORPHISM

For an elliptic curve with complex multiplication, Deuring proved a suggestion of Weil that the Frobenius endomorphism should be a Hecke character [13]. As Weil pointed out, this shows that the Hasse zeta function of the curve is a Hecke L-series, an interpretation which we shall indicate afterwards.

The reader interested at this stage principally in the generation of class fields by values of modular functions can omit this section and proceed directly to the Shimura reciprocity law, as a direct continuation of the preceding sections.

Throughout this section, we let A be an elliptic curve defined over a number field K, and we assume that A has complex multiplication. Let k be an imaginary quadratic field in $\mathbf{C}$, and let

$$\theta\colon k \to \mathrm{End}(A)_{\mathbf{Q}}$$

be a normalized isomorphism of k with the algebra of endomorphisms of A.

We recall that **normalized** means that for any differential form ω of the first kind of A, we have

$$\omega \circ \theta(\mu) = \mu\omega, \qquad \mu \in k.$$

Such a differential form can always be selected to be defined over K.

Remark. We have $k \subset K$ if and only if every element of $\mathrm{End}(A)$ is defined over K.

Proof. Let σ be an automorphism of the algebraic closure K_a over K. Then having chosen ω to be defined over K, we have

$$\omega \circ \theta(\mu)^\sigma = \mu^\sigma\omega.$$

Therefore $\theta(\mu)^\sigma = \theta(\mu)$ if and only if $\mu^\sigma = \mu$. From this our remark follows at once.

Recall that a **Hecke character** (or **quasi character**) is a continuous homomorphism

$$\chi\colon \mathbf{A}_K^* \to \mathbf{C}^*$$

from the ideles of K into the multiplicative group of complex numbers, which is trivial on K^*, i.e. such that $\chi(K^*) = 1$. [We do not require that the character has absolute value 1.] Such a character is said to be **unramified** at a prime $\mathfrak{p}$ if it is trivial on the local $\mathfrak{p}$-units (embedded at the $\mathfrak{p}$-component of the ideles, having all other components equal to 1). If this is the case, we define $\chi(\mathfrak{p}) = \chi(s)$, where s is an idele having component 1 except at $\mathfrak{p}$, and having $\mathfrak{p}$-component equal to an element of order 1 at $\mathfrak{p}$, thus

$$\chi(\mathfrak{p}) = \chi(\ldots, 1, s_\mathfrak{p}, 1, \ldots), \qquad \mathrm{ord}_\mathfrak{p} s_\mathfrak{p} = 1.$$

We consider first the case when $k \subset K$. Let $\mathfrak{p}$ be a prime of K where A has non-degenerate reduction $\bar{A} = \bar{A}(\mathfrak{p})$. Let $\mathfrak{o}$ be the order in k such that $\theta(\mathfrak{o}) = \mathrm{End}(A)$. Assume that $\mathfrak{p}$ is prime to the conductor of $\mathfrak{o}$. Let $\mathfrak{p}_o = \mathfrak{p} \cap \mathfrak{o}$ and $\mathfrak{p}_k = \mathfrak{p} \cap k$. Let $f = f(\mathfrak{p}/\mathfrak{p}_k)$ be the degree of the residue class field extension. Then

$$\mathbf{N}_k^K \mathfrak{p} = \mathfrak{p}_k^f.$$

The field K contains $k(j_A)$. By complex multiplication, and the elementary formalism of the Frobenius automorphism, it follows that the Artin symbol of $\mathfrak{p}_k^f$ on $k(j_A)$ is the identity. Therefore $\mathfrak{p}_k^f$ is principal, and there is an element $\mu \in \mathfrak{o}$ such that

$$\mathfrak{p}_k^f = \mu \mathfrak{o}_k.$$

Deuring's theorem asserts that we can select the generator $\mu = \mu(\mathfrak{p})$ in such a way that the endomorphism $\theta(\mu)$ reduces to the Frobenius endomorphism of $\bar{A}$, and that the values $\mu(\mathfrak{p})$ are the values of a Hecke character. For the proof of Deuring's theorem, we shall use the following idelized version as in Shimura, who also gives a generalization to abelian varieties [B12], 7.8.

As usual, we denote by A_{tor} the set of torsion points of A. We let $\bar{K}(\mathfrak{p})$ be the residue class field of K at $\mathfrak{p}$.

Theorem 8. *Assume that $k \subset K$. Let s be an idele of K. Let*

$$\varphi: \mathbf{C}/\mathfrak{a} \to A_{\mathbf{C}}$$

be an analytic parametrization of A. Then $K(A_{\mathrm{tor}})$ is abelian over K, and there is a unique element $\mu(s) \in k$, making the following diagram commutative.

$$\begin{array}{ccc} k/\mathfrak{a} & \xrightarrow{\varphi} & A_{\mathrm{tor}} \\ {\scriptstyle \mu(s)N_k^K(s^{-1})}\downarrow & & \downarrow{\scriptstyle (s,K)} \\ k/\mathfrak{a} & \xrightarrow[\varphi]{} & A_{\mathrm{tor}} \end{array}$$

Proof. Let σ be an automorphism of $K(A_{\mathrm{tor}})$ over K, inducing (s, K) on the maximal abelian subfield of $K(A_{\mathrm{tor}})$. The restriction of σ to k_{ab} is equal to $(N_k^K(s), k)$. Let

$$t = N_k^K(s).$$

According to the idelized formulation of complex multiplication, Theorem 3 of Chapter 10, §2, there is an analytic parametrization

$$\psi: \mathbf{C}/t^{-1}\mathfrak{a} \to A_{\mathbf{C}}$$

making the following diagram commutative, with $A^\sigma = A$.

$$\begin{array}{ccc} k/\mathfrak{a} & \xrightarrow{\varphi} & A_{\mathrm{tor}} \\ {\scriptstyle t^{-1}}\downarrow & & \downarrow{\scriptstyle \sigma} \\ k/t^{-1}\mathfrak{a} & \xrightarrow[\psi]{} & A_{\mathrm{tor}} \end{array}$$

Again, since $A^{\sigma} = A$, it follows that $t^{-1}\mathfrak{a} = \alpha\mathfrak{a}$ for some $\alpha \in k$. The map ψ is determined up to an automorphism of A. Changing ψ by a unit in End(A) corresponds to changing α by a root of unity in k, and we may thus get the diagram with φ at the bottom, and $\mu(s)N_k^K(s^{-1})$ on the left with some $\mu(s)$ in k, which is uniquely determined since φ and σ are isomorphisms. Suppose that the restriction of σ to the maximal abelian subfield of $K(A_{\mathrm{tor}})$ is the identity. Then we could have taken $s = 1$ and $\mu(s) = 1$. Since σ is determined by its effect on A_{tor}, it follows that $\sigma = \mathrm{id}$, and therefore $K(A_{\mathrm{tor}})$ is abelian over K. This proves Theorem 8.

We continue to assume that $k \subset K$. We note that the association

$$s \mapsto \mu(s)$$

in Theorem 8 is obviously a homomorphism, and we define a function $\chi_{A,K}$ on $\mathbf{A}_K^*$, by

$$\chi_{A,K}(s) = \mu(s)N_k^K(s^{-1})_\infty,$$

where t_∞ is the archimedean component of an idele $t \in \mathbf{A}_K^*$. As with χ, we say that μ is **unramified** at $\mathfrak{p}$ if μ is trivial on the local $\mathfrak{p}$-units, and in this case, we define $\mu(\mathfrak{p})$ as we defined $\chi(\mathfrak{p})$.

Theorem 9. *Assume $k \subset K$. The function $\chi = \chi_{A,K}$ is continuous and trivial on K^*. In other words, it is a Hecke character of the idele classes of K. If $\mathfrak{p}$ is a prime of K where A has non-degenerate reduction, then χ and μ are unramified at $\mathfrak{p}$, and $\chi(\mathfrak{p}) = \mu(\mathfrak{p})$. If we denote reduction* mod $\mathfrak{p}$ *by a bar, then*

$$\overline{\theta(\mu(\mathfrak{p}))}$$

is the Frobenius endomorphism of $\bar{A}$, over the residue class field $\bar{K}(\mathfrak{p})$.

Proof. It is clear that χ is a homomorphism. If $s \in K^*$, then $(s, K) = 1$, and we can take $\mu(s) = N_k^K(s)$, so that $\chi(s) = 1$. Hence χ is trivial on K^*. If $s_\mathfrak{p} = 1$ for all (non-archimedean) primes $\mathfrak{p}$, then we can take $\mu(s) = 1$, and

$$\chi(s) = N_k^K(s^{-1})_\infty,$$

so that χ is continuous on the archimedean part of the ideles. On the other hand, suppose that $s_\mathfrak{p}$ is very close to 1 at all $\mathfrak{p}$ dividing $\mathfrak{a}$ in Theorem 8. Then $N_k^K(s^{-1})$ is also close to 1, and

$$N_k^K(s^{-1})\mathfrak{a} = \mathfrak{a}.$$

Since we must also have

$$\mu(s)N_k^K(s^{-1})\mathfrak{a} = \mathfrak{a},$$

it follows that $\mu(s)$ is a root of unity in k. If in addition we select s such that (s, k) is the identity on the points of A of order N for large N, and such that s, whence $N_k^K(s^{-1})$, is close to 1 at primes dividing N, then multiplication by $N_k^K(s^{-1})$ on $(k/\mathfrak{a})_N$ is the identity. Consequently $\mu(s)$ must also be equal to 1.

This proves that the kernel of χ contains an open subgroup of the finite part of the ideles of K, and therefore that χ is continuous, whence a Hecke character.

Let $\mathfrak{p}$ be a prime of K at which A has non-degenerate reduction. By Theorem 1 of Chapter 9, §3 we know that $\mathfrak{p}$ is unramified in $K(A_N)$ for $\mathfrak{p} \nmid N$. Let s be an idele all of whose components are equal to 1, except for the $\mathfrak{p}$-component, where $\mathrm{ord}_{\mathfrak{p}} s_{\mathfrak{p}} = 1$. Note that $N_k^K(s^{-1})_\infty = 1$, and hence $\chi(s) = \mu(s)$. Then $\theta(\mu(s))$ is an endomorphism of A, and we shall prove that its reduction mod $\mathfrak{p}$ is the Frobenius endomorphism of $\bar{A}$, which we denote by

$$\pi_{\mathfrak{p}} = \pi : \bar{A} \to \bar{A}.$$

Let ℓ be a prime number not divisible by $\mathfrak{p}$. Since the ℓ-component of s is 1, we conclude that multiplication by $\mu(s)N_k^K(s^{-1})$ on the group of ℓ-primary elements $k_\ell/\mathfrak{a}_\ell$ is the same as multiplication by $\mu(s)$. Since θ is normalized, it follows that the following diagram is commutative.

$$\begin{array}{ccc} \mathbf{C} & \xrightarrow{\varphi} & A_{\mathbf{C}} \\ {\scriptstyle \mu(s)}\downarrow & & \downarrow{\scriptstyle \theta(\mu(s))} \\ \mathbf{C} & \xrightarrow{\varphi} & A_{\mathbf{C}} \end{array}$$

Let $A^{(\ell)}$ be the group of ℓ-primary points on A, i.e. the image of $k_\ell/\mathfrak{a}_\ell$ under φ. Then the commutative diagram of Theorem 8 shows that for any point $P \in A^{(\ell)}$ and $\sigma = (s, K)$ we have

$$\overline{\theta(\mu(s))}\,\bar{P} = \overline{P^\sigma} = \pi(\bar{P}).$$

Therefore $\overline{\theta(\mu(s))} = \pi$, because these two endomorphisms have the same value on $\bar{A}^{(\ell)}$.

If $u_{\mathfrak{p}}$ is a unit in $k_{\mathfrak{p}}$ and $s'_{\mathfrak{p}} = u_{\mathfrak{p}} s_{\mathfrak{p}}$, then from the above, we conclude that $\theta(\mu(s'))$ and $\theta(\mu(s))$ have the same reduction, namely π, and therefore that $\overline{\theta(\mu(u_{\mathfrak{p}}))} = \mathrm{id}$. Since reduction mod $\mathfrak{p}$ is injective on $\mathrm{End}(A)$, it follows that $\mu(u_{\mathfrak{p}}) = 1$, and hence that χ_A is unramified at $\mathfrak{p}$. This proves our theorem.

Remark. Deuring also proved that when A does not have good reduction at $\mathfrak{p}$, then the character ramifies. Today, one can use a result of Serre-Tate [27] to deduce this property at once from the ℓ-adic representations discussed later in Chapter 13. Indeed, the Serre-Tate result asserts that if ℓ is a prime number not divisible by $\mathfrak{p}$, and if $\mathfrak{p}$ is unramified in the extension $K(A^{(\ell)})$ generated over K by the points of ℓ-power order on A, then A has a model over K with non-degenerate reduction at $\mathfrak{p}$. Let $(k/\mathfrak{a})^{(\ell)}$ be the group of ℓ-power torsion points in $k/\mathfrak{a}$. By definitions, and the lemma applied to an idele $s_{\mathfrak{p}}$ having components equal to 1 except at $\mathfrak{p}$, we have a commutative diagram

$$\begin{array}{ccc} (k/\mathfrak{a})^{(\ell)} & \longrightarrow & A^{(\ell)} \\ {\scriptstyle \mu(s_{\mathfrak{p}})}\downarrow & & \downarrow{\scriptstyle (s_{\mathfrak{p}}, K),} \\ (k/\mathfrak{a})^{(\ell)} & \longrightarrow & A^{(\ell)} \end{array}$$

We see that the right-hand side depends only on the order of $s_\mathfrak{p}$ at $\mathfrak{p}$ if and only if $\mu(s_\mathfrak{p})$ depends only on the order of $s_\mathfrak{p}$ at $\mathfrak{p}$, i.e. if and only if the Hecke character is unramified at $\mathfrak{p}$. The Serre-Tate criterion shows that this occurs if and only if A has non-degenerate reduction at $\mathfrak{p}$.

Next we consider the case when k is not contained in the field of definition of A, again as in Deuring [13], (iv).

Theorem 10. *Let A be defined over a number field K_0 not containing k. Let $K = K_0 k$. Let $\mathfrak{p}_0$ be a prime of K_0 where A has non-degenerate reduction. Then $\mathfrak{p}_0$ is unramified in K. Let*

$$\rho : \xi \mapsto \xi'$$

be the automorphism of K over K_0, and let $\mathfrak{p}$, $\mathfrak{p}'$ be the primes of K above $\mathfrak{p}_0$. Let $\mu(\mathfrak{p}) = \chi_{A,K}(\mathfrak{p})$, and similarly for $\mathfrak{p}'$. Then

$$\mu(\mathfrak{p})' = \mu(\mathfrak{p}').$$

Let $\pi_0 = \pi_{\mathfrak{p}_0}$ be the Frobenius endomorphism of the reduction $\bar{A}(\mathfrak{p}_0)$ over $\bar{K}(\mathfrak{p}_0)$. Let $q_0 = \mathbf{N}\mathfrak{p}_0$.

Case 1. *$\mathfrak{p}_0$ remains prime in K, so $\mathfrak{p} = \mathfrak{p}'$. Then π_0 is not rational. We have*

$$\pi_0^2 = \pi_\mathfrak{p} \qquad \textit{and} \qquad \pi_0 = \pm\sqrt{-q_0}.$$

Case 2. *$\mathfrak{p}_0$ splits completely in K, so $\mathfrak{p} \neq \mathfrak{p}'$. Then $\pi_0 = \pi_\mathfrak{p}$. Furthermore,*

$$\pi_0 = \overline{\theta(\mu(\mathfrak{p}))} \qquad \textit{and} \qquad \pi_0' = \overline{\theta(\mu(\mathfrak{p}'))}.$$

Proof. By the remark at the beginning of this section, we know that there exists an endomorphism α of A defined over K but not over K_0, so that $\alpha^\rho \neq \alpha$. Suppose that $\mathfrak{p}_0$ ramifies in K. The effect of ρ on the residue class field is trivial, and consequently reducing mod $\mathfrak{p}$ yields

$$\overline{\rho\alpha} = \bar{\alpha},$$

contradicting the injectivity of the reduction map on End(A). This proves that $\mathfrak{p}_0$ is unramified in K.

We shall prove that

$$\mu(\mathfrak{p})' = \mu(\mathfrak{p}').$$

Let ℓ be a prime number relatively prime to $\mathfrak{p}_0$. Let $A^{(\ell)}$ be the torsion points of A whose order is a power of ℓ. Similarly for $(k/\mathfrak{a})^{(\ell)}$. Both $\mathfrak{p}$ and $\mathfrak{p}'$ are unramified in $K(A^{(\ell)})$. From the definitions and the lemma, we have a commutative diagram

$$\begin{array}{ccc} (k/\mathfrak{a})^{(\ell)} & \xrightarrow{\varphi} & A^{(\ell)} \\ {\scriptstyle \mu(\mathfrak{p})}\downarrow & & \downarrow{\scriptstyle (\mathfrak{p},K)} \\ (k/\mathfrak{a})^{(\ell)} & \xrightarrow[\varphi]{} & A^{(\ell)} \end{array}$$

and a similar one for $\mathfrak{p}'$. This means that $\theta(\mu(\mathfrak{p})) = (\mathfrak{p}, K)$ on $K(A^{(\ell)})$. Since $\mathfrak{p}' = \rho\mathfrak{p}$, we get similarly that

$$\theta(\mu(\mathfrak{p}')) = (\rho\mathfrak{p}, K) = \rho(\mathfrak{p}, K)\rho^{-1}$$

on $K(A^{(\ell)})$. Abbreviate $\theta(\mu(\mathfrak{p}))$ by λ. Extend ρ to $K(A^{(\ell)})$. From the formula

$$(\lambda(x))^\rho = \lambda^\rho(x^\rho),$$

we conclude that $\lambda^\rho = \rho(\mathfrak{p}, K)\rho^{-1}$ on $K(A^{(\ell)})$. This proves that

$$\theta(\mu(\mathfrak{p}')) = \theta(\mu(\mathfrak{p}))',$$

because these two endomorphisms have the same effect on $A^{(\ell)}$. It follows that $\mu(\mathfrak{p}') = \mu(\mathfrak{p})'$, from the definition of a normalized map θ, i.e.

$$\omega \circ \theta(\mu) = \mu\omega.$$

Consider the case that $\mathfrak{p} = \mathfrak{p}'$. Then $\mu(\mathfrak{p})$ is an element of k fixed under conjugation, whence is rational. It follows that $\pi_\mathfrak{p}$ is a rational (and therefore integral) multiple of the identity on $\bar{A}$, and hence $\pi_\mathfrak{p} = \pm q_0\delta$. This yields $\pi_0^2 = \pm q_0$. We contend that π_0 cannot be a trivial endomorphism. Otherwise, it commutes with all endomorphisms of $\bar{A}$. But if α is an endomorphism of A such that $\alpha^\rho \neq \alpha$, then $\bar{\alpha}^{\bar{\rho}} \neq \bar{\alpha}$ and $\bar{\alpha}$ is not defined over $\bar{K}(\mathfrak{p}_0)$. Consequently $\bar{\alpha}$ does not commute with π_0, whence π_0 is non-trivial. On the other hand, $\pi_0\pi_0' = q_0$, and since the map $\xi \mapsto \xi'$ is the automorphism of $\mathbf{Q}(\pi_0)$, it follows that $\pi_0 \neq \pm\sqrt{q_0}$. Hence

$$\pi_0 = \pm\sqrt{-q_0},$$

thereby proving the assertions in case 1.

Now suppose that $\mathfrak{p}_0$ splits completely in K. Then the residue class field extension has degree 1, and therefore $\pi_0 = \pi_\mathfrak{p}$. Hence π_0 is the reduction of $\theta(\mu(\mathfrak{p}))$ mod $\mathfrak{p}$, by Theorem 9. Furthermore

$$\theta(\mu(\mathfrak{p}))\theta(\mu(\mathfrak{p}))' = \nu(\theta(\mu(\mathfrak{p})))\delta.$$

Taking the bar (reduction mod $\mathfrak{p}$), and taking into account that

$$\nu(\theta(\mu(\mathfrak{p}))) = \nu(\pi_0) = q_0,$$

it follows that $\theta(\mu(\mathfrak{p}))'$ reduces mod $\mathfrak{p}$ to π_0'. This proves our theorem.

Remark. As with Theorem 9, we can apply the Serre-Tate result to prove Deuring's criterion:

> *If $\mathfrak{p}_0$ is unramified in K, and A has non-degenerate reduction at one prime $\mathfrak{p}$ of K extending $\mathfrak{p}_0$, then in fact A has non-degenerate reduction at $\mathfrak{p}_0$.*

Deuring had a rather hard time proving this in [13], (iv), and even comments that this is the "wesentliche Schwierigkeit" of his paper, as distinct from the rather formal arguments reproduced above for Theorem 10.

Theorems 9 and 10 were proved by Deuring to describe the zeta function of the elliptic curve as a Hecke L-function. We carry out the formalism.

Let $\mathbf{F}$ be a finite field with q elements and let $\bar{A}$ be an elliptic curve defined over $\mathbf{F}$. Let N be the number of rational points of $\bar{A}$ in $\mathbf{F}$. Let

$$\pi \colon \bar{A} \to \bar{A}$$

be the Frobenius endomorphism $\pi = \pi_q$. Then

$$N = \nu(\pi - \delta) = (\pi - \delta)(\pi' - \delta)$$
$$= q + 1 - \mathrm{Tr}(\pi),$$

where $\mathrm{Tr}(\pi) = \pi + \pi'$ is the trace. Let $\theta\colon \mathbf{Q}(\pi) \to \mathbf{C}$ be an embedding into the complex numbers such that $\theta(\mu) = \pi$. Then following Hasse, we define the **zeta function**

$$Z(\bar{A}, \mathbf{F}, X) = \frac{(1 - \mu X)(1 - \mu' X)}{(1 - X)(1 - qX)}.$$

Note that μ, μ' occur symmetrically, so that the numerator of the zeta function is often written

$$H(\bar{A}, \mathbf{F}, X) = (1 - \pi X)(1 - \pi' X).$$

Taking the logarithmic derivative, one sees by a trivial computation that

$$\frac{d}{dX} \log Z(\bar{A}, \mathbf{F}, X) = \sum_{d=1}^{\infty} N_d X^{d-1},$$

where N_d is the number of points of $\bar{A}$ in $\mathbf{F}_d$ (extension of $\mathbf{F}$ of degree d). Putting $X = q^{-s}$, it is then easily seen that the zeta function is equal to the usual expression

$$Z(\bar{A}, \mathbf{F}, q^{-s}) = \prod\left(1 - \frac{1}{\mathbf{N}\mathfrak{p}^s}\right)^{-1} = \sum \mathbf{N}\mathfrak{a}^{-s},$$

where the Euler product is taken over all primes $\mathfrak{p}$ of $\bar{A}$, rational over $\mathbf{F}$, and the sum is taken over all positive divisors (cycles) $\mathfrak{a}$ on $\bar{A}$, rational over $\mathbf{F}$. As already mentioned in Chapter 2, Hasse had determined the roots of the zeta function in this case as the eigenvalues of the Frobenius mapping. [This was generalized by Weil to arbitrary curves and abelian varieties, as is well known.]

Let again A be an elliptic curve defined over the number field K or K_0 as in the previous considerations. We define its **zeta function,** again following Hasse, by the product

$$\zeta(A, K, s) = \prod_{\mathfrak{p}} Z(\bar{A}(\mathfrak{p}), \bar{K}(\mathfrak{p}), \mathbf{N}\mathfrak{p}^{-s}),$$

taken over all $\mathfrak{p}$ where A has non-degenerate reduction. Then according to the above definitions, Theorem 9 implies that

$$\boxed{\zeta(A, K, s) = \zeta_K(s)\zeta_K(s - 1)[L(s, \chi_{A,K})L(s, \chi'_{A,K})]^{-1},}$$

while Theorem 10 implies that

$$\boxed{\zeta(A, K_0, s) = \zeta_K(s)\zeta_K(s - 1)L(s, \chi_{A,K})^{-1}.}$$

The function ζ_K is the Dedekind zeta function associated with the number field K, and $L(s, \chi)$ is the Hecke L-function

$$\prod_{\mathfrak{p}} (1 - \chi(\mathfrak{p})\mathbf{N}\mathfrak{p}^{-s})^{-1}$$

associated with a Hecke character of the idele classes of K.

Sometimes one wants to deal with a Hecke character of absolute value 1. Then one can define

$$\psi_{A,K}(\mathfrak{p}) = \chi_{A,K}(\mathfrak{p})\mathbf{N}\mathfrak{p}^{-\frac{1}{2}},$$

and then $L(s, \chi)$ is replaced by $L(s - \frac{1}{2}, \psi)$.

APPENDIX. A RELATION OF KRONECKER

The contents of this appendix will not be used anywhere else in the book and may be omitted. They are due to Kronecker. Cf. Weber, §115, 116.

We shall prove another property of the modular polynomial $\Phi_m(X, X)$, for an arbitrary positive integer m. Let $z \in \mathfrak{H}$ be imaginary quadratic. We are interested in the multiplicity of $j(z)$ as a root of $\Phi_m(X, X)$. This multiplicity may of course be 0. Write $z = z_1/z_2$ where z_1, z_2 lie in an imaginary quadratic field k, let $\mathfrak{a} = [z_1, z_2]$ and let $\mathfrak{o}$ be the order belonging to $\mathfrak{a}$ (or equivalently to $L_z = [z, 1]$). An element $\mu \in \mathfrak{o}$ is called **primitive** if it does not lie in $n\mathfrak{o}$ for any positive integer $n \neq 1$. If $\mu \in \mathfrak{o}$, then

$$\mu\begin{pmatrix} z_1 \\ z_2 \end{pmatrix} = \begin{pmatrix} a & b \\ c & d \end{pmatrix}\begin{pmatrix} z_1 \\ z_2 \end{pmatrix} = \alpha\begin{pmatrix} z_1 \\ z_2 \end{pmatrix}$$

with an integral matrix $\alpha = \alpha_\mu$ and we see that μ is primitive if and only if $(a, b, c, d) = 1$. Two elements of $\mathfrak{o}$ will be said to be **$\mathfrak{o}$-equivalent** if their quotient is a unit in $\mathfrak{o}$.

Theorem 11. *Let $z \in \mathfrak{H}$ be imaginary quadratic and let $\mathfrak{o}$ be the order of $L_z = [z, 1]$. Then the multiplicity of z as a root of $\Phi_m(X, X)$ is equal to the number of primitive $\mathfrak{o}$-equivalence classes of elements $\mu \in \mathfrak{o}$ such that* $\mathbf{N}\mu = m$.

Proof. Let $\{\alpha_i\}$ $(i = 1, \ldots, \psi(m))$ be representatives for the left cosets of Δ_m^* with respect to Γ. It is clear that $j(z)$ is a root of $\Phi_m(X, X)$ if and only if

$$j(z) = j(\alpha_i z)$$

for some α_i, and this is the case if and only if there exists $\gamma \in \Gamma$ such that $\gamma\alpha_i z = z$. Without loss of generality, we may assume that if $\alpha_i z$ and z lie in the same orbit of $SL_2(\mathbf{Z})$, then they are equal (multiply a representative α_i by a suitable element of $SL_2(\mathbf{Z})$).

Using the notation before the theorem, we see that the association

$$\mu \mapsto \alpha_\mu$$

induces a bijection of the primitive $\mathfrak{o}$-equivalence classes of elements $\mu \in \mathfrak{o}$ such that $\mathbf{N}\mu = m$, and those representatives α_i such that $\alpha_i z = z$. We are therefore reduced to proving that the number r of such representatives is precisely the multiplicity of $j(z)$ in $\Phi_m(X, X)$, i.e. that $\Phi_m(X, X)$ is exactly divisible by $(X - j(z))^r$. It will suffice to prove that

$$\lim_{\tau \to z} \frac{\Phi_m(j(\tau), j(\tau))}{(j(\tau) - j(z))^r} \neq 0, \infty$$

and therefore it will suffice to prove the following lemma.

Lemma. *Let $\alpha \in \Delta_m^*$ be such that $\alpha z = z$. Then*

$$\lim_{\tau \to z} \frac{j(\tau) - j(\alpha\tau)}{j(\tau) - j(z)} \neq 0, \infty.$$

Proof. We use the Taylor expansions:

$$j(\tau) - j(z) = \sum_{n=1}^{\infty} \frac{j^{(n)}(z)}{n!}(\tau - z)^n$$

and

$$j(\alpha\tau) - j(\tau) = \sum_{n=1}^{\infty} \frac{j^{(n)}(\tau)}{n!}(\alpha\tau - \tau)^n$$

$$= \sum_{n=1}^{\infty} \frac{j^{(n)}(z) + j^{(n+1)}(z)(\tau - z) + \cdots}{n!}(\alpha\tau - \tau)^n.$$

Since $\alpha\tau - \tau = 0$ has two distinct solutions (namely z and its complex conjugate) it follows at once that the above two expressions have the same order of zero at $\tau = z$, whence the lemma is proved.

Let $G_\mathfrak{o}$ be the group of proper $\mathfrak{o}$-ideal classes for an order $\mathfrak{o}$ in an imaginary quadratic field k. Let

$$H_\mathfrak{o}(X) = \prod_{\mathfrak{a} \in G_\mathfrak{o}} (X - j(\mathfrak{a})).$$

We know that all the numbers $j(\mathfrak{a})$ ($\mathfrak{a} \in G_\mathfrak{o}$) are conjugate over $\mathbf{Q}$. Hence $H_\mathfrak{o}(X)$ has integral coefficients, and is irreducible over $\mathbf{Q}$. Let $r(m, \mathfrak{o})$ be the number of primitive $\mathfrak{o}$-equivalence classes of elements $\mu \in \mathfrak{o}$ such that $\mathbf{N}\mu = m$. Then Theorem 8 shows that for a suitable constant c_m, we have

$$\Phi_m(X, X) = c_m \prod_{\mathfrak{o}} H_\mathfrak{o}(X)^{r(m,\mathfrak{o})}.$$

Counting up the degrees will yield the relation we are looking for. We make some more remarks concerning the degree of $\Phi_m(X, X)$ since the discussion in Chapter 5, §2 was kept brief for the limited purposes we had in mind then.

We have

$$\Phi_m(j, j) = \prod_{i=1}^{\psi(m)} (j - j \circ \alpha_i).$$

Suppose that α is in triangular form,

$$\alpha = \begin{pmatrix} a & b \\ 0 & d \end{pmatrix},$$

with $ad = m$, $a, d > 0$, and $0 \leqq b \leqq d - 1$. Also, α is primitive. The lowest term of the q-expansion of $j - j \circ \alpha$ is then:

i) $-e^{-2\pi ib/d} q^{-a/d}$ if $a > d$,

ii) q^{-1} if $a < d$,

iii) $(1 - e^{2\pi ib/d}) q^{-1}$ if $a = d$.

This third possibility occurs if and only if m is a square and had been disregarded before, but we must take it into account now. Note that in this last case, the coefficient of q^{-1} is $\neq 0$ since α is primitive, so that $b \neq 0$. The lowest term of the q-expansion of $\Phi_m(j, j)$ is therefore Cq^{-N}, where C is some non-zero constant, and

$$N = \sum_{a>d} \frac{a}{d} \frac{d}{e} \varphi(e) + \sum_{a<d} 1 \frac{d}{e} \varphi(e) + \varphi(\sqrt{m}),$$

As usual, we use the notation $e = (a, m/a)$, and by convention, $\varphi(\sqrt{m}) = 0$ if m is not a square, otherwise has the usual value of the Euler function. This yields:

Theorem 12. *The degree of $\Phi_m(X, X)$ is equal to*

$$N = 2 \sum_{\substack{a \mid m \\ a > \sqrt{m}}} \frac{a}{e} \varphi(e) + \varphi(\sqrt{m}),$$

and we have the Kronecker relation

$$\sum_{\mathfrak{o}} r(m, \mathfrak{o}) h_{\mathfrak{o}} = \deg \Phi_m(X, X),$$

taking the sum over all orders $\mathfrak{o}$, where $h_{\mathfrak{o}}$ is the number of elements in $G_{\mathfrak{o}}$.

From the elementary discussion about orders, we know that an order $\mathfrak{o}$ is of the form

$$\mathfrak{o} = \left[\frac{D + \sqrt{D}}{2}, 1\right],$$

where D is the discriminant of $\mathfrak{o}$, and $D \equiv 0$ or $1 \pmod 4$. We shall also write $r(m, \mathfrak{o}) = r(m, D)$ if $D = D(\mathfrak{o})$. We ask what is the largest possible value of

$|D|$ (for given m) such that $r(m, D) > 0$. The answer is given by the following theorem.

Theorem 13. *Let $m > 1$. The largest values of $|D|$ with $r(m, D) > 0$ are those for which D is equal to $-4m$ and $-4m + 1$. The corresponding values of $r(m, D)$ are*

$r(m, -4m) = 1$, and a representative primitive solution of $\mathbf{N}\mu = m$ is $\frac{\sqrt{D}}{2}$;

$r(m, -4m + 1) = 2$, and representative primitive solutions of $\mathbf{N}\mu = m$ are given by $\frac{1 + \sqrt{D}}{2}$ and $\frac{1 - \sqrt{D}}{2}$.

Proof. Write a primitive solution of $\mathbf{N}\mu = m$ in the form

$$\mu = \frac{x + y\sqrt{D}}{2}, \qquad \text{with } x, y \in \mathbf{Z},$$

Then $4m = x^2 - Dy^2$. If $y = 0$, then $x = \pm 2$, because μ is primitive. Hence $m = 1$, which is impossible. Hence $|y| > 0$. Therefore

$$|D| \leqq 4m.$$

Since $D \equiv 0$ or $1 \pmod 4$, the highest possible values for $|D|$ correspond to $D = -4m$ and $D = -4m + 1$. We now determine the multiplicities in these two cases.

Case 1. D = −4m. From the relation $4m = x^2 + 4my^2$ we conclude that $x = 0$ and $y = \pm 1$. Therefore $\mu = \frac{\sqrt{D}}{2}$, i.e.

$$\mu = -\frac{D}{2} + \frac{D + \sqrt{D}}{2}.$$

This is an element of $\mathfrak{o}$, and is primitive since $\frac{D + \sqrt{D}}{2}$ has coefficient 1. Thus we have found one solution, and its multiplicity is 1.

Case 2. D = −4m + 1. Then $4m = x^2 + 4my^2 - y^2$. Since μ is primitive, $y \neq \pm 2$, and also $y \neq 0$, so $y = \pm 1$. Hence $x = \pm 1$. Then

$$\mu = \frac{\pm 1 \pm \sqrt{D}}{2} = \frac{\pm 1 \mp D}{2} \pm \frac{D + \sqrt{D}}{2}$$

lies in $\mathfrak{o}$ and is primitive, because $D \equiv 1 \pmod 4$, D is odd, and

$$\frac{\pm 1 \mp D}{2}$$

is an integer, while the other term has coefficient 1. This shows that

$$r(m, -4m + 1) = 2,$$

with the two inequivalent primitive solutions

$$\frac{1 + \sqrt{D}}{2} \quad \text{and} \quad \frac{1 - \sqrt{D}}{2}$$

for $\mathbf{N}\mu = m$.

11 Shimura's Reciprocity Law

§1. RELATION BETWEEN GENERIC AND SPECIAL EXTENSIONS

Let F be the modular function field, studied in Chapter 6. We saw that F can be identified with the field of x-coordinates (or h-coordinates, h = Weber function) of division points of an elliptic curve A defined over $\mathbf{Q}(j)$, having invariant j. Let k be an imaginary quadratic field, and let $z \in k \cap \mathfrak{H}$. Then the fundamental theorem of complex multiplication tells us that the field $F(z)$ consisting of all values $f(z)$, $f \in F$, is k_{ab}. Let $\mathfrak{o}$ be the local ring in $k(j)$ of the place $f \mapsto f(z)$, with $f \in k(j)$. Let S be the integral closure of $\mathfrak{o}$ in F. Then every function $f \in S$ is defined at z, and we let $\mathfrak{M}$ be the kernel of the homomorphism $f \mapsto f(z)$, $f \in S$. We are now in a situation similar to that of the decomposition group, except that automorphisms of F do not necessarily leave $\mathbf{Q}(j)$ fixed. We want to determine in some fashion the decomposition group of this situation, which we shall see is isomorphic to $\mathrm{Gal}(k_{ab}/k)$, i.e. to the Galois group of the residue class field, so that our situation is essentially unramified, except in the two cases when z is equivalent to i or ρ under the modular group.

If $\xi \in k^*$, then there is a rational matrix $q(\xi) \in GL_2^+(\mathbf{Q})$ such that

$$\xi\begin{pmatrix} z \\ 1 \end{pmatrix} = \begin{pmatrix} \xi z \\ \xi 1 \end{pmatrix} = q(\xi)\begin{pmatrix} z \\ 1 \end{pmatrix}.$$

So we have an embedding

$$q_z = q: k^* \to GL_2^+(\mathbf{Q})$$

satisfying the above property. We note that z is a fixed point of $q(k^*)$. By continuity, we can extend q to an embedding

$$q_{z,p} = q_p: k_p^* \to GL_2(\mathbf{Q}_p),$$

whence to an embedding of the ideles, again denoted by $q = q_z$ (depending on z),

$$q: \mathbf{A}_k^* \to GL_2(\mathbf{A}),$$

although we shall use the homomorphism

$$q: \mathbf{A}_k^* \to GL_2(\mathbf{A}_f)$$

which drops the complex component, and otherwise is the same as above on the p-components. The image of $\mathbf{A}_k^*$ lies in the adelized GL_2 because for all p,

$$\mathbf{Q}_p \otimes k = k_p = \mathbf{Q}_p z \oplus \mathbf{Q}_p,$$

and for almost all p,

$$\mathfrak{o}_{k,p} = \mathbf{Z}_p z \oplus \mathbf{Z}_p.$$

Theorem 1. (**Shimura**) *Let s be an idele of k, and let (s^{-1}, k) be the Artin symbol on k_{ab}. Let $z \in k$ and $z \in \mathfrak{H}$. Let $\sigma = \sigma(q_z(s))$ be the automorphism of F of Chapter 7, §3. Then for every function $f \in F$ defined at z, we have*

$$f(z)^{(s^{-1},k)} = f^\sigma(z).$$

Proof. We shall first prove that the above relation holds for the Fricke functions f_a as in Chapter 7, and for j itself. After that, we shall give a formal decomposition group argument to show that the same relation holds for all f.

Write

$$q(s) = u\alpha$$

where

$$\alpha \in GL_2^+(\mathbf{Q}) \qquad \text{and} \qquad u \in \prod_p GL_2(\mathbf{Z}_p) = U,$$

as in Theorem 1 of Chapter 7, §1. Let $a \in \mathbf{Q}^2$, $a \notin \mathbf{Z}^2$. We recall the notation

$$a\begin{pmatrix} z \\ 1 \end{pmatrix} = a_1 z + a_2.$$

Locally, we have

$$\begin{aligned} s_p a\begin{pmatrix} z \\ 1 \end{pmatrix} &= a s_p\begin{pmatrix} z \\ 1 \end{pmatrix} = a q_p(s_p)\begin{pmatrix} z \\ 1 \end{pmatrix} \\ &= a u \alpha\begin{pmatrix} z \\ 1 \end{pmatrix} \\ &= a u \mu\begin{pmatrix} \alpha(z) \\ 1 \end{pmatrix} \end{aligned}$$

with some $\mu \in k$ (actually $\mu = cz + d$).

As usual, we write $L_z = [z, 1]$. What is sL_z? We contend that

$$sL_z = \mu L_{\alpha(z)}.$$

It suffices to verify this locally at each p. We have

$$\begin{aligned} s_p \mathbf{Z}_p L_z = \mathbf{Z}_p^2 s_p\begin{pmatrix} z \\ 1 \end{pmatrix} &= \mathbf{Z}_p^2 u_p \mu\begin{pmatrix} \alpha(z) \\ 1 \end{pmatrix} \\ &= \mathbf{Z}_p^2 \mu\begin{pmatrix} \alpha(z) \\ 1 \end{pmatrix}. \end{aligned}$$

This proves our contention.

We now have the usual diagram of complex multiplication,

$$\begin{array}{ccccc}
\mathbf{Q}^2/\mathbf{Z}^2 & \xrightarrow{\text{dot}} & \mathbf{Q}L_z/L_z & \longrightarrow & A^z_{\text{tor}} \\
 & & \downarrow s & & \downarrow \sigma = (s^{-1}, k) \text{ on } k^{\text{ab}} \\
\downarrow u & & \mathbf{Q}sL_z/sL_z & \longrightarrow & A^{\alpha(z)}_{\text{tor}} \\
 & & \downarrow \mu^{-1} & & \downarrow \text{id} \\
\mathbf{Q}^2/\mathbf{Z}^2 & \xrightarrow[\text{dot}]{} & \mathbf{Q}L_{\alpha(z)}/L_{\alpha(z)} & \longrightarrow & A^{\alpha(z)}_{\text{tor}}
\end{array}$$

Note that

$$j(z)^{(s^{-1},k)} = j(\alpha(z)),$$

so that we can select

$$A^{\alpha(z)} = (A^z)^\sigma.$$

We contend that multiplication by u on the far left makes the diagram commutative. This is trivially verified using our previous computation.

Therefore, if $\varphi_z \colon \mathbf{C}/L_z \to A^z_{\mathbf{C}}$ is our usual parametrization of A^z, we obtain, for some automorphism ε,

$$\varphi_z(a)^{(s^{-1},k)} = \varepsilon \circ \varphi_{\alpha(z)}(au),$$

if $q(s) = u\alpha$. Taking the Weber function, rewriting this in terms of the notation f_a where f_a is the Fricke function, and recalling that $q(s) = u\alpha$, we have

$$\boxed{\begin{aligned} f_a(z)^{(s^{-1},k)} &= f_a^\sigma(z) = f_{au}(\alpha(z)) \\ j(z)^{(s^{-1},k)} &= j(\alpha(z)) = j^\sigma(z). \end{aligned}}$$

Thus we have proved our theorem for the special functions f_a, j. Observe that we can take the Weber functions of any one of the three types, and these relations still hold.

Next we prove that the relation of the theorem holds true for all elements of F. Let S be the integral closure of the ring $R = k[j]$ in F. Since $j \circ \alpha$ is integral over $k[j]$, it follows that σ maps S on a ring which is integral over $k[j \circ \alpha]$, whence σ induces an automorphism of S. We let $\mathfrak{m}$ be the maximal ideal in $k[j]$, and $\mathfrak{M}$ the maximal ideal in S, which are the kernels of the homomorphism

$$f \mapsto \bar{f} = f(z).$$

If ρ is an automorphism of S which maps $\mathfrak{M}$ onto $\mathfrak{M}$, i.e. ρ lies in the isotropy group of $\mathfrak{M}$, then ρ induces an automorphism of the residue class field, denoted by

$$\bar{\rho} \colon \bar{S} \to \bar{S}.$$

We identify $\bar{S}$ as the set of all elements $\bar{f}, f \in S$.

We shall prove that there exists an element $\rho \in \operatorname{Aut}(F)$ such that:

i) ρ maps $\mathfrak{M}$ onto $\mathfrak{M}$.

ii) $\rho = \sigma$ on $k(j)$.

iii) $\bar{\rho} = (s^{-1}, k)$.

Let σ_α be the automorphism of F such that $\sigma_\alpha f = f \circ \alpha$. Then σ_α induces an automorphism of S (same argument as for σ). The formula

$$j(\alpha z) = j(z)^{(s^{-1},k)}$$

shows that $\sigma_\alpha \mathfrak{m} \subset \mathfrak{M}$, because σ_α leaves the constants fixed. Consequently $\mathfrak{m} \subset \sigma_\alpha^{-1}\mathfrak{M}$. By the ordinary Galois theory as in Chapter 8, §3 there exists an element $\tau \in \operatorname{Gal}(F/k(j))$ such that $\tau\mathfrak{M} = \sigma_\alpha^{-1}\mathfrak{M}$, whence we obtain

$$\sigma_\alpha \tau \mathfrak{M} = \mathfrak{M},$$

and $\sigma_\alpha \tau = \sigma$ on $k(j)$. This already achieves the first two of our desired conditions. Let $G = \operatorname{Gal}(F/k(j))$. The residue class field $R/\mathfrak{m} = \bar{R}$ is precisely $k(j(z))$. By the surjectivity of Proposition 4, Chapter 8, §3 we know that there exists an automorphism $\lambda \in G_{\mathfrak{M}}$ such that $\bar{\lambda}$ has any prescribed effect on the residue class field, in our case such that

$$\bar{\lambda}\, \overline{\sigma_\alpha \tau} = (s^{-1}, k).$$

We let $\rho = \lambda\sigma_\alpha\tau$. Then ρ satisfies all three of our requirements.

The automorphism $\rho^{-1}\sigma$ satisfies the condition

$$(\rho^{-1}\sigma f_a)(z) = f_a(z)$$

for all $a \neq 0$, $a \in \mathbf{Q}^2/\mathbf{Z}^2$, and leaves $\mathbf{Q}(j)$ fixed. By Theorems 2′ and 3′ of Chapter 9, §3 we conclude that $\rho^{-1}\sigma$ lies in the inertia group. Since the relations we want, i.e.

$$\overline{\rho f} = \bar{f}^{(s^{-1},k)}$$

are true for ρ and all $f \in S$, they are also true for σ, and this proves our theorem.

In his proof of the reciprocity law, as given in [B12], Shimura gave the arguments showing that the relation holds for j and the functions f_a. To extend this to all elements of F, he then went through a fairly elaborate discussion, even using the parametrizations of the models of the function fields F_N over $\mathbf{Q}$ from the upper half-plane. The difficulty concerning such a step had arisen before, in every treatment of complex multiplication. We have avoided the difficulty by a more direct usage of the formalism of decomposition groups, which follows the usual formalism of Galois extensions.

It is worthwhile also to describe the inertia group in the full group of automorphisms.

Theorem 2. *Let $z \in \mathfrak{H}$ be imaginary quadratic, and let $k = \mathbf{Q}(z)$. Let $\mathfrak{M}$ be the kernel of the place $f \mapsto f(z) = \bar{f}$ in the modular function field F. Let*

G be the group of automorphisms of F over k, and let $G_{\mathfrak{M}}$ be the isotropy group of $\mathfrak{M}$. Then the map

$$\sigma \mapsto \bar{\sigma}, \qquad \sigma \in G_{\mathfrak{M}},$$

is a homomorphism of $G_{\mathfrak{M}}$ onto $\mathrm{Gal}(k_{ab}/k)$, whose kernel consists of those elements σ_α, with $\alpha \in GL_2^+(\mathbf{Q})$ satisfying $\alpha z = z$.

Proof. By Theorem 6 of Chapter 7, §3 (the Shimura exact sequence), we can write an element $\sigma \in G_{\mathfrak{M}}$ in the form

$$\sigma = \sigma(u\alpha),$$

with $u \in \prod GL_2(\mathbf{Z}_p)$ and $\alpha \in GL_2^+(\mathbf{Q})$. Then $\sigma j = j \circ \alpha$. Suppose that $\bar{\sigma} = \mathrm{id}$. Then

$$j(\alpha z) = j(z),$$

whence $\alpha z = \gamma z$ for some $\gamma \in SL_2(\mathbf{Z})$. For any Fricke function f_a, we have $f_a^\sigma(z) = f_a(z)$, whence

$$f_{au}(\alpha z) = f_a(z).$$

But $f_{au}(\alpha z) = f_{au}(\gamma z) = f_{au\gamma}(z)$. By Theorem 1, 2′, 3′ of Chapter 9, §3, we conclude that $u\gamma = \gamma_1$ for some $\gamma_1 \in SL_2(\mathbf{Z})$ such that $\gamma_1 z = z$. Hence $\sigma = \sigma(\gamma_1\gamma^{-1}\alpha)$, and $\gamma_1\gamma^{-1}\alpha$ leaves z fixed. Conversely, if $\beta \in GL_2^+(\mathbf{Q})$ and $\beta z = z$, it is clear that $\sigma(\beta)$ lies in the kernel of $\sigma \mapsto \bar{\sigma}$. The surjectivity of our homomorphism on $\mathrm{Gal}(k_{ab}/k)$ comes from Theorem 1, thereby proving our theorem.

Corollary. *Let $\mathbf{A}_k^*$ be the group of ideles of k. Then $G_{\mathfrak{M}}$ is the image of $\mathbf{A}_k^*$ under the embedding q_z.*

Proof. Theorem 1 shows that the image of q_z is contained in $G_{\mathfrak{M}}$. Furthermore, if $\sigma \in G_{\mathfrak{M}}$, then

$$\bar{\sigma} = \overline{\sigma(q_z(s))}$$

for some idele s, whence σ and $\sigma(q_z(s))$ differ by an element of the kernel in Theorem 2, which we know is of type σ_α. If

$$\alpha = \begin{pmatrix} a & b \\ c & d \end{pmatrix},$$

we let $\mu = cz + d$, and identify μ with the idele having μ on each component. Then $\sigma(q_z(\mu)) = \alpha$, and our assertion follows.

§2. APPLICATION TO QUOTIENTS OF MODULAR FORMS

We shall proceed as in Shimura [B12]. If f is an automorphic function of a certain weight $2t$ as defined in Chapter 3, §2 we shall write f homogeneously, i.e.

$$f(\tau) = f\begin{pmatrix} \tau \\ 1 \end{pmatrix},$$

so that as a function of two variables, f is homogeneous of degree $-2t$, i.e.

$$f\left(\lambda\begin{pmatrix}\tau\\1\end{pmatrix}\right) = \lambda^{-2t} f\begin{pmatrix}\tau\\1\end{pmatrix}, \qquad \lambda \in \mathbf{C}.$$

Aside from the meromorphic conditions, the functional equation of an automorphic function with respect to $\Gamma = SL_2(\mathbf{Z})$ then reads in homogeneous notation

$$f\left(\gamma\begin{pmatrix}\tau\\1\end{pmatrix}\right) = f\begin{pmatrix}\tau\\1\end{pmatrix}, \qquad \text{all } \gamma \in \Gamma.$$

Theorem 3. *Let f, g be automorphic functions of the same weight, with rational Fourier coefficients, and let $\alpha \in M_2^+(\mathbf{Z})$,* det $\alpha = N$. *Let*

$$h(\tau) = \frac{f\left(\alpha\begin{pmatrix}\tau\\1\end{pmatrix}\right)}{g\begin{pmatrix}\tau\\1\end{pmatrix}}.$$

Then h is a modular function of level N. Furthermore:

i) *h is fixed under the group $\alpha^{-1}\Gamma\alpha \cap \Gamma$.*

ii) *Letting $U = \prod GL_2(\mathbf{Z}_p)$ as before, h is fixed under*

$$\alpha^{-1}U\alpha \cap U.$$

Proof. Let $\gamma \in \Gamma_N$, i.e. $\gamma \in \Gamma$ and $\gamma = I + N\beta$ for some integral matrix β. Then

$$\alpha\gamma\alpha^{-1} = I + N\alpha\beta\alpha^{-1}$$

is integral and has determinant 1. Hence

$$h(\tau) = \frac{f\left(\alpha\begin{pmatrix}\tau\\1\end{pmatrix}\right)}{g\begin{pmatrix}\tau\\1\end{pmatrix}} = \frac{f\left(\alpha\gamma\alpha^{-1}\alpha\begin{pmatrix}\tau\\1\end{pmatrix}\right)}{g\begin{pmatrix}\tau\\1\end{pmatrix}} = h(\gamma\tau),$$

which proves that h is modular of level N. Substituting an element $\alpha^{-1}\gamma\alpha$ which also lies in Γ into h leaves h invariant, as one sees at once.

The proof of the other assertion is slightly longer. We first make a reduction by diagonalizing α. There exist $\gamma, \delta \in SL_2(\mathbf{Z})$ such that

$$\alpha = \gamma\beta\delta, \qquad \beta = \begin{pmatrix}rm & 0\\0 & r\end{pmatrix}$$

with $r \in \mathbf{Q}$ and m equal to a positive integer. Then

$$h_\alpha(\tau) = h(\tau) = \frac{f\left(\gamma\beta\delta\begin{pmatrix}\tau\\1\end{pmatrix}\right)}{g\begin{pmatrix}\tau\\1\end{pmatrix}} = \frac{f\left(\beta\delta\begin{pmatrix}\tau\\1\end{pmatrix}\right)}{g\begin{pmatrix}\tau\\1\end{pmatrix}}$$

$$= h_\beta \circ \delta.$$

It suffices to prove that h_β is fixed under $\delta(\alpha^{-1}U\alpha \cap U)\delta^{-1}$, i.e. under $\beta^{-1}U\beta \cap U$, because $\delta\alpha^{-1} = \beta^{-1}\gamma^{-1}$, and $\gamma^{-1}U\gamma = U$. Thus it suffices to prove our assertion for

$$h(\tau) = \frac{f(\beta\tau)}{g(\tau)} = \frac{f(m\tau)}{g(\tau)},$$

which we have to show is invariant under $\beta^{-1}U\beta \cap U$, where

$$\beta = \begin{pmatrix} m & 0 \\ 0 & 1 \end{pmatrix}.$$

Let $u \in U$ and suppose $\beta^{-1}u\beta$ also lies in U. Write u_p as a matrix

$$u_p = \begin{pmatrix} a_p & b_p \\ c_p & d_p \end{pmatrix} \in GL_2(\mathbf{Z}_p).$$

Then

$$\beta^{-1}u_p\beta = \begin{pmatrix} a_p & b_p/m \\ mc_p & d_p \end{pmatrix}.$$

This lies in U_p if and only if $b_p = mb'_p$ for some p-adic unit b'_p. Consequently, we have proved:

Lemma. *If*

$$\alpha = \begin{pmatrix} m & 0 \\ 0 & 1 \end{pmatrix},$$

then $\alpha^{-1}U\alpha \cap U$ consists of all elements $v \in U$ such that

$$v_p = \begin{pmatrix} a_p & b_p \\ mc_p & d_p \end{pmatrix}, \qquad a_p, b_p, c_p, d_p \in \mathbf{Z}_p;$$

and $\alpha^{-1}\Gamma\alpha \cap \Gamma = \Gamma_0(m)$, i.e. *consists of all matrices*

$$\begin{pmatrix} a & b \\ mc & d \end{pmatrix} \in SL_2(\mathbf{Z}),$$

Reading mod m, our proof of Theorem 3 reduces to the following special case.

Theorem 4. *Let f, g be automorphic functions of the same weight with respect to $\Gamma = SL_2(\mathbf{Z})$, and with rational Fourier coefficients. Let m be a positive integer. Then the function*

$$h(\tau) = \frac{f(m\tau)}{g(\tau)}$$

has level m, and is fixed under the group of all automorphisms of F represented by the matrices

$$\begin{pmatrix} a & b \\ 0 & d \end{pmatrix} \in GL_2(\mathbf{Z}/m\mathbf{Z}). \tag{1}$$

Proof. Since h has rational Fourier coefficients, it is fixed under the automorphisms of F represented by matrices of the form

$$\begin{pmatrix} 1 & 0 \\ 0 & d \end{pmatrix}, \qquad d \in (\mathbf{Z}/m\mathbf{Z})^*.$$

Multiplying a matrix of type (1) by such a diagonal matrix, we are reduced to the case when the element of (1) lies in $SL_2(\mathbf{Z}/m\mathbf{Z})$. Such an element has a representative in $SL_2(\mathbf{Z})$ lying precisely in the group

$$\alpha^{-1}\Gamma\alpha \cap \Gamma = \Gamma_0(m)$$

as described in Lemma 1. This concludes the proof, in view of the first assertion in Theorem 3.

Corollary 1. *The function $f(m\tau)/g(\tau)$ lies in $\mathbf{Q}(j, j \circ m)$.*

Proof. The fixed field of the group of automorphisms of F_m in Theorem 4 is equal to $\mathbf{Q}(j, j \circ m)$, according to Theorem 5 of Chapter 6, §4, so our corollary is clear.

Corollary 2. *Let Δ be the usual discriminant function, and let*

$$\varphi_m(\tau) = \frac{\Delta(m\tau)}{\Delta(\tau)}.$$

Then $\mathbf{Q}(j, j \circ m) = \mathbf{Q}(j, \varphi_m)$.

Proof. Take $f = g = (2\pi)^{-12}\Delta$ in the theorem. Then we see that the function φ_m lies in $\mathbf{Q}(j, j \circ m)$, referring back to Theorem 5 of Chapter 6, §4. Looking at the conjugates of this function under the modular group, and the coefficients of the q-expansion at infinity, one sees that all $\psi(m)$ conjugates are distinct, and hence that we have an equality of fields as stated.

The arithmetic result corresponding to the function theoretic result of Corollary 2 will be proved in Chapter 21, §1, Theorem 3.

The next theorem is Excercise 6.37 of Shimura's book [B12]. In his book Shimura proves the result only for ideals of the ring of all algebraic integers. It is needed explicitly in general for certain applications.

Theorem 5. (***Shimura***). *Let f, g be automorphic functions of the same weight with respect to Γ, and with rational Fourier coefficients. Let $\alpha \in M_2^+(\mathbf{Z})$,* $\det \alpha = N$. *Let*

$$h(\tau) = \frac{f\left(\alpha\begin{pmatrix}\tau \\ 1\end{pmatrix}\right)}{g\begin{pmatrix}\tau \\ 1\end{pmatrix}}.$$

Let k be an imaginary quadratic field. Let s be an idele of k such that $s_p = 1$ for all $p|N$. Let $L = [z_1, z_2]$ be a lattice in k with $z = z_1/z_2 \in \mathfrak{H}$. There exists $\eta \in GL_2^+(\mathbf{Q})$ such that:

i) $\eta\begin{pmatrix} z_1 \\ z_2 \end{pmatrix}$ *is a basis for* $s^{-1}L$.

ii) $\alpha\eta\alpha^{-1} \in GL_2(\mathbf{Z}_p)$ *for all* $p|N$.

Assume that $f \circ \alpha$ *and* g *are defined at* z. *For any* η *satisfying* (i) *and* (ii), *we have*

$$h(z)^{(s,k)} = h(\eta z).$$

Proof. We first prove the existence of η satisfying the desired conditions. It is trivial to find some η satisfying (i). We need only prove that there exists $\gamma \in \Gamma$ such that replacing η by $\gamma\eta$ satisfies (ii), i.e.

$$\alpha\gamma\eta\alpha^{-1} \in GL_2(\mathbf{Z}_p)$$

for all $p|N$. We have

$$s_p^{-1}L_p = \mathbf{Z}_p^2\eta\begin{pmatrix} z_1 \\ z_2 \end{pmatrix}.$$

Hence

$$q_p(s_p^{-1}) = u_p\eta,$$

with some $u_p \in GL_2(\mathbf{Z}_p) = U_p$. Hence $u_p = \eta^{-1}$ for $p|N$, and $q(s^{-1}) = u\eta$. Note that multiplication on the right by the element $u_p \in GL_2(Z_p)$ yields an automorphism of $\mathbf{Z}_p^2$, and hence we have an isomorphism

$$\mathbf{Z}^2/\mathbf{Z}^2\alpha \approx \coprod \mathbf{Z}_p^2/\mathbf{Z}_p^2\alpha u_p.$$

There exists a sublattice M of $\mathbf{Z}^2$ such that $M_p = \mathbf{Z}_p^2\alpha u_p$ for all p, and then

$$\mathbf{Z}^2/\mathbf{Z}^2\alpha \approx \mathbf{Z}^2/M.$$

By elementary divisor theory, there exists $\gamma \in \Gamma$ such that $\mathbf{Z}^2\alpha\gamma = M$, or in other words

$$\mathbf{Z}_p^2\alpha\gamma = \mathbf{Z}_p^2\alpha u_p$$

for all p. Hence

$$\alpha\gamma u_p^{-1}\alpha^{-1} \in U_p$$

for all p. In particular, for $p|N$, we know that $u_p^{-1} = \eta$, whence the existence of the desired γ follows.

To give the effect of (s, k), we use $q(s^{-1}) = u\eta$ and apply Theorem 1. By Theorem 1, we find

$$h(z)^{(s,k)} = h^\sigma(z) = h^{\sigma(u)}(\eta z),$$

and it suffices to prove that $h^{\sigma(u)} = h$. By Theorem 3(ii), we need only show that $u \in \alpha^{-1}U\alpha \cap U$, i.e. $\alpha u\alpha^{-1} \in U$. We check this at each prime. If $p|N$, then this amounts to the second hypothesis on η. If $p\nmid N$, then α itself lies in U_p, so this is clear. This completes the proof of Theorem 5.

Let $\mathfrak{o}$ be an order in k. If $\mathfrak{b}$ is a proper $\mathfrak{o}$-ideal prime to the conductor of $\mathfrak{o}$, we recall the notation of Chapter 10, §3, where $(\mathfrak{b}, k) = (\mathfrak{b}_k, k)$, $\mathfrak{b}_k = \mathfrak{b}\mathfrak{o}_k$, and

$(\mathfrak{b}_k, k)$ is the Artin automorphism on any abelian extension of k in which all the prime factors of $\mathfrak{b}_k$ are unramified (in particular $k(j(\mathfrak{o}))$).

Theorem 6. *Let f, g be automorphic functions of the same weight with respect to Γ, and with rational Fourier coefficients. Assume that they are holomorphic on $\mathfrak{H}$, and g does not vanish on $\mathfrak{H}$. Let $\mathfrak{o}$ be an order of k, and let $\mathfrak{a}, \mathfrak{b}$ be proper $\mathfrak{o}$-ideals, with $\mathfrak{b}$ prime to the conductor of $\mathfrak{o}$. For any proper $\mathfrak{o}$-ideal $\mathfrak{c}$, define*

$$h_{\mathfrak{a}}(\mathfrak{c}) = \frac{f(\mathfrak{a}\mathfrak{c})}{g(\mathfrak{c})}.$$

Then $h_{\mathfrak{a}}(\mathfrak{o})$ lies in $k(j(\mathfrak{o}))$, and

$$h_{\mathfrak{a}}(\mathfrak{o})^{(\mathfrak{b},k)} = h_{\mathfrak{a}}(\mathfrak{b}^{-1}).$$

Proof. Let $\mathfrak{o} = [z, 1]$ and let α be an integral matrix such that

$$\alpha\begin{pmatrix} z \\ 1 \end{pmatrix}$$

is a basis of $\mathfrak{a}$. Let $\det \alpha = N$. Assume first that $\mathfrak{b}$ is also prime to N. Let s be an idele such that $s_p = 1$ for all $p|N$, and such that $s_p\mathfrak{o}_p = \mathfrak{b}_p$ for all p. Let h be as in Theorem 5. We find:

$$h(z) = h_{\mathfrak{a}}(\mathfrak{o}) = \frac{f(\mathfrak{a})}{g(\mathfrak{o})} = \frac{f\left(\alpha\begin{pmatrix} z \\ 1 \end{pmatrix}\right)}{g\begin{pmatrix} z \\ 1 \end{pmatrix}},$$

and

$$h(\eta z) = \frac{f\left(\alpha\eta\begin{pmatrix} z \\ 1 \end{pmatrix}\right)}{g\left(\eta\begin{pmatrix} z \\ 1 \end{pmatrix}\right)}.$$

We shall now prove that $\alpha\eta\begin{pmatrix} z \\ 1 \end{pmatrix}$ is a basis for $\mathfrak{b}^{-1}\mathfrak{a}$, and we check this for each prime p. If $p|N$, then $\mathfrak{b}_p = \mathfrak{o}_p$ and

$$\mathbf{Z}_p^2\alpha\eta\begin{pmatrix} z \\ 1 \end{pmatrix} = \mathbf{Z}_p^2\alpha\eta\alpha^{-1}\alpha\begin{pmatrix} z \\ 1 \end{pmatrix} = \mathfrak{a}_p = \mathfrak{b}_p^{-1}\mathfrak{a}_p$$

using the definition of α. On the other hand, if $p \nmid N$, then $\alpha_p = \mathfrak{o}_p$ and $\mathbf{Z}_p^2\alpha = \mathbf{Z}_p^2$, so that

$$\mathbf{Z}_p^2\alpha\eta\begin{pmatrix} z \\ 1 \end{pmatrix} = \mathbf{Z}_p^2\eta\begin{pmatrix} z \\ 1 \end{pmatrix} = \mathfrak{b}_p^{-1} = \mathfrak{b}_p^{-1}\mathfrak{a}_p.$$

This gives us the desired basis for $\mathfrak{b}^{-1}\mathfrak{a}$.

By Theorem 5 we conclude that

$$h_{\mathfrak{a}}(\mathfrak{o})^{(s,k)} = h_{\mathfrak{a}}(\mathfrak{b}^{-1}) = \frac{f(\mathfrak{a}\mathfrak{b}^{-1})}{g(\mathfrak{b}^{-1})}.$$

Let $K = k(j(\mathfrak{o}), h_{\mathfrak{a}}(\mathfrak{o}))$. Let S be a finite set of primes containing all primes dividing N, all primes dividing the conductor of $\mathfrak{o}$, and all primes which may ramify in K. Assume that $\mathfrak{b}$ is also prime to S. We apply the above relation to $\mathfrak{b}$, and select s such that $s_p = 1$ for all $p \in S$. Then

$$h_{\mathfrak{a}}(\mathfrak{o})^{(s,k)} = h_{\mathfrak{a}}(\mathfrak{o})^{(\mathfrak{b},k)} = \frac{f(\mathfrak{a}\mathfrak{b}^{-1})}{g(\mathfrak{b}^{-1})}.$$

Suppose that (s, k) leaves $j(\mathfrak{o})$ fixed. The Kronecker congruence relation tells us that

$$j(\mathfrak{o})^{(s,k)} = j(\mathfrak{o})^{(\mathfrak{b},k)} = j(\mathfrak{b}^{-1}).$$

Consequently $\mathfrak{b} = \lambda\mathfrak{o}$ is principal, and the above formula shows that $h_{\mathfrak{a}}(\mathfrak{o})$ is also kept fixed by (s, k). It follows that $h_{\mathfrak{a}}(\mathfrak{o})$ lies in $k(j(\mathfrak{o}))$.

If we start with a proper $\mathfrak{o}$-ideal $\mathfrak{b}$ which is prime to the conductor of $\mathfrak{o}$ (and hence contains only prime factors in k which are unramified in $k(j(\mathfrak{o}))$, then there exists $\lambda \in k$ such that $\lambda\mathfrak{b}$ is an $\mathfrak{o}$-ideal prime to S, by Theorem 5 of Chapter 8, §1. On $k(j(\mathfrak{o}))$ the two symbols $(\mathfrak{b}, k)$ and $(\lambda\mathfrak{b}, k)$ have the same effect. This reduces our theorem to the case already treated, and concludes the proof.

Corollary. *The values* $\Delta(\mathfrak{a}\mathfrak{o})/\Delta(\mathfrak{o})$ *lie in* $k(j(\mathfrak{o}))$, *and we have*

$$\left(\frac{\Delta(\mathfrak{a}\mathfrak{o})}{\Delta(\mathfrak{o})}\right)^{(\mathfrak{b},k)} = \frac{\Delta(\mathfrak{a}\mathfrak{b}^{-1})}{\Delta(\mathfrak{b}^{-1})}.$$

Proof. Take $f = g = (2\pi)^{-12}\Delta$ in Theorem 6.

12 The Function $\Delta(\alpha\tau)/\Delta(\tau)$

§1. BEHAVIOR UNDER THE ARTIN AUTOMORPHISM

In this section we give an example for the Shimura theorem concerning the quotient of automorphic functions. Throughout this section we let k be an imaginary quadratic field and we let

$$\mathfrak{o}_k = [z, 1], \qquad z \in \mathfrak{H}.$$

We consider the special case when $\alpha = \begin{pmatrix} m & 0 \\ 0 & 1 \end{pmatrix}$, so that $\alpha(z) = mz$, and

$$\mathfrak{o} = [mz, 1]$$

is the order with conductor m. If $\mathfrak{b}$ is a proper $\mathfrak{o}$-ideal prime to m, we recall the notation of Chapter 10, §3, where $(\mathfrak{b}, k) = (\mathfrak{b}_k, k)$, $\mathfrak{b}_k = \mathfrak{b}\mathfrak{o}_k$, and $(\mathfrak{b}_k, k)$ is the Artin automorphism.

Theorem 1. *The value $\Delta(\mathfrak{o})/\Delta(\mathfrak{o}_k)$ lie in $k(j(\mathfrak{o}))$, and for any proper $\mathfrak{o}$-ideal $\mathfrak{b}$ prime to the conductor m of $\mathfrak{o}$, we have*

$$\left(\frac{\Delta(\mathfrak{a})}{\Delta(\mathfrak{a}_k)}\right)^{(\mathfrak{b},k)} = \frac{\Delta(\mathfrak{b}^{-1}\mathfrak{a})}{\Delta(\mathfrak{b}^{-1}\mathfrak{a}_k)}$$

for any proper $\mathfrak{o}$-ideal $\mathfrak{a}$ prime to m.

Proof. We shall first prove the special formula

$$\left(\frac{\Delta(\mathfrak{o})}{\Delta(\mathfrak{o}_k)}\right)^{(\mathfrak{b},k)} = \frac{\Delta(\mathfrak{b}^{-1})}{\Delta(\mathfrak{b}_k^{-1})}.$$

It will be a special case of Theorem 5, taking $f = g = (2\pi)^{-12}\Delta$. Note that the power of 2π cancels in the quotient, so that the condition on rational Fourier coefficients is satisfied. We let η be as in the theorem, such that

$$\eta\begin{pmatrix} z \\ 1 \end{pmatrix}$$

is a basis of $\mathfrak{b}_k^{-1}$, and $\alpha\eta\alpha^{-1} \in GL_2(\mathbf{Z}_p)$ for all $p|m$. Theorem 5 of Chapter 11 gives us

$$\left(\frac{\Delta(\mathfrak{o})}{\Delta(\mathfrak{o}_k)}\right)^{(\mathfrak{b},k)} = \frac{\Delta\left(\alpha\begin{pmatrix} z \\ 1 \end{pmatrix}\right)}{\Delta\begin{pmatrix} z \\ 1 \end{pmatrix}} = \frac{\Delta(\mathfrak{b}^{-1})}{\Delta(\mathfrak{b}_k^{-1})},$$

which is what we want, provided we check that

$$\alpha\eta\begin{pmatrix} z \\ 1 \end{pmatrix}$$

is a basis of $\mathfrak{b}^{-1}$. We do this at each prime p. If $p\nmid m$, then α is locally invertible at p, and the matter is clear. If $p|m$, we write

$$\alpha\eta\begin{pmatrix} z \\ 1 \end{pmatrix} = \alpha\eta\alpha^{-1}\alpha\begin{pmatrix} z \\ 1 \end{pmatrix}. \tag{1}$$

By hypothesis, $\alpha\eta\alpha^{-1}$ is p-integral and its determinant is a p-unit. Hence locally at p, the local lattice whose basis is given by (1) is precisely $\mathfrak{o}_p$, which is also $\mathfrak{b}_p^{-1}$ since $\mathfrak{b}$ is assumed prime to m.

From the special formula, we get the general one of the theorem, simply by applying the special case to $(\mathfrak{b}\mathfrak{a}^{-1}, k)$ instead of $(\mathfrak{b}, k)$, and using the special result twice.

Finally, to see that the desired values lie in the ring class field $k(j(\mathfrak{o}))$, let σ be any automorphism of the ray class field k_m over k which leaves $j(\mathfrak{o})$ fixed. Select an $\mathfrak{o}_k$-ideal $\mathfrak{b}_k$ prime to m such that $\sigma = (\mathfrak{b}_k, k_m/k)$. Since the restriction of σ to the ring class field $k(j(mz))$ is the identity, it follows from Theorem 5 of Chapter 10, §3 that the proper $\mathfrak{o}$-ideal $\mathfrak{b} = \mathfrak{b}_k \cap \mathfrak{o}$ must be principal. The formula giving the effect of $(\mathfrak{b}, k)$ and the homogeneity property of Δ now show that $(\mathfrak{b}_k, k)$ is also the identity on $\Delta(\mathfrak{o})/\Delta(\mathfrak{o}_k)$, which is therefore contained in $k(j(\mathfrak{o}))$. This proves all of Theorem 1.

Corollary. *The numbers $j(\mathfrak{o}), j(\mathfrak{o}_k), \Delta(\mathfrak{o})/\Delta(\mathfrak{o}_k)$ are real, and*

$$j(\mathfrak{o}_k), \frac{\Delta(\mathfrak{o})}{\Delta(\mathfrak{o}_k)} \in \mathbf{Q}(j(\mathfrak{o})).$$

Proof. The reality assertion is clear, because from the original series, say for g_2 and g_3, and any lattice L in k, with complex conjugate L', we have

$$g_2(L') = g_2(L)' \quad \text{and} \quad g_3(L') = g_3(L)',$$

whence $\Delta(L') = \Delta(L)'$ and $j(L') = j(L)'$. Since $\mathfrak{o}' = \mathfrak{o}$, the reality assertion follows, and so does the corollary, because $\mathbf{Q}(j(\mathfrak{o}))$ is the maximal real subfield of $k(j)\mathfrak{o}))$.

It will be proved in Chapter 21, §1, Theorem 3, that the numbers $j(\mathfrak{o}_k)$, $\Delta(\mathfrak{o})/\Delta(\mathfrak{o}_k)$ actually generate $k(j(\mathfrak{o}))$ over k.

§2. PRIME FACTORIZATION OF ITS VALUES

We want to describe the prime factorization of the values

$$\frac{\Delta(\alpha z)}{\Delta(z)}$$

for imaginary quadratic z, and $\alpha \in M_2^+(\mathbf{Z})$. This was done completely by Hasse, but we shall work out here only the most important special case, and refer to Deuring [B1], §22, p. 43 for the general tables. We begin with some integral properties similar to those for the j function. Let α have determinant n, also denoted by $|\alpha|$. Define

$$\varphi_\alpha(\tau) = |\alpha|^{12} \frac{\Delta\left(\alpha\begin{pmatrix}\tau\\1\end{pmatrix}\right)}{\Delta\begin{pmatrix}\tau\\1\end{pmatrix}}.$$

For any $\gamma \in SL_2(\mathbf{Z})$ we have $\varphi_{\gamma\alpha} = \varphi_\alpha$, so the value of φ depends only on the left coset of α, which we may assume primitive. We may then also assume that α is in triangular form

$$\alpha = \begin{pmatrix}a & b\\0 & d\end{pmatrix}, \tag{2}$$

as in Chapter 5, §1. Then

$$\varphi_\alpha(\tau) = |\alpha|^{12} d^{-12} \frac{\Delta(\alpha\tau)}{\Delta(\tau)}.$$

Theorem 2. *The function φ_α is integral over $\mathbf{Z}[j]$.*

Proof. Let $\alpha_1, \ldots, \alpha_{\psi(n)}$ be representatives of the left cosets of primitive matrices in $M_2(\mathbf{Z})$ having determinant n, with respect to the modular group, and take these representatives in triangular form as above. We use the same method as in Chapter 5, §2. The q-expansion for Δ is of the form

$$\Delta = (2\pi)^{12} q(1 + A(q))$$

where $A(q)$ is a power series with integer coefficients, and does not vanish on $\mathfrak{H}$. Each φ_{α_i} is holomorphic on $\mathfrak{H}$, and has a $q^{1/n}$-expansion at infinity. Each φ_α in fact has level n, as one sees by the same argument that we used for j (Theorem 4, Chapter 6, §4). The symmetric functions of the φ_{α_i} are therefore modular functions of level 1, and being holomorphic on $\mathfrak{H}$, they lie in $\mathbf{C}[j]$.

To get them in $\mathbf{Z}[j]$ we use the q-expansion. For α in triangular form as in (1), we see that under the transformation

$$q \mapsto q^{a/d}\zeta_d^b$$

the q-expansion for Δ transforms in such a way that

$$\varphi_\alpha(\tau) = n^{12}d^{-12}\frac{\Delta(\alpha\tau)}{\Delta(\tau)} = a^{12}\zeta_d^b\frac{1 + A(q^{a/d}\zeta_d^b)}{1 + A(q)}. \tag{3}$$

The Fourier coefficients of this expression lie in $\mathbf{Z}[\zeta_d]$. An automorphism σ_s sending ζ_d on ζ_d^s (with $(s, n) = 1$) extends to the power series field in $q^{1/n}$, and permutes the expansions of the $\psi(n)$ functions φ_{α_i} $(i = 1, \ldots, \psi(n))$. Hence the elementary symmetric functions of $\varphi_{\alpha_1}, \ldots, \varphi_{\alpha_{\psi(n)}}$ are invariant under these automorphisms, and therefore have coefficients in $\mathbf{Z}$. Together with the fact that each φ_α is holomorphic on $\mathfrak{H}$, this proves our theorem.

Actually if one analyses the proofs of Theorem 2, one finds that they are valid in the more general context of a quotient of automorphic functions having the same weight, under the following conditions.

Theorem 3. *Let f, g be automorphic functions of the same weight $-m$ with respect to $SL_2(\mathbf{Z})$. Assume that:*

i) *both f, g have Fourier coefficients in $\mathbf{Z}$.*

ii) *both f, g are holomorphic on $\mathfrak{H}$ and g is not zero anywhere on $\mathfrak{H}$.*

iii) *the function g has a q-expansion of the form*

$$g = q^\nu(1 + qB(q))$$

where ν is some integer, and $B(q)$ is a power series in q with integer coefficients. Then the function

$$h_\alpha(\tau) = |\alpha|^m \frac{f\left(\alpha\begin{pmatrix}\tau\\1\end{pmatrix}\right)}{g\begin{pmatrix}\tau\\1\end{pmatrix}}$$

is integral over $\mathbf{Z}[j]$.

Theorem 4. *For imaginary quadratic z, the values $\varphi_\alpha(z)$ are algebraic integers, which divide $|\alpha|^{12}$.*

Proof. Since the values $j(z)$ are algebraic integers, and since $\varphi_\alpha(z)$ is integral over $\mathbf{Z}[j(z)]$, we conclude that $\varphi_\alpha(z)$ is also an algebraic integer. To get the divisibility, we let α' be the integral matrix such that

$$\alpha'\alpha = |\alpha|I_2 = \begin{pmatrix}|\alpha| & 0\\0 & |\alpha|\end{pmatrix},$$

and we consider the product

$$|\alpha|^{12}|\alpha'|^{12}\frac{\Delta\left(\alpha'\alpha\begin{pmatrix}z\\1\end{pmatrix}\right)\Delta\left(\alpha\begin{pmatrix}z\\1\end{pmatrix}\right)}{\Delta\left(\alpha\begin{pmatrix}z\\1\end{pmatrix}\right)\quad\Delta\begin{pmatrix}z\\1\end{pmatrix}}.$$

We cancel a numerator and denominator, and use $\alpha\alpha' = |\alpha|I_2$ with the homogeneity of Δ, of degree -12, to get

$$|\alpha|^{12}|\alpha'|^{12}|\alpha|^{-12}\frac{\Delta\begin{pmatrix}z\\1\end{pmatrix}}{\Delta\begin{pmatrix}z\\1\end{pmatrix}} = |\alpha|^{12}.$$

Knowing that $\varphi_\alpha(z)$ is an algebraic integer proves our theorem.

We use the following notation. If ξ is an algebraic integer, and $\mathfrak{a}$ is some $\mathfrak{o}_k$-ideal, we write

$$\xi \approx \mathfrak{a}$$

to mean that $\xi\mathfrak{o}_K = \mathfrak{a}\mathfrak{o}_K$ in some large number field K. Similarly, if ξ_1, ξ_2 are algebraic numbers, we write

$$\xi_1 \approx \xi_2$$

to mean that ξ_1/ξ_2 is a unit. We then say that ξ_1, ξ_2 are **associated.**

Let $\mathfrak{o}$ be an order in k and let $\mathfrak{a}$, $\mathfrak{b}$ be proper $\mathfrak{o}$-ideals. Let $\mathfrak{b} = [z_1, z_2]$ and let $\alpha \in M_2^+(\mathbf{Z})$ be such that

$$\alpha\begin{pmatrix}z_1\\z_2\end{pmatrix}$$

is a basis of $\mathfrak{a}\mathfrak{b}$. We denote by $\mathbf{N}\mathfrak{a}$ the index $(\mathfrak{o} : \mathfrak{a})$, and it is clear (say from elementary divisors) that $\mathbf{N}\mathfrak{a} = |\alpha|$. Thus we use the notation

$$\varphi_{\mathfrak{a}}(\mathfrak{b}) = \varphi_\alpha(z) = \mathbf{N}\mathfrak{a}^{12}\frac{\Delta(\mathfrak{a}\mathfrak{b})}{\Delta(\mathfrak{b})}.$$

Theorem 5. *Let p be a prime number which splits completely in k, and does not divide the conductor of $\mathfrak{o}$. Let $p\mathfrak{o} = \mathfrak{p}\mathfrak{p}'$ be its factorization in $\mathfrak{o}$, $\mathfrak{p} \neq \mathfrak{p}'$. Then for any proper $\mathfrak{o}$-ideal $\mathfrak{a}$,*

$$\varphi_{\mathfrak{p}}(\mathfrak{a}) = p^{12}\frac{\Delta(\mathfrak{p}\mathfrak{a})}{\Delta(\mathfrak{a})} \approx \mathfrak{p}'^{12}.$$

Proof. Let $\mathfrak{b}$ be a proper $\mathfrak{o}$-ideal prime to p such that $\mathfrak{b}\mathfrak{p}$ is principal, say $\mathfrak{b}\mathfrak{p} = \lambda\mathfrak{o}$. Then

$$\begin{aligned}\mathbf{N}\mathfrak{b}^{12}\frac{\Delta(\mathfrak{b}\mathfrak{p}\mathfrak{a})}{\Delta(\mathfrak{p}\mathfrak{a})}\, p^{12}\frac{\Delta(\mathfrak{p}\mathfrak{a})}{\Delta(\mathfrak{a})} &= \mathbf{N}\mathfrak{b}^{12}\, p^{12}\frac{\Delta(\lambda\mathfrak{a})}{\Delta(\mathfrak{a})}\\ &= \mathbf{N}\mathfrak{b}^{12}\, p^{12}\,\lambda^{-12}.\end{aligned}$$

By Theorem 3, the first factor on the left divides $\mathbf{N}\mathfrak{b}^{12}$, which is prime to p, and the second factor divides p^{12}. On the other hand, in the prime factorization of $\lambda\mathfrak{o}$, we know that $\mathfrak{p}$ appears with multiplicity 1, and $\mathfrak{p}'$ does not appear since $\mathfrak{b}$ is prime to p. Hence in the prime factorization of the right-hand side, $\mathfrak{p}'^{12}$ is the precise contribution of $\mathfrak{p}'$. This proves our theorem.

Corollary. *For any proper* $\mathfrak{o}$*-ideal* $\mathfrak{b}$ *the number*

$$\mathbf{N}\mathfrak{b}^{12}\frac{|\Delta(\mathfrak{b})|^2}{|\Delta(\mathfrak{o})|^2}$$

is a unit.

Proof. Let

$$\varepsilon(\mathfrak{b}) = \mathbf{N}\mathfrak{b}^{6}\frac{|\Delta(\mathfrak{b})|}{|\Delta(\mathfrak{o})|}.$$

For any $\lambda \in k$, $\lambda \neq 0$ we see that $\varepsilon(\lambda\mathfrak{b}) = \varepsilon(\mathfrak{b})$, i.e. $\varepsilon(\mathfrak{b})$ depends only on the class of the ideal $\mathfrak{b}$. We can always find some λ such that $\lambda\mathfrak{b}$ is equal to a prime $\mathfrak{p}$ of degree 1, i.e.

$$(p) = \mathfrak{p}\mathfrak{p}'$$

with $\mathfrak{p} \neq \mathfrak{p}'$, and p is a prime number not dividing the conductor. (We are using here the existence of primes in generalized arithmetic progressions from class field theory. The theorem for $\mathfrak{o}_k$, combined with Theorem 5 of Chapter 8, §1 gives us what we want.) Replacing $\mathfrak{b}$ by such a prime ideal $\mathfrak{p}$, and taking the product of the expression in Theorem 5 with its conjugate yields the corollary.

We prove one other statement which is occasionally useful in applications.

Theorem 6. *Let p be a prime number, let z be imaginary quadratic, and let $\alpha_i \in M_2^+(\mathbf{Z})$ $(i = 0, \ldots, p)$ be representatives for the left cosets of matrices with determinant p, with respect to $SL_2(\mathbf{Z})$. Then*

$$\prod_{i=1}^{p+1} \varphi_{\alpha_i} = (-1)^{p-1}p^{12}.$$

Proof. We know that $\psi(p) = p + 1$, and that representative matrices can be selected as

$$\alpha_i = \begin{pmatrix} 1 & i \\ 0 & p \end{pmatrix}, \qquad i = 0, \ldots, p - 1$$

$$\alpha_p = \begin{pmatrix} p & 0 \\ 0 & 1 \end{pmatrix}.$$

Hence we get the q-expansions for the φ_{α_i} from the q-expansions as given in (3). The leading term of φ_{α_i} for $i = 0, \ldots, p - 1$ is

$$\zeta_p^i q^{1/p-1}.$$

The leading term of φ_{α_p} is $p^{12}q^{p-1}$. From this it follows that the product has a q-expansion beginning with the constant

$$\zeta_p^{1+2+\cdots+p}\,p^{12} = (-1)^{p-1}p^{12}.$$

Since this product is a modular function having no pole at infinity, it is constant, thereby proving our theorem.

Corollary. *Let $\mathfrak{a}$ be a proper $\mathfrak{o}$-ideal, and let p be a prime number, prime to the conductor, and splitting completely in k, so that $p\mathfrak{o} = \mathfrak{p}\mathfrak{p}'$, $\mathfrak{p} \neq \mathfrak{p}'$. Let $\mathfrak{a} = [z_1, z_2]$ and let P, P' be matrices of determinant p such that*

$$P\begin{pmatrix} z_1 \\ z_2 \end{pmatrix} \quad \textit{and} \quad P'\begin{pmatrix} z_1 \\ z_2 \end{pmatrix}$$

are bases of $\mathfrak{p}\mathfrak{a}$ and $\mathfrak{p}'\mathfrak{a}$ respectively. If $\alpha \in M_2^+(\mathbf{Z})$ has determinant p and α does not lie in the orbit of P or P' under $SL_2(\mathbf{Z})$, then $\varphi_\alpha(z)$ is a unit, where $z = z_1/z_2$.

Proof. The contributions to the product in Theorem 6 coming from the two terms $\varphi_{\mathfrak{p}}(\mathfrak{a})$ and $\varphi_{\mathfrak{p}'}(\mathfrak{a})$ arising from Theorem 5 will already contribute p^{12} to the factorization of the product. Since there can be no other prime factor contribution by Theorem 6, we conclude that all other terms in the product must be units, because they are all algebraic integers by Theorem 3.

To find the values of $\varphi_\alpha(z)$ in general, one can use an inductive procedure. For suppose $\alpha = \beta\gamma$, with $\beta, \gamma \in M_2^+(\mathbf{Z})$. Then for any lattice $L = [z_1, z_2]$ with $z = z_1/z_2 \in \mathfrak{H}$ we have

$$\frac{\Delta\left(\alpha\begin{pmatrix} z_1 \\ z_2 \end{pmatrix}\right)}{\Delta\begin{pmatrix} z_1 \\ z_2 \end{pmatrix}} = \frac{\Delta\left(\beta\gamma\begin{pmatrix} z_1 \\ z_2 \end{pmatrix}\right)}{\Delta\left(\gamma\begin{pmatrix} z_1 \\ z_2 \end{pmatrix}\right)} \frac{\Delta\left(\gamma\begin{pmatrix} z_1 \\ z_2 \end{pmatrix}\right)}{\Delta\begin{pmatrix} z_1 \\ z_2 \end{pmatrix}},$$

in other words

$$\boxed{\varphi_{\beta\gamma}(z) = \varphi_\beta(\gamma z)\varphi_\gamma(z).}$$

Given a lattice L and a sublattice M we can find a chain of lattices

$$L = L_0 \supset L_1 \supset L_2 \supset \cdots \supset L_r = M$$

such that $(L_i : L_{i+1}) = p_i$ is a prime number. Now, if $(L : M) = p$, then M has a basis

$$P\begin{pmatrix} z_1 \\ z_2 \end{pmatrix}$$

with a matrix P such that $|P| = p$. So in principle, the values of φ_α are reduced to computing values φ_P where P has prime determinant.

§3. ANALYTIC PROOF FOR THE CONGRUENCE RELATION OF j

For the convenience of the reader, we shall reproduce here the classical proof of Hasse [19], [B1] for the congruence relation of the j-function.

Theorem 7. *Let $\mathfrak{o}$ be an order in k and let p be a prime not dividing the conductor of $\mathfrak{o}$, such that $p\mathfrak{o} = \mathfrak{p}\mathfrak{p}'$, $\mathfrak{p} \neq \mathfrak{p}'$. Let $\mathfrak{a}$ be a proper ideal of $\mathfrak{o}$. Let K be a finite Galois extension of k containing all the numbers $j(\mathfrak{c})$, where $\mathfrak{c}$ ranges over the proper ideals of $\mathfrak{o}$. Then*

$$j(\mathfrak{a})^p \equiv j(\mathfrak{p}'\mathfrak{a}) \pmod{\mathfrak{p}\mathfrak{o}_K}.$$

Proof. Without loss of generality we can extend K to a bigger finite Galois extension of k to contain other algebraic numbers which will occur in the proof, e.g. values $\varphi_\alpha(z)$ where α is a primitive integral matrix with determinant p. We select the same representatives as before for the left cosets of such matrices with respect to $SL_2(\mathbf{Z})$, namely

$$\alpha_i = \begin{pmatrix} 1 & i \\ 0 & p \end{pmatrix} \qquad \text{for} \quad i = 0, \ldots, p-1$$

$$\alpha_p = \begin{pmatrix} p & 0 \\ 0 & 1 \end{pmatrix}.$$

If f is a function on the upper half-plane, we write $f^*(q)$ for its powers series in q (or $q^{1/p} = e^{2\pi i\tau/p}$). We agree that

$$f^*(q) \equiv 0 \pmod{\mathfrak{p}}$$

means that all coefficients of the power series lie in $\mathfrak{p}$. We also write a congruence mod p, or $1 - \zeta$, where ζ is a primitive p-th root of unity, to mean that the coefficients lie in the ideal generated by these elements.

We consider the polynomial in two variables

$$F(X, Y) = \sum_{i=0}^{p} (X - j \circ \alpha_i)(Y - \varphi_{\alpha_1}) \cdots \widehat{(Y - \varphi_{\alpha_i})} \cdots (Y - \varphi_{\alpha_p})$$

the factor $Y - \varphi_{\alpha_i}$ being omitted from the product on the right. The above polynomial has coefficients which are functions on $\mathfrak{H}$, modular of level p. The permutation induced by $j \circ \alpha_i \mapsto j \circ \alpha_i\gamma$ for $\gamma \in \Gamma$ is the same as $\varphi_{\alpha_i} \mapsto \varphi_{\alpha_i\gamma}$. Hence $F(X, Y)$ has coefficients which are invariant under Γ. Furthermore, if $(d, p) = 1$, the automorphism σ_d on roots of unity such that $\sigma_d\zeta = \zeta^d$ has the effect

$$j \circ \alpha_i \mapsto j \circ \alpha_{di} \qquad \text{and} \qquad \varphi_{\alpha_i} \mapsto \varphi_{\alpha_{di}}$$

for $i = 0, \ldots, p-1 \pmod p$, while leaving $j \circ \alpha_p$ and φ_{α_p} fixed. Therefore the coefficients of $F(X, Y)$ are modular functions invariant under Γ, and with rational Fourier coefficients. We may therefore write

$$F(X, Y) = F(X, Y, j) \in \mathbf{Z}[X, Y, j]$$

as a polynomial in X, Y, j with integer coefficients.

Observe that if

$$f^*(q) = A(q) \in \mathbf{Z}((q))$$

is a power series in q with coefficients in $\mathbf{Z}$, then for $i = 0, \ldots, p-1$ we have

$$\begin{aligned}(f \circ \alpha_i)^*(q) &= A(q^{1/p}\zeta^i) \\ &\equiv A(q^{1/p}) \qquad (\mathrm{mod}\ 1 - \zeta).\end{aligned}$$

On the other hand

$$(f \circ \alpha_p)^*(q) = A(q^p) \equiv A(q)^p \quad (\mathrm{mod}\ p).$$

Here ζ is a primitive p-th root of unity, and the congruences mean that all coefficients are divisible by $1 - \zeta$ in the ring of algebraic integers in K, taken sufficiently large to contain the p-th roots of unity. Applying this to the first p terms, and to the functions j and $(2\pi)^{-12}\Delta$, we conclude that these first p terms are all congruent to each other mod $1 - \zeta$.

The last term involves $(X - j^*(q)^p)$ as a factor. If we substitute j^p for X, then this factor becomes $\equiv 0\ (\mathrm{mod}\ p)$. Therefore

$$F(j^*(q)^p, Y, j^*(q)) \equiv 0 \quad (\mathrm{mod}\ 1 - \zeta),$$

i.e. this expression lies in $(1 - \zeta)\mathbf{Z}[\zeta]((q^{1/p}))[Y]$. Since the Fourier coefficients are integers, we conclude that

$$F(j^*(q)^p, Y, j^*(q)) \equiv 0 \quad (\mathrm{mod}\ p).$$

Therefore

$$F(j^p, Y, j) \in p\mathbf{Z}[Y, j].$$

Let $\mathfrak{a} = [z_1, z_2]$, with $z = z_1/z_2 \in \mathfrak{H}$. Since $(\mathfrak{a} : \mathfrak{p}\mathfrak{a})$ and $(\mathfrak{a} : \mathfrak{p}'\mathfrak{a})$ have index p, we can find two of the matrices α_i, say P and P', such that

$$P\begin{pmatrix} z_1 \\ z_2 \end{pmatrix} \quad \text{and} \quad P'\begin{pmatrix} z_1 \\ z_2 \end{pmatrix}$$

are bases of $\mathfrak{p}\mathfrak{a}$ and $\mathfrak{p}'\mathfrak{a}$ respectively. Substitute $j(\mathfrak{a})$ for j and $\varphi_{P'}(z)$ for Y. We find that

$$F(j(\mathfrak{a})^p, \varphi_{P'}(z), j(\mathfrak{a})) \equiv 0 \quad (\mathrm{mod}\ \mathfrak{p}).$$

On the other hand, in the original sum defining $F(X, Y)$, all the terms become equal to 0 except one, and we find

$$(j(\mathfrak{a})^p - j(\mathfrak{p}'\mathfrak{a})) \prod_{\alpha_i \neq P'} (\varphi_{P'}(z) - \varphi_{\alpha_i}(z)) \equiv 0 \qquad (\mathrm{mod}\ \mathfrak{p}).$$

From the preceding section, we know that $\varphi_{P'}(z) \approx \mathfrak{p}^{12}$. We also know that $\varphi_{\alpha_i}(z)$ is a unit for $\alpha_i \neq P$ or P'. This proves our theorem.

In his paper, Hasse gave a further slightly elaborate argument to show that $k(j(\mathfrak{o}))$ is abelian over k and that the Frobenius automorphism gives, for almost all $\mathfrak{p}$, the effect

$$\sigma_{\mathfrak{p}} j(\mathfrak{a}) = j(\mathfrak{p}'\mathfrak{a}).$$

Deuring [B1] observed that this now follows trivially. Indeed, disregard the finite number of primes dividing all differences $j(\mathfrak{a}_\nu) - j(\mathfrak{a}_\mu)$ and the differences of their conjugates, if $\mathfrak{a}_\nu$ represent the different proper $\mathfrak{o}$-ideal classes. By general properties of the Frobenius automorphism (cf. the end of Chapter 8, §3) we see that the precise equality

$$\sigma_{\mathfrak{P}} j(\mathfrak{a}) = j(\mathfrak{p}'\mathfrak{a})$$

does hold for any $\mathfrak{P}$ dividing $\mathfrak{p}$ in K. Since multiplication of proper $\mathfrak{o}$-ideal classes is abelian, it follows that the map

$$\mathfrak{P} \mapsto \sigma_{\mathfrak{P}}$$

is a homomorphism from the free abelian group generated by almost all primes into the Galois group of the smallest Galois extension of k containing $j(\mathfrak{a})$. Using now theorems concerning the existence of primes with given Frobenius element, one concludes that this extension is abelian, and that the proper $\mathfrak{o}$-ideal class group is isomorphic to the Galois group under the map induced both by the Frobenius element on almost all primes, and also by the property

$$\sigma_{\mathfrak{B}} j(\mathfrak{A}) = j(\mathfrak{B}^{-1}\mathfrak{A})$$

for proper $\mathfrak{o}$-ideal classes $\mathfrak{A}$ and $\mathfrak{B}$.

13 The ℓ-adic and p-adic Representations of Deuring

It was first proved by Hasse that even in characteristic $p > 0$, if N is an integer prime to p, then the points of order N on an elliptic curve A form a cyclic group of type $\mathbf{Z}/N\mathbf{Z} \times \mathbf{Z}/N\mathbf{Z}$. On the other hand, Hasse also discovered that there may not be points of period p, and if there are some, then the group of points of order p^r is then cyclic. Essentially one sees this from the representation of the endomorphism

$$N\delta : a \mapsto Na,$$

whose degree is N^2. If we represent this endomorphism on the local tangent space at the origin, or equivalently on the differential forms, we see that it must be separable if $(N, p) = 1$, and must be inseparable if p divides N. Thus in characteristic $p > 0$, there cannot exist two points of period p linearly independent over $\mathbf{Z}/p\mathbf{Z}$. Therefore either

$$A_p = 0 \qquad \text{or} \qquad A_p \approx \mathbf{Z}/p\mathbf{Z}.$$

The first case is called **supersingular**. The second case is called **singular** or **generic** according as the j-invariant is transcendental over the prime field or not. Hasse also discovered that over finite fields the algebra of endomorphisms must be either an imaginary quadratic field, or a division algebra of rank 4 over $\mathbf{Q}$, depending on the two cases.

Using ℓ-adic and p-adic representations, Deuring [4] gave a more comprehensive theory, and especially determined what happens to the ring of endomorphism of an elliptic curve under reduction mod p. We shall closely follow Deuring's paper, except that as usual we use the projective limit of the groups A_{p^r}, forming the Tate vectors and Tate module $T_p(A)$, which gives a natural representation of the endomorphisms over the p-adic integers. Except as specified above, the results of this chapter are due to Deuring.

§1. THE ℓ-ADIC SPACES

Let, therefore, A be an elliptic curve defined over a field of characteristic p. Points of A are taken in a fixed algebraic closure. For any prime number ℓ, define the **ℓ-adic module** $T_\ell(A)$ to be the set of infinite vectors

$$(a_1, a_2, \ldots)$$

with $a_i \in A_{\ell^i}$ (that is $\ell^i a_i = 0$) and $\ell a_{i+1} = a_i$. Addition is defined componentwise so $T_\ell(A)$ is a group. It is clear that $T_\ell(A)$ is a module over the ℓ-adic integers $\mathbf{Z}_\ell$. To define the multiplication of a vector by an ℓ-adic number, we define it componentwise. On the i-th component, we approximate the ℓ-adic number by an integer mod ℓ^i, and multiply the i-th component by this integer. It is immediately verified that this multiplication is well defined, and gives an operation of $\mathbf{Z}_\ell$ on $T_\ell(A)$.

Theorem 1. *If $\ell \neq p$, then $T_\ell(A)$ is a free $\mathbf{Z}_\ell$-module of dimension 2. On the other hand, $T_p(A) = 0$, or is a free module of dimension 1 over $\mathbf{Z}_p$, according as we are in the supersingular or singular case.*

Proof. Take first $\ell \neq p$. Let x_1, x_2 be elements of $T_\ell(A)$ whose first components $a_{1,1}, a_{2,1}$ are linearly independent over the field $\mathbf{Z}/\ell\mathbf{Z}$. Then these vectors x_1, x_2 are linearly independent over $\mathbf{Z}_\ell$, for if we had a relation of linear dependence over $\mathbf{Z}_\ell$, we could assume that not all coefficients are divisible by ℓ, and hence the projection of this relation on the first component would contradict the hypothesis made on x_1, x_2.

I contend that x_1, x_2 form a basis of $T_\ell(A)$ over $\mathbf{Z}_\ell$. We are going to prove this by an inductive argument. Suppose that we can write every element w of $T_\ell(A)$ as a linear combination

$$w \equiv z_1 x_1 + z_2 x_2 \quad \bmod \ell^n T_\ell(A) \tag{1}$$

with integers $z_j \in \mathbf{Z}$. Let $w = (b_1, \ldots, b_n, b_{n+1}, \ldots)$. By definition, we have for the first $n + 1$ components,

$$z_1(a_{1,1}, \ldots, a_{1,n+1}) + z_2(a_{2,1}, \ldots, a_{2,n+1}) = (b_1, \ldots, b_n, b_{n+1}) + (0, \ldots, 0, c_{n+1})$$

for some point c_{n+1} of order ℓ. By the very choice of the vectors x_i, there exist integers d_1, d_2 such that

$$c_{n+1} = d_1 \ell^n a_{1,n+1} + d_2 \ell^n a_{2,n+1}.$$

If we replace z_1, z_2 by $z_1 + d_1\ell^n$, $z_2 + d_2\ell^n$, we see that we have extended the congruence (1) from n to $n + 1$. That gives us what we wanted.

If $\ell = p$, then one verifies at once that A_{p^i} is cyclic of order p^i in the singular case, and that $T_p(A)$ is free over $\mathbf{Z}_p$ in this case (easier than for $\ell \neq p$ in this instance). If there is no points of order p, then $T_p(A) = 0$.

From now on, we use ℓ to denote a prime number other than p.

Let $\lambda: A \to B$ be a homomorphism of elliptic curves. Then λ induces a homomorphism also denoted by λ,

$$\lambda: T_\ell(A) \to T_\ell(B),$$

and similarly for T_p. Its effect on a vector $(a_1, a_2, \ldots)$ is given by

$$\lambda(a_1, a_2, \ldots) = (\lambda a_1, \lambda a_2, \ldots).$$

Theorem 2. *If $\lambda_1, \ldots, \lambda_r$ are endomorphisms of A which are linearly independent over $\mathbf{Z}$, then as endomorphisms of $T_\ell(A)$, they are linearly independent over $\mathbf{Z}_\ell$.*

Proof. Say $c_1\lambda_1 + \cdots + c_r\lambda_r = 0$ with $c_i \in \mathbf{Z}_\ell$. It will suffice to prove that all c_i are divisible by ℓ (then cancel the ℓ and start over again to get an impossibility unless all $c_i = 0$). Write

$$c_i = m_i + \ell d_i,$$

with $d_i \in \mathbf{Z}_\ell$ and $m_i \in \mathbf{Z}$. It suffices to prove that $\ell | m_i$ for all i. The endomorphism

$$\lambda = m_1\lambda_1 + \cdots + m_r\lambda_r = -\ell(d_1\lambda_1 + \cdots + d_r\lambda_r)$$

lies in End(A). Acting on A, we see that λ kills A_ℓ. Hence λ factors through $\ell\delta$, i.e. $\lambda = \ell\alpha$ for some $\alpha \in$ End(A). But $\lambda_1, \ldots, \lambda_r$ generate a space $\mathbf{Q}\lambda_1 + \cdots + \mathbf{Q}\lambda_r$ over $\mathbf{Q}$, and

$$(\mathbf{Q}\lambda_1 + \cdots + \mathbf{Q}\lambda_r) \cap \mathrm{End}(A)$$

is a lattice of rank r in this subspace. Without loss of generality, it suffices to prove that a basis of this lattice is linearly independent over $\mathbf{Z}_\ell$, i.e. we can assume that $\lambda_1, \ldots, \lambda_r$ themselves form a basis of this lattice. But then it follows that α lies in $\mathbf{Z}\lambda_1 + \cdots + \mathbf{Z}\lambda_r$, whence $\ell | m_i$ for all i, as desired.

The above theorem shows that our representation of End(A) on $T_\ell(A)$ corresponds to tensoring with $\mathbf{Z}_\ell$, i.e. we get an injection

$$\mathbf{Z}_\ell \otimes_{\mathbf{Z}} \mathrm{End}(A) \to \mathrm{End}_{\mathbf{Z}_\ell}(T_\ell(A)).$$

We denote by $A^{(\ell)}$ the set of points of A whose order is a power of ℓ.

Let $V_\ell(A)$ be the set of vectors

$$(a_0, a_1, a_2, \ldots)$$

with $a_0 \in A^{(\ell)}$ any point of order a power of ℓ, and satisfying

$$\ell a_{i+1} = a_i.$$

It is clear that

$$V_\ell \approx \mathbf{Q}_\ell \otimes_{\mathbf{Z}_\ell} T_\ell.$$

In fact for any point x in V_ℓ we can find a power ℓ^s such that $\ell^s x$ has its first component equal to 0. Identifying the vectors

$$(0, a_1, a_2, \ldots)$$

such that $\ell a_1 = 0$ in V_ℓ with elements of T_ℓ, we get an exact sequence

$$0 \to T_\ell(A) \to V_\ell(A) \to A^{(\ell)} \to 0,$$

the mapping on the right being projection on the first component.

Of course we have a similar sequence with T_p and V_p; however, when $T_p = 0$, we do not get a faithful representation of End(A). We do if $T_p \neq 0$, because an isogeny has finite kernel.

For an arbitrary ℓ we get a faithful representation

$$\mathbf{Q} \otimes_{\mathbf{Z}} \operatorname{End}(A) = \operatorname{End}(A)_{\mathbf{Q}} \to \operatorname{End}_{\mathbf{Q}_\ell}(V_\ell).$$

Since $\dim_{\mathbf{Q}_\ell}(V_\ell) = 2$ it follows that $\dim_{\mathbf{Q}_\ell} \operatorname{End}_{\mathbf{Q}_\ell}(V_\ell) = 4$. Hence:

Theorem 3. *In any characteristic,* $\dim_{\mathbf{Q}} \operatorname{End}(A)_{\mathbf{Q}} \leqq 4$ *and*

$$\dim_{\mathbf{Z}} \operatorname{End}(A) \leqq 4.$$

This gives a proof for a result mentioned previously.

We already know that every element of End(A) is invertible in $\operatorname{End}(A)_{\mathbf{Q}}$, as discussed in Chapter 2, §2. Hence $\operatorname{End}(A)_{\mathbf{Q}}$ is a division algebra of dimension $\leqq 4$ over $\mathbf{Q}$. The only possibilities are that it has dimension 1, 2 in which case it is commutative, or dimension 4. In that case, it cannot be commutative, because we have an injection

$$\mathbf{Q}_\ell \otimes \operatorname{End}(A)_{\mathbf{Q}} \to M_2(\mathbf{Q}_\ell)$$

in the representation on V_ℓ.

§2. REPRESENTATIONS IN CHARACTERISTIC p

We first give a proof in arbitrary characteristic for the following fact, using only the involutive property of $\alpha \mapsto \alpha'$.

Let $\alpha \in \operatorname{End}(A)$ *be a non-trivial endomorphism. Then* $\mathbf{Q}(\alpha)$ *is quadratic imaginary.*

Proof. Since $\mathbf{Q}(\alpha)$ is a commutative subfield of a division algebra of dimension 4 over $\mathbf{Q}$, it follows that $[\mathbf{Q}(\alpha) : \mathbf{Q}] = 2$, so α is quadratic. The mapping

$$\lambda \mapsto \lambda'$$

on $\mathbf{Q}(\alpha)$ (where λ' is the endomorphism such that $\lambda\lambda' = \nu(\lambda)\delta$ as discussed in Chapter 2) defines an automorphism of $\mathbf{Q}(\alpha)$, and is not the identity, for otherwise

$$\lambda^2 = \lambda\lambda' = \nu(\lambda) \in \mathbf{Z}$$

for all $\lambda \in \mathbf{Z}[\alpha]$ which is patently false. Hence $\lambda \mapsto \lambda'$ is *the* non-trivial automorphism of $\mathbf{Q}(\alpha)$. Furthermore $\mathbf{Q}(\alpha)$ must be imaginary (in any embedding in $\mathbf{C}$) because

$$N_{\mathbf{Q}(\alpha)/\mathbf{Q}}(\lambda) = \lambda\lambda' = \nu(\lambda) > 0$$

for all $\lambda \in \mathbf{Z}[\alpha]$, $\lambda \neq 0$.

Theorem 4. *Let A be defined over the finite field with q elements, and let π_q be its Frobenius endomorphism. If $\pi_q \in \mathbf{Z}$, then $T_p = 0$. So if $T_p \neq 0$, then π_q is a non-trivial endomorphism.*

Proof. Let $q = p^r$. We know that π_q has degree q. If $\pi_q = n\delta$, then

$$q = \nu(\pi_q) = n^2,$$

whence $n = p^m$ for some integer m. But π_q is purely inseparable, and therefore $p^m\delta$ has kernel 0, whence $T_p = 0$.

Theorem 5. *Let A be an elliptic curve over a finite field* $\mathbf{F}$ *of characteristic p, and assume $T_p(A) \neq 0$. Then:*

i) $\mathrm{End}(A)_{\mathbf{Q}} = k$ *is a quadratic imaginary field, and* $\mathrm{End}(A) = \mathfrak{o}$ *is an order in k.*

ii) *The prime p does not divide the conductor c of $\mathfrak{o}$.*

iii) *The prime p splits completely in k.*

Proof. Theorem 4 shows that there exists a non-trivial endomorphism of A, namely π_q. The representation of $\mathrm{End}(A)_{\mathbf{Q}}$ on V_p (or of $\mathrm{End}(A)$ on T_p) is faithful, and therefore gives rise to an embedding of $\mathrm{End}(A)$ in $\mathbf{Z}_p$, which shows that $\mathrm{End}(A) = \mathfrak{o}$ is commutative. It follows that $\mathrm{End}(A)$ has dimension 2 over $\mathbf{Z}$, whence $\mathrm{End}(A)_{\mathbf{Q}} = k$ is a quadratic field. Since k admits an embedding in $\mathbf{Q}_p$, it follows that p splits completely in k. There remains only to prove that p does not divide the conductor c of $\mathfrak{o}$. We know that $\mathfrak{o} = \mathbf{Z} + c\mathfrak{o}_k$. There is an integer m such that

$$\pi_q = m + c\alpha \quad \text{and} \quad \pi'_q = m + c\alpha'$$

for some $\alpha \in \mathfrak{o}_k$. We get

$$q\delta = \pi_q\pi'_q \equiv m^2 \ (\mathrm{mod}\ c\mathfrak{o}_k).$$

Viewing $\mathfrak{o}_k$ as embedded in $\mathbf{Z}_p$, from the representation on $T_p(A)$, we conclude that p divides m, whence from the representation on $T_p(A)$ it follows that π_q kills the points of order p on A. This is a contradiction, since π_q is purely inseparable, and our theorem is proved.

Corollary. *Let $q = p^d$ be the number of elements of* $\mathbf{F}$. *Let $\pi = \pi_q$ be the Frobenius endomorphism. If $p\mathfrak{o} = \mathfrak{p}\mathfrak{p}'$ is the factorization of p in $\mathfrak{o} =$* $\mathrm{End}(A)$, *then*

$$\pi\mathfrak{o} = \mathfrak{p}^d \quad \text{or} \quad \pi\mathfrak{o} = \mathfrak{p}'^d,$$

and any other generator of $\pi\mathfrak{o}$ is $\pm\pi$.

Proof. Since $\pi\pi' = q\delta$, in the unique factorization in $\mathfrak{o}$ (with primes not dividing the conductor), only divisors of p can occur as divisors of π and π'. Since p does not divide π (because π is purely inseparable, and $p\delta$ has a non-trivial kernel), it follows that there is a positive integer m such that, after permuting $\mathfrak{p}$ and $\mathfrak{p}'$ if necessary, we get

$$\pi\mathfrak{o} = \mathfrak{p}^m \qquad \text{and} \qquad \pi'\mathfrak{o} = \mathfrak{p}'^m.$$

Therefore $\pi\pi'\mathfrak{o} = p^d\mathfrak{o}$, and $m = d$. Since the curve is not supersingular, the only automorphisms are $\pm\delta$ (according to the tabulation of Appendix 1), and π is uniquely determined up to ± 1. This proves our corollary.

Next we consider the supersingular case. We observe that if $T_p(A) = 0$ and B is isogenous to A then $T_p(B) = 0$ also (obvious).

Theorem 6. *Let A be defined over a field of characteristic p. If $T_p(A) = 0$, then $j_A = j_A^{p^2}$.*

Proof. If $T_p = 0$, then $p\delta$ must be purely inseparable of degree p^2. Hence there is an isomorphism

$$\lambda\colon A \to \pi_{p^2}(A),$$

whence $j_A = j_A^{p^2}$. We are using the fact in characteristic p that j_A is the invariant of isomorphism classes of elliptic curves.

In particular, we see that $j_A \in \mathbf{F}_{p^2}$ must lie in the field with p^2 elements if $T_p(A) = 0$, and *there is only a finite number of isomorphism classes of elliptic curves A in characteristic p such that $T_p(A) = 0$.*

Corollary. *Assume that A is supersingular, with invariant j, and that A is defined over $\mathbf{F}_p(j) = \mathbf{F}$. Then for $p \neq 2, 3$ we have:*

$$\pi_p^2 = -p\delta \qquad \text{if } j \in \mathbf{F}_p$$

$$\pi_{p^2} = \pm p\delta \qquad \text{if } j \notin \mathbf{F}_p.$$

Proof. Suppose first that $\mathbf{F} = \mathbf{F}_p$, i.e. that $j \in \mathbf{F}_p$. Let $\pi = \pi_p$. Since $p\delta$ and π^2 have the same degree p^2, it follows that they differ by an automorphism of A. Since $p \neq 2, 3$ and the curve is supersingular, it follows from Appendix 1 that the only automorphisms of A are $\pm\delta$, whence $\pi^2 = -p\delta$, and the first formula is proved.

Secondly, suppose that j is of degree 2, so $\mathbf{F} = \mathbf{F}_p(j) = \mathbf{F}_{p^2}$, and $q = p^2$. Then π_q and $p\delta$ have the same degree p^2, so that they differ by an automorphism of A. Again by Appendix 1, it follows that in the supersingular case the only possible automorphisms are $\pm\delta$, whence $\pi = \pm p\delta$. The second formula is then obvious.

Remark. In characteristic 2 or 3 there are slight variations on the formulas of the corollary. Take for instance characteristic 2. The curve A defined by

$$y^2 + y = x^3 + x$$

over $\mathbf{F}_2$ has 5 rational points (counting the point at infinity). If N is the number of rational points, then

$$\begin{aligned} N = \nu(\pi - \delta) &= (\pi - \delta)(\pi' - \delta) \\ &= q + 1 - (\pi + \pi'). \end{aligned}$$

In the present case,

$$5 = 2 + 1 - (\pi + \pi')$$

whence $\mathrm{Tr}(\pi) = -2$. Thus $\pi = -1 \pm i$ and $\pi^2 = \pm 2i$. In general, when we have

$$\pi^2 = p\varepsilon,$$

with some unit ε, one must take a power of this expression to get rid of the non-rational unit.

Theorem 7. *If* $T_p(A) = 0$, *then* $\mathrm{End}(A)_{\mathbf{Q}}$ *is a division algebra of dimension* 4 *over* $\mathbf{Q}$.

Proof. First we prove that there exist non-trivial endomorphisms. Let $\ell_1, \ell_2, \ldots$ be a sequence of distinct primes unequal to p, and let $a_1, a_2, \ldots$ be a sequence of points such that a_i has order ℓ_i. Let (a_i) be the cyclic group generated by a_i. Each factor curve $A/(a_i)$ has no point of order p. By Theorem 6 we must have an isomorphism

$$A/(a_i) \approx A/(a_j)$$

for some $i \neq j$. Consider the composite homomorphisms

$$A \to A/(a_i) \approx A/(a_j) \to A,$$

where the first is the canonical homomorphism λ_i of degree ℓ_i, and the last is a homomorphism of degree ℓ_j, say λ_j'. We then obtain an endomorphism of A of degree $\ell_i \ell_j$, which cannot be of type $n\delta$ for $n \in \mathbf{Z}$ because its degree is not a square. Hence we have obtained non-trivial endomorphisms of A.

Next we prove that $\mathrm{End}(A)_{\mathbf{Q}}$ cannot be a quadratic field. Suppose it is a quadratic field k. Let $p_1, p_2, \ldots$ be a sequence of primes $\neq p$ which remain prime in k, and let a_i be a point of order p_i on A. We consider the factor curves

$$A/(a_1), \qquad A/(a_1, a_2), \qquad A/(a_1, a_2, a_3), \ldots$$

none of which has a point of period p. Hence by Theorem 6 we have an isomorphism

$$A/(a_1, \ldots, a_r) \approx A/(a_1, \ldots, a_r, a_{r+1}, \ldots, a_s)$$

for some pair of integers $r < s$. Let $B = A/(a_1, \ldots, a_r)$, and let $b_1, \ldots, b_s$ be the images in B of $a_{r+1}, \ldots, a_{r+s}$ under the canonical map. We have an endomorphism of B,

$$\lambda: B \to B/(b_1, \ldots, b_s) \approx B$$

of degree $p_1 \cdots p_s$. Let

$$\lambda \mathfrak{o}_k = \mathfrak{q}_1 \cdots \mathfrak{q}_m$$

be the prime factorization in $\mathfrak{o}_k$. Then

$$\nu(\lambda) = \lambda\lambda' = \mathbf{N}\mathfrak{q}_1 \cdots \mathbf{N}\mathfrak{q}_m = p_1 \cdots p_s.$$

Hence the prime ideals $\mathfrak{q}_j$ must be the prime ideals of $\mathfrak{o}_k$ dividing $p_1, \ldots, p_s$ and must occur to the first power. Since p_i remains prime in k for all i, it follows that $\mathbf{N}\mathfrak{q}_j$ is the square of a rational prime, a contradiction which proves the theorem.

Theorem 8. *Let $T_p(A) = 0$, and let D be the division algebra* $\mathrm{End}(A)_{\mathbf{Q}}$. *Then D splits at all primes $\ell \neq p$.*

Proof. If $\ell \neq p$, then D is represented as a ring of endomorphisms of $V_\ell(A)$, in a way which we know is the tensor product with $\mathbf{Q}_\ell$, and $V_\ell(A)$ has dimension 2 over $\mathbf{Q}_\ell$. Hence locally at ℓ, we must have $\mathbf{Q}_\ell \otimes D \approx M_2(\mathbf{Q}_\ell)$.

For the reader who knows the Hasse theorem on simple algebra, we now see that D ramifies at p and also at infinity (i.e. becomes the ordinary quaternion algebra over $\mathbf{R}$) because the sum of its invariants is equal to 0, and D cannot split everywhere, otherwise D is a matrix algebra globally, which is not the case.

Theorem 9. *If $T_p(A) = 0$, then* $\mathrm{End}(A)$ *is a maximal order in* $\mathrm{End}(A)_{\mathbf{Q}}$.

Proof. We shall omit this proof, which the reader can look up in Deuring [4], and which depends on a counting argument, considering left ideals. The result will not be used in this book.

For the further properties of supersingular invariants, we refer the reader to Deuring's basic paper [4], and more recently to Manin's fairly comprehensive survey [30]. Observe that in characteristic p, if the group of automorphisms of an elliptic curve has order >2, then $j = 0$ or 12^3 (as you can verify from the tables in the Appendix). If the curve is supersingular in addition, then we necessarily have $j = 0$. Connections with the Hasse invariant are discussed in an appendix.

§3. REPRESENTATIONS AND ISOGENIES

We continue to suppose that $\ell \neq p$ where p is the characteristic of the field over which A is defined.

We want to see how the modules T_ℓ correspond under isogenies. In many ways these modules play the same role as a lattice L in $\mathbf{C}$. Let

$$\lambda: A \to B$$

be an isogeny. Tensoring $\mathrm{Hom}(A, B)$ with $\mathbf{Q}$ to get $\mathrm{Hom}(A, B)_{\mathbf{Q}}$, we can find an inverse λ^{-1} in $\mathrm{Hom}(B, A)_{\mathbf{Q}}$.

First we have a simple lemma, giving us a criterion of ℓ-integrality for an element $\alpha \in \mathrm{End}(A)_{\mathbf{Q}}$ in terms of its representation on $T_\ell(A)$.

Lemma 1. *Let S_ℓ be the multiplicative monoid of positive integers prime to ℓ, let $\mathfrak{o} = \mathrm{End}(A)$, and let $\mathfrak{o}_{(\ell)} = S_\ell^{-1}\mathfrak{o}$ be the localization of $\mathfrak{o}$ at ℓ. Let $\alpha \in \mathrm{End}(A)_{\mathbf{Q}}$. Then $\alpha T_\ell \subset T_\ell$ if and only if $\alpha \in \mathfrak{o}_{(\ell)}$. In other words, $\mathfrak{o}_{(\ell)}$ is the set of $\alpha \in \mathrm{End}(A)_{\mathbf{Q}}$ such that $\alpha T_\ell \subset T_\ell$.*

Proof. If $\alpha \in \mathfrak{o}_{(\ell)}$, it is clear that $\alpha T_\ell \subset T_\ell$. Conversely, suppose that $\alpha T_\ell \subset T_\ell$. There exists $\lambda \in \mathfrak{o}$ such that $m\ell^r\alpha = \lambda$ for some integer m prime to ℓ. Then

$$m\ell^r\alpha T_\ell \subset \ell^r T_\ell,$$

whence $\lambda T_\ell \subset \ell^r T_\ell$, and $\lambda = \ell^r\beta$ for some $\beta \in \mathfrak{o}$. It follows that

$$m\ell^r\alpha = \ell^r\beta,$$

and therefore $m\alpha = \beta$. This proves that $\alpha \in S_\ell^{-1}\mathfrak{o}$, α is ℓ-integral, as desired.

Lemma 2. *Let $\lambda: A \to B$ be an isogeny, and let M_ℓ be the set of vectors $(a_0, a_1, \ldots)$ in $V_\ell(A)$ such that $a_0 \in \mathrm{Ker}\,\lambda$. Then $\lambda M_\ell = T_\ell(B)$.*

This is clear from the definition of T_ℓ, and therefore gives us some description of the inverse image of $T_\ell(B)$ under λ, in $V_\ell(A)$.

Theorem 10. *Let $\lambda: A \to B$ be an isogeny, and let $\alpha \in \mathrm{End}(A)_{\mathbf{Q}}$. Let M_ℓ be the inverse image of $T_\ell(B)$ in $V_\ell(A)$ under λ. We have $\lambda\alpha\lambda^{-1} \in \mathrm{End}(B)$ if and only if $\alpha M_\ell \subset M_\ell$, for all ℓ.*

Proof. Assume first that $p \nmid \nu(\lambda)$. Suppose that $\lambda\alpha\lambda^{-1} \in \mathrm{End}(B)$. Then for all ℓ, we get

$$\alpha M_\ell = \lambda^{-1}\lambda\alpha\lambda^{-1}\lambda M_\ell \subset \lambda^{-1}T_\ell(B) \subset M_\ell.$$

Conversely, assume that $\alpha M_\ell \subset M_\ell$ for all ℓ. Then

$$\lambda\alpha\lambda^{-1}T_\ell(B) = \lambda\alpha M_\ell \subset \lambda M_\ell \subset T_\ell(B).$$

By Lemma 1, we conclude that $\lambda\alpha\lambda^{-1}$ is ℓ-integral for each ℓ. There remains to prove that $\lambda\alpha\lambda^{-1}$ is also p-integral. Suppose that

$$\lambda\alpha\lambda^{-1} = p^{-r}\beta$$

for some $\beta \in \mathrm{End}(B)$. Let $n = \nu(\lambda)$. Then

$$n\beta = p^r\lambda\alpha n\lambda^{-1} = p^r\gamma,$$

for some $\gamma \in \operatorname{End}(B)$. So

$$\lambda\alpha\lambda^{-1} = \frac{1}{n}\gamma.$$

But

$$\frac{1}{n}\gamma T_\ell(B) \subset T_\ell(B)$$

for all $\ell | n$, and by Lemma 1 we conclude that $\gamma = n\gamma'$ for some $\gamma' \in \operatorname{End}(B)$. Therefore

$$\lambda\alpha\lambda^{-1} = \gamma' \in \operatorname{End}(B),$$

thus proving our theorem in the present case $p \nmid \nu(\lambda)$.

Next we have a result which will be used to deal with the remaining case, but is of interest in itself, so we state it separately as a theorem.

Theorem 11. *Let $\lambda\colon A \to B$ be an isogeny and $\nu(\lambda) = p^r$. The map*

$$\alpha \mapsto \lambda\alpha\lambda^{-1}$$

is an isomorphism between $\operatorname{End}(A)$ *and* $\operatorname{End}(B)$.

Proof. We may decompose λ into a composition of isogenies each of whose degrees is p, and it will therefore suffice to prove the theorem under the assumption that $\nu(\lambda) = p$, which we now make. It will suffice to prove that if $\alpha \in \operatorname{End}(A)$, then $\lambda\alpha\lambda^{-1} \in \operatorname{End}(B)$, for then we get an inverse mapping using λ', and the fact that

$$\lambda'\lambda\alpha\lambda^{-1}\lambda'^{-1} = p\alpha p^{-1} = \alpha.$$

Let $\alpha \in \operatorname{End}(A)$. Suppose that λ is separable. Let $\lambda\lambda' = p\delta$. Then λ' is purely inseparable, and $\lambda^{-1} = p^{-1}\lambda'$. Suppose that

$$\lambda\alpha\lambda^{-1} = \frac{1}{p}\beta,$$

with some $\beta \in \operatorname{End}(B)$. Then

$$\lambda\alpha\lambda^{-1} = \frac{1}{p}\lambda\alpha\lambda' = \frac{1}{p}\beta,$$

whence $\lambda\alpha\lambda' = \beta$. But λ' is purely inseparable, λ is separable, hence Ker β contains the point of period p. Hence $\beta = p\gamma$ with some $\gamma \in \operatorname{End}(B)$, so

$$\lambda\alpha\lambda^{-1} = \gamma \in \operatorname{End}(B),$$

proving our theorem in this case.

If λ is purely inseparable, then $\lambda = \varepsilon\pi$ with some isomorphism ε, and then

$$\lambda^{-1} = \pi^{-1}\varepsilon^{-1},$$

so that

$$\lambda\alpha\lambda^{-1} = \varepsilon\pi\alpha\pi^{-1}\varepsilon^{-1}.$$

For any point $x \in \pi(A)$ we have

$$\pi\alpha\pi^{-1}(x) = \pi(\alpha(x^{(1/p)})) = \alpha^{(p)}(x),$$

where $\alpha^{(p)}$ is the image of α under the automorphism $c \mapsto c^p$ of the universal domain. Hence we find

$$\pi\alpha\pi^{-1} = \alpha^{(p)} \in \mathrm{End}(\pi A),$$

whence $\varepsilon\pi\alpha\pi^{-1}\varepsilon^{-1} \in \mathrm{End}(B)$, thereby proving Theorem 11.

Returning to the proof of Theorem 10, we decompose an arbitrary isogeny into a product of an isogeny whose degree is prime to p, and an isogeny whose degree is p^r for some r. The theorem follows at once.

§4. REDUCTION OF THE RING OF ENDOMORPHISMS

We now investigate the relationship between elliptic curves in characteristic 0 and curves in characteristic p, and consider especially how the ring of endomorphisms reduces.

First we look at the ℓ-adic spaces. Suppose that A is an elliptic curve defined over a number field. Let $\mathfrak{P}$ be a place of the algebraic numbers ${}^a\mathbf{Q}$ (algebraic closure of $\mathbf{Q}$ in $\mathbf{C}$) with values in an algebraic closure of the finite field with p elements, denoted by ${}^a\mathbf{F}_p$. On each number field finite over $\mathbf{Q}$, the place, denoted by

$$x \mapsto \bar{x} = x(\mathfrak{P})$$

induces a discrete valuation ring. Suppose that A has non-degenerate reduction mod $\mathfrak{P}$. Again we use ℓ for a prime number unequal to p. Then we know that we have an isomorphism

$$A^{(\ell)} \xrightarrow{\approx} \bar{A}^{(\ell)},$$

where $A^{(\ell)}$ denotes the group of points of A whose order is a power of ℓ, in the given algebraic closure ${}^a\mathbf{Q}$. Consequently we have an isomorphism

$$T_\ell(A) \xrightarrow{\approx} T_\ell(\bar{A}).$$

If we want to specify the $\mathfrak{P}$ in the notation, we also write

$$\bar{A} = A(\mathfrak{P}).$$

On the other hand, we only have a homomorphism

$$T_p(A) \to T_p(\bar{A}).$$

If $T_p(\bar{A}) \neq 0$, then the kernel of this homomorphism is a 1-dimensional module over $\mathbf{Z}_p$.

We observe that a result like that of Theorem 10 allows us to test for integrality both upstairs and downstairs, i.e. on A and $\bar{A}$.

Theorem 12. *Let A be an elliptic curve over a number field, with* $\mathrm{End}(A) \approx \mathfrak{o}$, *where $\mathfrak{o}$ is an order in an imaginary quadratic field k. Let $\mathfrak{P}$ be a place of ${}^a\mathbf{Q}$ over a prime number p, where A has non-degenerate reduction $\bar{A}$. The curve $\bar{A}$ is supersingular if and only if p has only one prime of k above it (p ramifies or remains prime in k). Suppose that p splits completely in k. Let c be the conductor of $\mathfrak{o}$, and write $c = p^r c_0$, where $p \nmid c_0$. Then:*

i) $\mathrm{End}(\bar{A}) = \mathbf{Z} + c_0\mathfrak{o}_k$ *is the order in k with conductor c_0.*

ii) *If $p \nmid c$, then the map $\lambda \mapsto \bar{\lambda}$ is an isomorphism of* $\mathrm{End}(A)$ *onto* $\mathrm{End}(\bar{A})$.

Proof. Suppose that p splits completely in k, say $p\mathfrak{o}_k = \mathfrak{p}\mathfrak{p}'$, $\mathfrak{p} \neq \mathfrak{p}'$, and $\mathfrak{P} \cap \mathfrak{o}_k = \mathfrak{p}$. To prove that $\bar{A}$ has a point of period p, it suffices to do it for any elliptic curve isogenous to $\bar{A}$. By changing A with an isogeny over some number field, we may assume without loss of generality that we have a normalized embedding

$$\theta \colon k \to \mathrm{End}(A)_{\mathbf{Q}}$$

such that $\theta(\mathfrak{o}_k) = \mathrm{End}(A)$. Let m be a positive integer such that $\mathfrak{p}^m$ and $\mathfrak{p}'^m$ are principal, say

$$\mathfrak{p}^m = \mu\mathfrak{o}_k \qquad \text{and} \qquad \mathfrak{p}'^m = \mu'\mathfrak{o}_k.$$

Then $\mu\mu' = p^m$. Note that $\mu' \notin \mathfrak{p}$, and since θ is a normalized embedding, it follows that $\overline{\theta(\mu')}$ is separable because the reduction of $\mu'\omega$ (for a differential form of first kind ω) mod $\mathfrak{P}$ is not 0. Since $\theta(\mu')$ has degree a power of p, so does its reduction mod $\mathfrak{P}$, and hence $\bar{A}$ has a non-trivial point of order p, thus proving that $\bar{A}$ is not supersingular.

On the other hand, if p does not split completely in k, we know from Theorem 9 of Chapter 10, §4 that there is some element $\mu \in \mathfrak{o}_k$ such that $\theta(\mu)$ reduces to a Frobenius endomorphism. Since $p\mathfrak{o}_k = \mathfrak{p}^m$, with only one prime $\mathfrak{p}$, and since $\mu\mu'$ is equal to a power of p, it follows that μ' differs from μ by a unit in $\mathfrak{o}_k$, and that $\theta(\mu)\theta(\mu') = q\delta$, where q is a power of p. This implies that $q\delta$ is purely inseparable, whence $\bar{A}$ is supersingular.

Let us assume now that p splits completely in k, and that quite generally $\mathrm{End}(A) \approx \mathfrak{o}$, where $\mathfrak{o}$ is an order in k with conductor $c = p^r c_0$, and $p \nmid c_0$. We want to determine $\mathrm{End}(\bar{A})$.

We know from general reduction theory that the reduction map

$$\mathrm{End}(A) \to \mathrm{End}(\bar{A})$$

is an injection. So $\mathrm{End}(\bar{A})$ contains at least $\overline{\mathrm{End}(A)}$. Theorem 5 of §2 puts some limitation on how much more there is in $\text{End}(\bar{A})$, for we know that $\text{End}(\bar{A})_{\mathbf{Q}}$ is an imaginary quadratic field. Hence at least we get an isomorphism

$$\text{End}(A)_{\mathbf{Q}} \xrightarrow{\approx} \text{End}(\bar{A})_{\mathbf{Q}}$$

induced by reduction.

Now suppose that p does not divide the conductor of $\mathfrak{o} = \text{End}(A)$. We have an isomorphism

$$T_\ell(A) \xrightarrow{\approx} T_\ell(\bar{A})$$

for every prime $\ell \neq p$, and we use Lemma 1 of §3, which tells us that $\text{End}(A)$ and $\text{End}(\bar{A})$ have the same localizations at ℓ. On the other hand, $\mathfrak{o}_{k,(p)} = \mathfrak{o}_{(p)}$ if p does not divide the conductor, and therefore $\mathfrak{o}_{(p)}$ is integrally closed, hence must coincide with the localization at p of $\text{End}(\bar{A})$. This proves that

$$\text{End}(A) \approx \text{End}(\bar{A})$$

because they have the same localizations at all primes.

If p divides the conductor, the argument is similar. We see that $\text{End}(A)$ and $\text{End}(\bar{A})$ have the same localizations at $\ell \neq p$. Theorem 5 of §2 tells us that p does not divide the conductor of $\text{End}(\bar{A})$. This proves our theorem.

Let $\mathcal{J}$ be the set of all invariants j_A of elliptic curves A over the complex numbers with non-trivial endomorphisms. If $j \in \mathcal{J}$, we let k_j be the quadratic imaginary field isomorphic to the endomorphism algebra corresponding to the given invariant. We know that $\mathcal{J}$ is contained in the integral closure of $\mathbf{Z}$ in the field of algebraic numbers, and we denote this integral closure by ${}^a\mathbf{Z}$.

For each prime number p we let $\mathcal{J}_p$ be the set of $j \in \mathcal{J}$ such that p splits completely in k_j, and p does not divide the conductor of the ring $\mathfrak{o}_j$ of endomorphisms of an elliptic curve A with invariant j. We shall sometimes use Ihara's notation, and write

$$\left(\frac{\mathfrak{o}}{p}\right) = 1$$

if p splits completely in the field k and does not divide the conductor of the order $\mathfrak{o}$ in k.

Let $\mathfrak{P}$ be a place of ${}^a\mathbf{Q}$, lying above p. We get a map

$$\mathcal{J}_p \to {}^a\mathbf{F}_p$$

denoted by the usual bar,

$$j \mapsto \bar{j},$$

into the set of singular (and not supersingular) invariants in characteristic p, according to Theorem 12. One of Deuring's major results is:

Theorem 13. *The map $\mathcal{J}_p \to {}^a\mathbf{F}_p$ is a bijection of $\mathcal{J}_p$ with the set of singular invariants in characteristic p.*

Proof. We first prove that the map is injective.

Suppose that $j_{\bar{A}} = \bar{j}_A = \bar{j}_B = j_{\bar{B}}$. We know by Theorem 12 that k_j is preserved under the reduction map, whence A and B have the same field k. Hence there exists an isogeny $\lambda: A \to B$, giving rise to a reduced isogeny

$$\bar{\lambda}: \bar{A} \to \bar{B}.$$

Also there exists an isomorphism $\varepsilon: \bar{B} \to \bar{A}$. By Theorem 12 again, we know that $\mathrm{End}(\bar{A}) = \overline{\mathrm{End}(A)}$. Hence there exists $\alpha \in \mathrm{End}(A)$ such that

$$\bar{\alpha} = \varepsilon \circ \bar{\lambda}.$$

Let C be the image of the map $\lambda \times \alpha: A \times A \to B \times A$. Then $\bar{C}$ is the image of $\bar{\lambda} \times \bar{\alpha}$. The projection of $\bar{C}$ on each factor induces an isomorphism of $\bar{C}$ on its projection, i.e. has degree 1. By general reduction theory, this must also be true for C, and therefore C is the graph of an isomorphism between A and B. It follows that $j_A = j_B$, thereby proving the injectivity.

The surjectivity will be proved in the next section by a method different from that which we have been using, also as in Deuring's paper [4]. In fact, somewhat more is proved, since one shows that given an elliptic curve in characteristic p, and some endomorphism, then they can both be lifted to characteristic 0. Given a singular elliptic curve $\bar{A}$ in characteristic p, we then select an endomorphism $\bar{\alpha}$ such that $\mathrm{End}(\bar{A}) = [\bar{\alpha}, 1]$ and lift back, to an endomorphism α of an elliptic curve A. It follows that the reduction of $\mathrm{End}(A)$ is precisely equal to $\mathrm{End}(\bar{A})$ (it is contained in $\mathrm{End}(\bar{A})$ and cannot be bigger).

§5. THE DEURING LIFTING THEOREM

Theorem 14. *Let A_0 be an elliptic curve in characteristic p, with an endomorphism α_0 which is not trivial. Then there exists an elliptic curve A defined over a number field, an endomorphism α of A, and a non-degenerate reduction of A at a place $\mathfrak{P}$ lying above p, such that A_0 is isomorphic to $\bar{A}$, and α_0 corresponds to $\bar{\alpha}$ under the isomorphism.*

Proof. We shall give the proof only in cases which imply the surjectivity of Theorem 13. It is a little simpler than the proof of the general theorem, on which we shall make technical comments at the end.

First we can assume that $\nu(\alpha_0)$ is prime to p, by considering $\alpha_0 + n\delta$ with suitable n, namely such that

$$\nu(\alpha_0 + n\delta) = \alpha_0\alpha_0' + n(\alpha_0 + \alpha_0') + n^2$$

is prime to p, which we can obviously do. Indeed, if we can lift $\alpha_0 + n\delta$, we can lift α_0, since the trivial endomorphisms lift in a trivial way.

We can also assume that α_0 is cyclic, for otherwise, factor out any multiple of the identity. Let $n = \nu(\alpha_0)$. Let $A(j)$ be an elliptic curve with transcendental invariant j over $\mathbf{Q}$. Let $Z_1, \ldots, Z_{\psi(n)}$ be the cyclic subgroups of $A(j)$, of order n. Let

$$\lambda_i : A(j) \to A(j_i)$$

be a homomorphism with kernel Z_i, $i = 1, \ldots, \psi(n)$. Let R be the integral closure of

$$\mathbf{Z}[j, j_1, \ldots, j_{\psi(n)}]$$

in a suitable finite extension of $\mathbf{Q}(j)$.

Let $\bar{j} \in {}^a\mathbf{F}_p$ be the invariant of A_0. There exists a homomorphism

$$\mathbf{Z}[j] \to {}^a\mathbf{F}_p$$

whose kernel contains p and sending j on $\bar{j}$. We can extend this homomorphism to R, say $R \to \bar{R}$, because all the j_i are integral over $\mathbf{Z}[j]$. We can select models for the $A(j_i)$ so that they have non-degenerate reduction at the local ring of the homomorphism

$$R \to \bar{R}.$$

Without loss of generality, we can select $A_0 = \overline{A(j)} = \bar{A}$ since they have the same invariant $\bar{j}$. For one of the indices i, say $i = 1$, the kernel $\bar{Z}_1$ of $\bar{\lambda}_1$ will be the kernel of α_0. Therefore

$$\bar{A} \approx \bar{A}/\bar{Z}_1 \approx \overline{A(j_1)}.$$

Let $\mathfrak{M}$ be the kernel of the homomorphism $R \to \bar{R}$. We have the inclusions

$$R \supset \mathfrak{M} \supset (p, j - j_1).$$

Let $\mathfrak{q}_R$ be a minimal prime containing $(j - j_1)$. Then $\mathfrak{q}_R$ is of dimension 1 (geometrically speaking, $\mathfrak{q}_R$ defines a component of the hypersurface $j = j_1$). Then $\mathfrak{q}_R \cap \mathbf{Z} = \{0\}$, for if $q \in \mathfrak{q}_R$ is a rational prime, then $\mathfrak{q}_R$ contains q and $j - j_1$, whence would be of dimension 2, which is impossible.

Let $\mathfrak{q}$ be an extension of $\mathfrak{q}_R$ to a prime ideal in the integral closure ${}^aR$ of R in the algebraic closure of $\mathbf{Q}(j)$. We reduce mod $\mathfrak{q}$. Then $j - j_1$ goes to 0, and $A(j_1)$ reduces to an elliptic curve $A(j_1)_{\mathfrak{q}}$, while A reduces to $A_{\mathfrak{q}}$, and we have an isomorphism

$$A(j_1)_{\mathfrak{q}} \approx A_{\mathfrak{q}}.$$

We have an isogeny

$$\lambda_{\mathfrak{q}} : A_{\mathfrak{q}} \to A(j_1)_{\mathfrak{q}},$$

whose kernel is $Z_{1\mathfrak{q}}$, and therefore $A_{\mathfrak{q}}$ admits an endomorphism α whose kernel is $Z_{1\mathfrak{q}}$. Reducing further mod $\mathfrak{M}$, we conclude that $\bar{A}$ has the endomorphism $\bar{\alpha}$ whose kernel is $\bar{Z}_1$, which is the same kernel as α_0.

If $\bar{A} = A_0$ has no automorphisms other than ± 1, we have now completed the proof, because two endomorphisms with the same kernel differ by ± 1.

This suffices for our purposes of lifting singular, i.e. not supersingular, invariants. Indeed, in characteristic > 0, if an elliptic curve admits automorphisms other than ± 1, then one sees from Appendix 1 that *either* the characteristic is $\neq 2, 3$ and the curve is not supersingular, and is definable by an ordinary equation,

$$y^2 = x^3 - x \quad \text{or} \quad y^2 = x^3 - 1,$$

whence the ring of endomorphisms obviously lifts; *or* the characteristic is 2 or 3, in which case the curve is *necessarily* supersingular, and actually $j = 0$!

So, for our purposes, we are done.

Observe the compatibility of the present situation with the general system of Theorem 12, say. If A is an elliptic curve over a number field with ring of endomorphisms $\mathbf{Z}[i]$, then its reduction mod 2 or 3 must be supersingular, because 2 ramifies in $\mathbf{Q}(i)$ and 3 remains prime in $\mathbf{Q}(i)$. Similarly, if A admits $\mathbf{Z}[\rho]$ as endomorphisms, then its reduction mod 2 or 3 must be supersingular, because 3 ramifies in $\mathbf{Q}(\rho)$ and 2 remains prime in $\mathbf{Q}(\rho)$.

14 Ihara's Theory

One can reduce the modular function field mod p and obtain an infinite extension of $\mathbf{F}_p(j)$, with j transcendental over $\mathbf{F}_p$. Igusa determined the Galois group [22], pointing out that it has the same SL_2 part as in characteristic zero, and that the part acting on the roots of unity is just that generated by the Frobenius element, i.e. those matrices having determinant a power of p. Ihara had the idea of lifting back singular values $\bar{j}$ of j in the algebraic closure ${}^a\mathbf{F}_p$ by the Deuring lifting, and to represent the Frobenius automorphism in the decomposition group of the modular function field in characteristic p by an element of the isotropy group of the point $z \in \mathfrak{H}$ such that $\overline{j(z)} = \bar{j}$, with a suitable place of the algebraic numbers, denoted also by a bar. This led him to deep conjectures concerning non-abelian extensions of the rational field $\mathbf{F}_p(j)$, for which we refer to his original treatise [B6].

However, as pointed out in [28], one can use some of Ihara's ideas in the context of extensions of $\mathbf{Z}[j]$ in characteristic 0, also allowing for the possibility of studying extensions of number fields generated by coordinates of point of finite order on elliptic curves without complex multiplication. The ideas used by Ihara for his proofs could be extended to this context, and we shall follow here the exposition of [28].

§1. DEURING REPRESENTATIVES

As in Ihara, we start with Deuring's canonical bijection,

$$\mathscr{I}_p \to \overline{\mathscr{I}}_p,$$

from singular invariants in number fields to singular invariants in ${}^a\mathbf{F}_p$, **with respect to a fixed place $\mathfrak{P}$ of ${}^a\mathbf{Q}$ into ${}^a\mathbf{F}_p$**. We take ${}^a\mathbf{Q}$ as the algebraic closure of $\mathbf{Q}$ in $\mathbf{C}$.

The elements of $\mathscr{J}_p$ are values $j(z)$, such that the order $\mathfrak{o}$ of $[z, 1]$ has conductor not divisible by p, and p splits completely in the imaginary quadratic field $k = \mathbf{Q}(z)$. We shall abbreviate these two conditions by $(\mathfrak{o}/p) = 1$.

The association

$$j(z) \mapsto \overline{j(z)}$$

gives a bijection between $\mathscr{J}_p$ and the set of singular invariants in ${}^a\mathbf{F}_p$, by Theorem 13 of the preceding chapter. A point $z \in \mathfrak{H}$ such that $\overline{j(z)} = \bar{j}$ will be called a **Deuring representative** of $\bar{j}$ in $\mathfrak{H}$.

We consider such a point z, let the order $\mathfrak{o}$ be as above. We let

$$\mathfrak{p} = \mathfrak{P} \cap \mathfrak{o},$$

so that $p\mathfrak{o} = \mathfrak{p}\mathfrak{p}'$. Note that $\mathfrak{p}$ is determined by our original place $\mathfrak{P}$.

Theorem 1. *Having fixed the place $\mathfrak{P}$ of ${}^a\mathbf{Q}$, let z be a Deuring representative for $\bar{j} \in {}^a\mathbf{F}_p$, let $\mathfrak{o}$ be the order of $[z, 1]$, and let $\mathfrak{p} = \mathfrak{P} \cap \mathfrak{o}$. Then the period D of $\mathfrak{p}$ in the proper ideal class group of $\mathfrak{o}$ is equal to the degree of $\bar{j}$ over $\mathbf{F}_p$. Furthermore, letting $\mathfrak{a} = [z, 1]$, the elements*

$$\overline{j(\mathfrak{a})}, \quad \overline{j(\mathfrak{p}\mathfrak{a})}, \quad \ldots, \quad \overline{j(\mathfrak{p}^{D-1}\mathfrak{a})}$$

form a complete set of conjugates of $\bar{j}$ over $\mathbf{F}_p$.

Proof. The Kronecker congruence relation

$$\overline{j(\mathfrak{p}^{-1}\mathfrak{a})} = \overline{j(\mathfrak{a})}^p,$$

together with the fact that the elements listed above are distinct (no repetition because of the injectivity in Deuring's reduction mapping on $\mathscr{J}_p$), implies that these elements form a complete set of conjugates over $\mathbf{F}_p$, and also that D is the degree of

$$\bar{j} = \overline{j(\mathfrak{a})}$$

over $\mathbf{F}_p$. This proves the theorem.

We denote by $M^p = M_2^p(\mathbf{Z})$ the set of 2×2 rational integral matrices whose determinant is equal to a power of p. Then M^p operates on $\mathfrak{H}$. We let M_z^p be the isotropy set of z, i.e. the subset of matrices $\alpha \in M^p$ such that $\alpha(z) = z$.

Theorem 2. *Let z be in the upper half plane, let $\mathfrak{o}$ be the order of $[z, 1]$, and assume $(\mathfrak{o}/p) = 1$. Then there exist two elements α, α' of M_z^p such that M_z^p is a disjoint union of two direct products*

$$M_z^p = \{\alpha\} \times p^{\mathbf{N}} \times T \qquad \cup \qquad \{\alpha'\} \times p^{\mathbf{N}} \times T,$$

where $\{\alpha\}$, $\{\alpha'\}$ are the positive powers of α, α' respectively, $p^{\mathbf{N}}$ consists of all powers of p with natural numbers, and T is isomorphic to the group of units in the order $\mathfrak{o}$ of $[z, 1]$.

Proof. Let the notation be as in Theorem 1. Let $\mathfrak{p}^D = \mu\mathfrak{o}$. Then there exists a unique matrix α in M^p such that

$$\alpha\begin{pmatrix} z \\ 1 \end{pmatrix} = \mu\begin{pmatrix} z \\ 1 \end{pmatrix},$$

and $\alpha(z) = z$, i.e. z is a fixed point of α. Let $L_z = [z, 1]$, and let A^z be, as in complex multiplication, an elliptic curve whose j-invariant is $j(z)$, and having non-degenerate reduction mod $\mathfrak{P}$. We identify $k = \mathbf{Q}(z)$ as $\mathrm{End}(A^z)_{\mathbf{Q}}$ in the normalized way. Let

$$\varphi_z : \mathbf{Q}^2 \to \mathbf{Q}L_z/L_z \to A^z$$

be our usual coordinatization, as in Chapter 7, §2. By the lemma of Chapter 7, §2, if

$$\alpha = \begin{pmatrix} * & * \\ c & d \end{pmatrix}$$

so that $\mu = cz + d$, then there exists an isogeny $\lambda : A^{\alpha(z)} \to A^z$ such that the following diagram is commutative.

$$\begin{array}{ccccc} \mathbf{Q}^2 & \to & \mathbf{Q}L_{\alpha(z)} & \to & A^{\alpha(z)} \\ {\scriptstyle\alpha}\downarrow & & \downarrow & & \downarrow{\scriptstyle\lambda} \\ \mathbf{Q}^2 & \to & \mathbf{Q}L_z & \to & A^z \end{array}$$

In other words,

$$\lambda \circ \varphi_{\alpha(z)}(a) = \varphi_z(a\alpha).$$

In the present case, $\alpha(z) = z$. Reducing mod $\mathfrak{P}$, we obtain

$$\bar{\lambda}\, \overline{\varphi_z(a)} = \overline{\varphi_z(a\alpha)}.$$

Furthermore, since $\mu \in \mathfrak{p}$, it follows that $\bar{\lambda}$ is purely inseparable. Hence $\bar{\lambda}$ differs from the Frobenius map π_{p^D} by an automorphism ε of $\bar{A}^z$, and consequently we get the relation

$$\overline{\varphi_z(a\alpha)} = \varepsilon\, \overline{\varphi_z(a)}^{p^D}.$$

The matrix α has infinite period modulo $p^{\mathbf{N}} \times T$ (T = torsion) because μ does not lie in $p\mathfrak{o}$ (not divisible by the conjugate $\mathfrak{p}'$).

Let $\beta \in M_z^p$. Dividing out a positive power of p, we may assume that β is primitive. Then

$$\beta\begin{pmatrix} z \\ 1 \end{pmatrix} = \mu_1\begin{pmatrix} z \\ 1 \end{pmatrix}$$

with some $\mu_1 \in \mathfrak{o}$, because z is a fixed point of β. But $\mu_1 \notin p\mathfrak{o}$. Hence if $\mathfrak{p}'$ is the conjugate of $\mathfrak{p}$, then

$$\mu_1\mathfrak{o} = \mathfrak{p}^m \quad \text{or} \quad \mu_1\mathfrak{o} = \mathfrak{p}'^m$$

for some positive integer m. Since D is the period of $\mathfrak{p}$ in the proper ideal class group of $\mathfrak{o}$, we must have $D|m$. Hence

$$\mu_1 = \mu^{m/D}\zeta \quad \text{or} \quad \mu_1 = \mu'^{m/D}\zeta,$$

where ζ is a unit of $\mathfrak{o}$. Hence $\beta = \alpha^{m/D}\gamma$, where γ has finite period, and corresponds to a unit of $\mathfrak{o}$, or $\beta = \alpha'^{m/D}\gamma$, where α' relates to μ' as α relates to μ. This proves Theorem 2.

Observe that the distinction between α and α' was due to the determination of $\mathfrak{p}$ as $\mathfrak{P} \cap \mathfrak{o}$. We call α a **$\mathfrak{p}$-generator of** $M_z^{\mathfrak{p}}$. It is well-defined modulo T, and is characterized as being that matrix such that if D is the period of $\mathfrak{p}$ in the proper ideal class group of $\mathfrak{o}$, and $\mathfrak{p}^D = \mu\mathfrak{o}$, then

$$\alpha\begin{pmatrix} z \\ 1 \end{pmatrix} = \mu\begin{pmatrix} z \\ 1 \end{pmatrix}.$$

§2. THE GENERIC SITUATION

Let j be the modular function. Let $F_1 = \mathbf{Q}(j)$ and F_N the field of modular functions of level N. As usual we let F be the union of all F_N. We let $R_1 = \mathbf{Z}[j]$, and let R be the integral closure of R_1 in F.

Theorem 3. *Let $z \in \mathfrak{H}$ be imaginary quadratic, and let $k = \mathbf{Q}(z)$. Let $\mathfrak{P}$ be a place of k_{ab}, denoted by a bar, and lying above p. Let $\mathfrak{o}$ be the order of $[z, 1]$ and assume $(\mathfrak{o}/p) = 1$. For $f \in R$, let $\bar{f} = \overline{f(z)}$, and let $\mathfrak{M}$ be the kernel of the bar mapping in R. Let $\mathfrak{p} = \mathfrak{P} \cap \mathfrak{o}$. Let α be a $\mathfrak{p}$-generator of $M_z^{\mathfrak{p}}$. Then a Frobenius automorphism $(\mathfrak{M}, F/F_1)$ restricted to those subfields F_N with $p \nmid N$ is given by the automorphism*

$$f_a \mapsto f_{a\alpha}$$

on the Fricke functions f_a with $a \in (\mathbf{Q}^2/\mathbf{Z}^2)_N$, $p \nmid N$.

Proof. Let $\mathfrak{p}^D = \mu\mathfrak{o}$ as before, and let s be the idele

$$s = (\ldots, \mu, \mu, 1, \mu, \mu, \ldots)$$

having $\mathfrak{p}$-component equal to 1, and all other components equal to μ. For any prime $\ell \neq p$, the embedding $q_\ell(s)$ in $GL_2(\mathbf{Z}_\ell)$ is simply the matrix α itself. By Shimura's reciprocity law in Chapter 11, we know that for any function $f \in F$ defined at z, we have

$$f(z)^{(s^{-1},k)} = f^\sigma(z),$$

where $\sigma = \sigma(q(s))$. Note that $\alpha = u_\ell$. So the right-hand side of the above relation gives us the desired effect on functions. As for the left-hand side, (s^{-1}, k) is the same automorphism as (r, k) where r is the idele

$$r = (\ldots, 1, 1, \mu, 1, 1, \ldots)$$

with μ in the $\mathfrak{p}$-component and 1 everywhere else. Now one knows by local class field theory (cf. *Algebraic Number Theory*, Chapter XI, §4) that (r, k) lies in the decomposition group of $\mathfrak{P}$, and has precisely the effect $(\mathfrak{P}, k_{ab}/k)$, modulo the inertia group of $\mathfrak{P}$. Consequently we find that

$$\overline{f^{\sigma}(z)} = \overline{f(z)}^{p^D},$$

because μ has order D at $\mathfrak{p}$. This proves our theorem.

Remark 1. On the subfield of F which is the union of all F_N with $p \nmid N$, it is clear that the inertia group of $\mathfrak{M} \cap F_N$ is precisely T as in Theorem 2.

Remark 2. From the argument in Theorem 3, we also get some description of the Frobenius automorphism in the p-part of F. Indeed, it is the matrix $q_p(s_p)$, where $s_p = (\mu, 1)$ with μ at $\mathfrak{p}$ and 1 at $\mathfrak{p}'$.

§3. SPECIAL SITUATIONS

Let F again be the modular function field, and let R_1, R be as in §2. Let $\bar{j}$ be a singular value in ${}^a\mathbf{F}_p$ and let $\mathfrak{m}$ be the kernel of the homomorphism

$$R_1 = \mathbf{Z}[j] \to \mathbf{F}_p[\bar{j}]$$

in R_1. Let $\mathfrak{M}$ be a maximal ideal of R lying above $\mathfrak{m}$. If $\mathfrak{q}$ is a prime of dimension 1 in $\mathfrak{m}$, and $\mathfrak{Q}$ is a prime in $\mathfrak{M}$ lying above $\mathfrak{q}$, then we can reduce mod $\mathfrak{Q}$. Let $G = \text{Gal}(F/F_1)$. Those elements of $G_{\mathfrak{M}}$ which leave $\mathfrak{Q}$ invariant then induce a Frobenius automorphism of $R/\mathfrak{Q}$ over $R_1/\mathfrak{q}$. In this way we can recover Ihara's theorem in characteristic p, if we select $\mathfrak{q}$ to be the ideal generated by p, and make use of Igusa's irreducibility theorem, which says that the modular function field reduces mod p in a non-degenerate way [23].

We can, however, take a prime $\mathfrak{q}$ which yields extensions of a number field. We start with a value $z \in \mathfrak{H}$ such that $j(z)$ is algebraic, and that an elliptic curve with invariant $j(z)$ does *not* have complex multiplication. Let us give ourselves again a place $\mathfrak{P}$ of ${}^a\mathbf{Q}$, and assume that $j(z)$ is $\mathfrak{P}$-integral. Let $\mathfrak{P}_z = \mathfrak{P} \cap \mathbf{Q}(j(z))$. Suppose that $\overline{j(z)}$ is not supersingular, and let

$$\bar{j} = \overline{j(z)}.$$

Let $F(z)$ be the field of all values $f(z)$ with $f \in F$, f defined at z. The Galois group of $F(z)$ over $\mathbf{Q}(j(z))$ is a factor group of the decomposition group of the place

$$f \mapsto f(z).$$

Let $\mathfrak{M}_z$ be the maximal ideal in R which is the kernel of the map

$$f \mapsto \overline{f(z)}.$$

We can find a Deuring representative z' for $\bar{j}$, and we let $\mathfrak{M}_{z'}$ be the kernel of the map

$$f \mapsto \overline{f(z')}$$

in R. Both $\mathfrak{M}_z$ and $\mathfrak{M}_{z'}$ lie above $\mathfrak{m}$, and the Frobenius automorphisms

$$(\mathfrak{M}_z, F/F_1) \qquad \text{and} \qquad (\mathfrak{M}_{z'}, F/F_1)$$

are conjugate to each other (as are the ideals $\mathfrak{M}_z$ and $\mathfrak{M}_{z'}$). We can then apply Theorem 3 to z' to get a description of the Frobenius automorphism in $F(z)$.

Thus we obtain a correspondence from certain non-abelian extensions of $\mathbf{Q}(j(z))$ to abelian extensions of $\mathbf{Q}(z', j(z'))$. In some sense, the study of the non-abelian Frobenius automorphism can be thrown back to the study of an abelian one, which, however, varies with p. Thus it becomes a major problem to determine the distribution laws of this variation with p, having fixed z. This concerns both the distribution of $z'_{\mathfrak{P}}$ and of the values $j(z'_{\mathfrak{P}})$. For instance, one may start with a given integer $j_0 \in \mathbf{Z}$, such that an elliptic curve with invariant j_0 does *not* have complex multiplication. One then asks for the distribution of values $j(z'_p)$ with Deuring representatives z'_p such that

$$j(z'_p) \equiv j_0 \pmod{p},$$

and $j_0 \pmod{p}$ is not supersingular. One can conjecture that the set of p for which $j_0 \pmod{p}$ has a given quadratic imaginary field k as algebra of endomorphisms must have density 0, but is infinite. Hale Trotter and I have made extensive computations about this problem, and a more precise discussion will appear, with the data, in a forthcoming joint paper [*Frobenius Distributions in GL_2-extensions*, Springer Lecture Notes 504, 1976]. For supersingular reduction, Serre has proved that the density is 0, cf. [35], 3.4 and 4.3.

One can also recall a problem which I had encountered many years ago, for abelian class field theory over finitely generated rings over $\mathbf{Z}$, namely describe an appropriate equivalence among the maximal ideals to determine which ones have the same Artin symbol in an abelian extension. It turns out here that we are studying a non-abelian situation of Kronecker dimension 2, i.e. a situation where both p and j vary, not only with fixed j, variable p as in ordinary complex multiplication, or fixed p, variable j, as in Ihara's work. In this way, complex multiplication seems to have a much wider range of applicability than thought of previously, since it affects the most general non-abelian situation.

Part Three

Elliptic Curves with Non-Integral Invariants

The preceding part studied elliptic curves with singular invariants, having complex multiplication from an imaginary quadratic field. We now study a case, which is both special and generic, of elliptic curves with invariant which is not integral at a given place, and find that there is a very convenient way to parametrize them, as shown by Tate, over a field with a non-archimedean valuation. Actually, as pointed out in [28], one also can work over complete local rings such that if j is the invariant of the curve, then $1/j$ lies in the maximal ideal, and this allows us to treat the generic case as well, since we can always send a transcendendental j to infinity.

For the higher dimensional theory, the reader is referred to:

H. MORIKAWA, "On theta functions and abelian varieties over valuation fields of rank one," I and II, *Nagoya Math. Jour.* **20** (1962), pp. 1–27 and 231–250.

D. MUMFORD, "An analytic construction of degenerating curves over complete local rings," *Compositio Math.* **24,** Fasc. 2, (1972), pp. 129–174.

15 The Tate Parametrization

§1. ELLIPTIC CURVES WITH NON-INTEGRAL INVARIANTS

In this section, we have essentially copied an unpublished manuscript of Tate. For an exposition of Tate's results which is more complete we refer to Roquette [B9]. We have done essentially what is needed to prove the isogeny theorem afterwards.

Consider the formal series in variables q, w given by

$$g_2 = \frac{1}{12}\left[1 + 240 \sum_{n=1}^{\infty} \frac{n^3 q^n}{1 - q^n}\right]$$

$$g_3 = \frac{1}{6^3}\left[-1 + 504 \sum_{n=1}^{\infty} \frac{n^5 q^n}{1 - q^n}\right]$$

$$j = \frac{1}{q} + 744 + \cdots$$

$$x(w) = \frac{1}{12} + \sum_{m \in \mathbf{Z}} \frac{q^m w}{(1 - q^m w)^2} - 2\sum_{n=1}^{\infty} \frac{nq^n}{1 - q^n}$$

$$y(w) = \sum_{m \in \mathbf{Z}} \frac{q^m w(1 + q^m w)}{(1 - q^m w)^3}$$

The denominators involving the primes 2, 3 are a slight blemish on these series, and so we make a transformation which gets rid of them.

First we get rid of the 4 in $4x^3$ by letting $y \mapsto y/2$. Next we get rid of the $1/12$ by letting $x \mapsto x - 1/12$. Finally we make a translation on y, to give us new variables X, Y whose relations to the original x, y are

$$X = x - \frac{1}{12}, \qquad Y = \frac{y}{2} + \frac{1}{2}\left(x - \frac{1}{12}\right).$$

Then the Weierstrass equation is transformed into the **Tate equation**

$$Y^2 - XY = X^3 - h_2 X - h_3$$

where

$$h_2 = 5 \sum_{n=1}^{\infty} \frac{q^n}{1 - q^n}$$

$$h_3 = \sum_{n=1}^{\infty} \frac{5n^3 + 7n^5}{12} \frac{q^n}{1 - q^n}$$

$$X(w) = \sum_{n \in \mathbf{Z}} \frac{q^n w}{(1 - q^n w)^2} - 2 \sum_{n=1}^{\infty} \frac{nq^n}{1 - q^n} \tag{1X}$$

$$Y(w) = \sum_{n \in \mathbf{Z}} \frac{(q^n w)^2}{(1 - q^n w)^3} - \sum_{n=1}^{\infty} \frac{nq^n}{1 - q^n}. \tag{1Y}$$

By expanding the square of the geometric series, one sees that the last term can be rewritten in the form

$$\sum_{n=1}^{\infty} \frac{nq^n}{1 - q^n} = \sum_{n=1}^{\infty} \frac{q^n}{(1 - q^n)^2}.$$

The reader will find both expressions in the literature.

We shall see that the series $(1X)$ and $(1Y)$ parametrize the elliptic curve A defined by the Tate equation over any field k complete under a non-archimedean absolute value, in any characteristic, under the following conditions.

Let q be an element of k such that $0 < |q| < 1$. Consider the series $X(w)$ in $(1X)$ where w is a variable in k^*. Using the identity

$$\frac{w}{(1 - w)^2} = \frac{1}{w + w^{-1} - 2} = \frac{w^{-1}}{(1 - w^{-1})^2},$$

we can rewrite the series in the form

$$X(w) = \frac{w}{(1 - w)^2} + \sum_{n=1}^{\infty} \left(\frac{q^n w}{(1 - q^n w)^2} + \frac{q^n w^{-1}}{(1 - q^n w^{-1})^2} - 2 \frac{q^n}{(1 - q^n)^2} \right) \tag{2X}$$

which shows, by comparison with the geometric series $\sum q^n$, that the convergence is absolute for all $w \in k^*$ and is uniform for w in an annulus

$$0 < r_1 \leqq |w| \leqq r_2.$$

We get the functional equations

$$X(qw) = X(w) = X(w^{-1}), \tag{3X}$$

trivially from $(1X)$ and $(2X)$ respectively. In the restricted range

$$|q| < |w| < |q|^{-1}$$

we have $|q^n w| < 1$ and $|q^n w^{-1}| < 1$ for all positive integers n, and hence we can expand the fractions under the summation signs in $(2X)$ to obtain

$$(4X)\qquad X(w) = \frac{w}{(1-w)^2} + \sum_{m=1}^{\infty}\sum_{n=1}^{\infty}(nq^{mn}w^n + nq^{mn}w^{-n} - 2nq^{mn})$$
$$= \frac{1}{w + w^{-1} - 2} + \sum_{n=1}^{\infty}\frac{nq^n}{1-q^n}(w^n + w^{-n} - 2)$$

for $|q| < |w| < |q|^{-1}$.

Similar to $(2X)$ we have the analogous expression for the other coordinate, namely for all $w \in k^*$ we have

$$(2Y)\qquad Y(w) = \frac{w^2}{(1-w)^3} + \sum_{n=1}^{\infty}\left[\frac{q^{2n}w^2}{(1-q^n w)^3} - \frac{q^n w^{-1}}{(1-q^n w^{-1})^3} - \frac{q^n}{(1-q^n)^2}\right].$$

Trivial rearrangements of the defining series show that Y satisfies the functional equation

$$(3Y)\qquad Y(qw) = Y(w) \quad\text{and}\quad Y(w^{-1}) + Y(w) = -X(w).$$

The series giving h_2 and h_3 converge, because the coefficients are integers, so of absolute value $\leqq 1$.

As usual, we let

$$\Delta = h_3 + h_2^2 + 72h_2h_3 - 432h_3^2 + 64h_2^3$$
$$= q - 24q^2 + 252q^3 + \cdots,$$

the polynomial in h_2, h_3 being simply obtained from the formal relation

$$g_2^3 - 27g_3^2 = \left(4h_2 + \frac{1}{12}\right)^3 - 27\left(4h_3 - \frac{1}{3}h_2 - \frac{1}{216}\right)^2.$$

We have $\Delta \neq 0$ because $\Delta \equiv q \pmod{q^2}$ (non-archimedean absolute value!). Therefore we have the absolute invariant

$$j = \frac{(12g_2)^3}{\Delta} = \frac{(1 + 48h_2)^3}{\Delta} = \frac{1 + 240q + 2160q^2 + \cdots}{q - 24q^2 + 252q^3 + \cdots}$$

$$j = \frac{1}{q}(1 + 744q + 196884q^2 + \cdots)$$

as expected. The Tate equation defines an elliptic curve, called the **Tate curve.**

Theorem 1. *Let $q^{\mathbf{Z}}$ be the infinite cyclic group generated by q in k^*. Let A be the Tate curve. Let*

$$\varphi(w) = (X(w), Y(w)) \quad \text{if } w \notin q^{\mathbf{Z}}$$
$$\varphi(w) = 0 \quad \text{if } w \in q^{\mathbf{Z}},$$

where 0 is the origin (point at infinity) on A. The map φ is a homomorphism of k^* into A_k with kernel $q^{\mathbf{Z}}$.

Proof. We prove first that φ maps k^* into A. Since $0 \in A_k$, this amounts to proving that the points $\varphi(w)$ for $w \notin q^{\mathbf{Z}}$ satisfy the equation of the curve. Because the functions $X(w)$ and $Y(w)$ have multiplicative period q, it is enough to consider values of w such that $|q| < |w| \leqq 1$ and $w \neq 1$. In this range we can use formula ($2X$) which expressed X as a power series in q with coefficients which are rational functions of w, and similarly for Y. Our first task will be completed if we can show that the Tate equation is a formal identity when we interpret X, Y, h_2, h_3 as formal power series in q with coefficients which are rational functions of an indeterminate w. In fact, the coefficients of the formal power series in question are expressed as elements of the ring

$$\mathbf{Z}[w, w^{-1}, (1 - w)^{-1}].$$

The canonical homomorphism $\mathbf{Z} \to k$ extends to a homomorphism of this ring into $k(w)$. Hence the formal identity we are trying to establish is a "universal" one, and will hold in any characteristic provided it holds in characteristic 0.

From the classical theory over the complex numbers we know that the point $\varphi(w)$ satisfies the Tate equation if we substitute any pair of complex numbers $w \neq 1, q \neq 0$ such that

$$|q| < |w| < |q|^{-1}.$$

Fixing first w such that $|w| < 1$ and letting q vary, we conclude that the resulting power series in q with complex coefficients are equal coefficient-wise. Then letting w vary, we conclude that the coefficients are formally equal as rational functions of an indeterminate, as was to be shown.

Next we prove that our map is a homomorphism. Given $w_1, w_2 \in k^*$, let $w_3 = w_1 w_2$. We must prove

$$\varphi(w_1 w_2) = \varphi(w_1) + \varphi(w_2). \tag{5}$$

Let $P_i = \varphi(w_i)$, $i = 1, 2, 3$. In view of the periodicity

$$\varphi(qw) = \varphi(w),$$

we can restrict our considerations to values of w_1 and w_2 in the range

$$|q| < |w_1| \leqq 1 \quad \text{and} \quad 1 \leqq |w_2| < |q|^{-1}.$$

Then

$$|q| < |w_3| < |q|^{-1},$$

so that all three w_i are within the domain of convergence of the power series expressions for X and Y considered before.

Since $\varphi(1) = 0$ by definition, (5) holds trivially if $w_1 = 1$ or $w_2 = 1$. The algebraic addition formula derived for the $\wp$-function yields an addition formula for points on the Tate curve.

Let $P_i = (X_i, Y_i)$, $i = 1, 2, 3$. If P_1, P_2 are on the curve, then $P_1 + P_2 = 0$ if and only if

$$X_1 = X_2 \quad \text{and} \quad Y_1 + Y_2 = -X_1. \tag{6}$$

From this we see that (5) holds if $w_1 w_2 = 1$.

In general, suppose that all three points P_i are different from 0. If $X_1 \neq X_2$ then the addition formula for the $\wp$-function yields at once an addition formula for points on the Tate curve, which reads

(7) $(X_1 - X_2)^2 X_3 = (Y_1 - Y_2)^2 + (Y_1 - Y_2)(X_1 - X_2) - (X_1 - X_2)^2(X_1 + X_2)$

(8) $(X_1 - X_2) Y_3 = -(X_1 - X_2)(Y_1 + X_3) + (Y_1 - Y_2)(X_1 - X_3).$

Now we can argue just as in the proof that $\varphi(w)$ lies on the curve. Relations (7) and (8) hold in the classical case. Hence they are identities in the ring of formal power series in q with coefficients in

$$\mathbf{Z}[w_1, w_1^{-1}, w_2, w_2^{-1}, (1 - w_1)^{-1}, (1 - w_2)^{-1}, (1 - w_1 w_2)^{-1}],$$

and (5) is therefore a functional identity in any complete field k. The remaining case $X_1 = X_2$ can be taken care of also by an explicit formula or by a continuity argument.

If $w \in q^{\mathbf{Z}}$ then $X(w)$ and $Y(w)$ lie in k, so $\varphi(w) \neq 0$. Hence the kernel of φ is $q^{\mathbf{Z}}$. Tate has also shown that φ maps k^* onto A_k. For this and a description of the function field in terms of the functional equation, we refer to the exposition of Roquette [B8].

Theorem 2. *Let $A(q)$ be the Tate curve corresponding to a choice of $q \in k$ with $|q| < 1$. For any positive integer N, the curves $A(q)$ and $A(q^N)$ are isogenous.*

Proof. Let $\Phi_N(T, j) = 0$ be the modular equation of order N. Write $j(q)$ for the q-expansion of j. Then from the complex theory we know that we have a formal power series relation

$$\Phi_N(j(q^N), j(q)) = 0.$$

Hence this relation is valid for $q \in k^*$ and $|q| < 1$. This proves the theorem in characteristic 0 by Theorem 5 of Chapter 5, §3. Actually the theorem is valid in general, and we again refer to Roquette's exposition for this.

It was convenient to give the above proof here, but of course it is also natural to see the theorem from the general theory. The group $q^{\mathbf{Z}}$ plays the role of a lattice, and in this analogy, any sublattice gives rise to an isogeny in a natural way.

Suppose given an element $j \in k^*$ such that $|j| > 1$. Then the formal q-expansion for the modular function can be inverted, to give

$$q = \frac{1}{j} + f\left(\frac{1}{j}\right),$$

where f is a power series with coefficients in $\mathbf{Z}$. Hence we can define q in k^* and get a Tate curve having the given invariant, chosen to be non-integral.

§2. ELLIPTIC CURVES OVER A COMPLETE LOCAL RING

Throughout this section, let R be a complete local ring, Noetherian, without divisors of zero, and integrally closed, with maximal ideal $\mathfrak{m}$, *and quotient field K.*

Let $j \in K$ be such that $j^{-1} \in \mathfrak{m}$. Then we can get an element $q \in \mathfrak{m}$ such that

$$q = \frac{1}{j} + f\left(\frac{1}{j}\right)$$

where f is the power series at the end of the last section. Conversely, given $q \in \mathfrak{m}$ the series $j(q)$ converges in R.

We can always find a discrete valuation on K which induces the topology on R such that the powers of $\mathfrak{m}$ form a fundamental system of neighborhoods of 0. For instance if R is regular, for any element $a \in R$ we define

$$\text{ord } a$$

to be the largest exponent r such that $a \in \mathfrak{m}^r$, and extend the order function to the quotient field so as to make it a homomorphism. In general, we use the Cohen structure theorem, which states that a ring R as above is always a finite module over a subring R_0, satisfying the same conditions, and in addition regular. We can then put a discrete valuation on the quotient field of R_0 as above, and extend it to the quotient field of R. It serves our purposes. Such a valuation will be called **admissible.**

Alternatively, one could also use the procedure known by geometers as blowing up the point corresponding to $\mathfrak{m}$ in spec(R), and one way of doing it is to take generators $\mathfrak{m} = (a_1, \ldots, a_m)$. For at least one of the a_i, say a_1, the ideal

$$(a_1, \ldots, a_m, a_2/a_1, \ldots, a_m/a_1)$$

is not the unit ideal in $R[a_2/a_1, \ldots, a_m/a_1]$. Let S be the integral closure of $R[a_2/a_1, \ldots, a_s/a_1]$ in K, and let $\mathfrak{p}$ be a minimal prime ideal containing the ideal Sa_1. Then $a_i \to 0$ under the canonical homomorphism $S \to S/\mathfrak{p}$. The local ring $S_{\mathfrak{p}}$ is a discrete valuation ring whose maximal ideal induces $\mathfrak{m}$ in R.

Geometrically, the above construction amounts to the following. We have a morphism spec(S) $\to$ spec(R), and we intersect S with the hypersurface $a_1 = 0$. Then all components of this intersection have dimension dim spec(S) $- 1$, and since S is integrally closed, these components are non-singular divisors on spec(S). One of them lies above the point in spec(R), thus giving rise to the discrete valuation. Cf. Zariski, *A simple analytical proof of a fundamental property of birational transformations*, Proc. Nat. Acad. Sci. USA (1949), pp. 62–66.

For a formal reference to the commutative algebra used above, you can always look up Grothendieck's *EGA*, Chapter IV, 7.8.3 and 7.8.6. The point is that starting with a certain type of ring called excellent, and including **Z**, a field, or a complete Noetherian local ring, then the rings obtained by taking completions, localizing, taking finitely generated extension rings, or taking integral closure, will have all desirable properties. For instance, we used the fact that the integral closure of $R[b_1, \ldots, b_m]$ is finite over this ring ($b_i = a_i/a_1$). We shall continue to assume such basic results from commutative algebra. For another reference, the reader can look up Matsumura's *Commutative Algebra*, Benjamin, Reading, Mass. 1970, Chapter XIII.

Let A be an elliptic curve defined over K, with invariant $j = j(q)$. Let $D_q = q^{\mathbf{Z}}$. We denote by $D_q^{1/N}$ the subgroup of K^* consisting of all elements whose N-th power lies in D_q. This subgroup is generated by the N-th roots of unity, and any N-th root of q, say $q^{1/N}$. The factor group

$$D_q^{1/N}/D_q$$

is isomorphic to a direct product of cyclic groups of order N, generated respectively by a primitive ζ_N and $q^{1/N} \bmod q^{\mathbf{Z}}$, if the characteristic of K does not divide N.

Theorem 3. *Let A have invariant $j(q)$ as above, and q lie in the maximal ideal $\mathfrak{m}$ of R. Let R_N be the integral closure of R in $K_N = K(\zeta_N, q^{1/N})$. Then the Tate mapping defined by the same formulas as in Theorem 1 converges in R_N and induces a homomorphism of $D_q^{1/N}$ into A_N. If N is prime to the characteristic of K, it induces a Galois isomorphism of $D_q^{1/N}/D_q$ onto A_N, and*

$$K(A_N) = K(\zeta_N, q^{1/N}).$$

Proof. Let $w = \zeta q^{s/N}$ where ζ is an N-th root of unity, and s is an integer. The series giving $X(w)$ and $Y(w)$ in the preceding section are seen to converge in R_N, and even in $R[\zeta_N, q^{1/N}]$, to yield elements in $K(\zeta_N, q^{1/N})$. Formulas $(2X)$ and $(2Y)$ exhibit the desired convergence. Note that a finite number of terms are rational functions in q, w, but that all but a finite number of terms lie in the maximal ideal of R_N, and tend to 0. We then see that the mapping is a homomorphism of $D_q^{1/N}$ either by repeating the arguments of Theorem 1, or by reducing the present situation to the preceding one by means of a discrete valuation v as constructed above. We get an injective homomorphism of $D_q^{1/N}/D_q$ into A_N. If N is not divisible by the characteristic of K, the homomorphism must be surjective since A_N has order N^2.

Let G be the Galois group of $K(A_N)$ over K. Then G operates in a manner compatible with the Tate parametrization, i.e. that for $\sigma \in G$ we have

$$X(\sigma w) = X(w)^\sigma \quad \text{and} \quad Y(\sigma w) = Y(w)^\sigma$$

if $w \in D_q^{1/N}$. This is clear by continuity. It is then clear that

$$K(A_N) = K(\zeta_N, q^{1/N}),$$

thereby proving our theorem.

Example. Let A be an elliptic curve with transcendental invariant j over $\mathbf{Q}$. We consider the ring $\mathbf{Z}[1/j]$, and its completion at the maximal ideal generated by $(p, 1/j)$. Let R be this completion, so that actually

$$R = R_1 = \mathbf{Z}_p[[1/j]] = \mathbf{Z}_p[[q]].$$

Let K_1 be its quotient field and let $K_N = K_1(A_N)$. Then

$$K_N = K_1(\zeta_N, q^{1/N}).$$

The Galois group of this extension is easily determined. Over $K_1(\zeta_N)$ it is a Kummer extension, whose Galois group is generated by the map

$$q^{1/N} \mapsto \zeta_N q^{1/N}.$$

Let $\varphi = (X, Y)$ be the Tate mapping, and let $P_1 = \varphi(\zeta_N)$, $P_2 = \varphi(q^{1/N})$. Then the above element in the Galois group is represented by the matrix

$$\begin{pmatrix} 1 & 1 \\ 0 & 1 \end{pmatrix}.$$

On the other hand, suppose for simplicity that $p \nmid N$. Then ζ_N generates an unramified extension of $\mathbf{Z}_p[[q^{1/N}]]$, whose Galois group is generated by the Frobenius automorphism such that

$$\zeta_N \mapsto \zeta_N^p,$$

represented on the points of period N by the matrix

$$\begin{pmatrix} p & 0 \\ 0 & 1 \end{pmatrix}.$$

The full Galois group $\mathrm{Gal}(K_N/K_1)$ is the subgroup of $GL_2(\mathbf{Z}/N\mathbf{Z})$ generated by the above two elements (when $p \nmid N$). When $p|N$, then the root of unity ramifies, but the group is again easily determined, since K_N over the quotient field of $\mathbf{Z}_p[[q^{1/N}]]$ has the same Galois group as $\mathbf{Q}_p(\zeta_N)$ over $\mathbf{Q}_p$.

Observe that taking the union of all fields K_N yields a field which we denote by K. The group of all matrices

$$\begin{pmatrix} 1 & b \\ 0 & 1 \end{pmatrix}, \qquad b \in \prod_\ell \mathbf{Z}_\ell$$

is contained in the inertia group of a maximal ideal $\mathfrak{M}$ lying above (p, q) in R_1. If we restrict this group to the subfield obtained as the union of all K_N such that $p \nmid N$, then it is the inertia group of this subfield, since the N-th roots of unity for $p \nmid N$ generate an unramified extension.

Igusa was the first to recognize the presence of such unipotent elements in the Galois group in the case of bad reduction [25].

16 The Isogeny Theorems

Throughout this chapter we let K denote a field of characteristic 0.

§1. THE GALOIS p-ADIC REPRESENTATIONS

We return to p-adic representations. Let A be an elliptic curve defined over K. We take points of A in a fixed algebraic closure K_a. We have the p-adic spaces

$$T_p(A) \quad \text{and} \quad V_p(A)$$

over $\mathbf{Z}_p$ and $\mathbf{Q}_p$ respectively. We recall that $T_p(A)$ consists of all vectors

$$(a_1, a_2, \ldots), \qquad a_i \in A$$

such that $p^i a_i = 0$, $pa_{i+1} = a_i$; and $V_p(A)$ consists of all vectors

$$(a_0, a_1, a_2, \ldots)$$

such that a_0 is an arbitrary point of order a power of p, and $pa_{i+1} = a_i$. We know that $T_p(A)$ (resp. $V_p(A)$) is free of dimension 2 over $\mathbf{Z}_p$ (resp. $\mathbf{Q}_p$).

The Galois group $\text{Gal}(K_a/K)$, also denoted by G_K, operates continuously on both $T_p(A)$ and $V_p(A)$ in the obvious way. If $\sigma \in G_K$, then

$$\sigma(a_1, a_2, \ldots) = (\sigma a_1, \sigma a_2, \ldots).$$

Thus we get a representation

$$\rho: G_K \to GL_2(\mathbf{Z}_p)$$

if a basis of $T_p(A)$ over $\mathbf{Z}_p$ has been selected, and without such a selection, into $\text{Aut}_{\mathbf{Z}_p}(T_p(A))$.

For simplicity, we shall write T_p, V_p, omitting the A if the reference to A is fixed throughout a discussion. We call the above representations the **p-adic (Galois) representations associated with A over K.**

If $\lambda: A \to B$ is an isogeny defined over K, then λ induces a G_K-isomorphism

$$V_p(\lambda): V_p(A) \to V_p(B),$$

but of course only an injection of $T_p(A)$ into $T_p(B)$. Indeed, if λ is defined over K and $\sigma \in G_K$, then for any point a of A in the algebraic closure of K, we have

$$\lambda(a)^\sigma = \lambda^\sigma(a^\sigma) = \lambda(a^\sigma).$$

It is then clear that the induced map on $V_p(A)$ commutes with the action of the Galois group. For simplicity, we also write $V_p(\lambda) = \lambda$.

It is a major problem to prove the converse over fields which are of arithmetic interest, and the first progress in this direction was made by Serre [B11], whose results and methods we reproduce in this chapter.

Remark 1. We observe that all the results will be such that they allow us to pass to open subgroups of the Galois group over the field K. Thus whenever we want to prove an isogeny theorem, it suffices to do it over a finite extension of K, which we select at our convenience. We can also do it over a finitely generated extension, because the Galois group of a Galois extension does not change when we lift this extension over a purely transcendental extension of K.

Remark 2. The Galois representation of G_K on V_p factors through the Galois group leaving $K(A^{(p)})$ fixed, where $A^{(p)}$ is the group of points on A having p-power order. Hence we are really concerned with the representation of the Galois group of $K(A^{(p)})$ over K. In particular, if A, A' are two elliptic curves defined over K, and $V_p(A)$, $V_p(A')$ are G_K-isomorphic, then $K(A^{(p)}) = K(A'^{(p)})$.

There is a converse to the preceding remark in certain cases.

Theorem 1. *Let A, A' be elliptic curves defined over K, and assume that $K(A^{(p)}) = K(A'^{(p)})$. Let G be the Galois group of $K(A^{(p)})$ over K, and assume that the representations of G on $T_p(A)$ and $T_p(A')$ map G onto open subgroups of $SL_2(\mathbf{Z}_p)$. Then $V_p(A)$ and $V_p(A')$ are G_E-isomorphic for some finite extension E of K.*

The theorem follows from the next lemma.

Lemma 1. *Let G be an open subgroup of $SL_2(\mathbf{Z}_p)$ and let*

$$\rho_1 : G \to SL_2(\mathbf{Z}_p) \qquad \text{and} \qquad \rho_2 : G \to SL_2(\mathbf{Z}_p)$$

be continuous injective representations. Then there exists $g \in GL_2(\mathbf{Q}_p)$ such that $g^{-1}\rho_2 g = \rho_1$ on an open subgroup of G.

Proof. Without loss of generality we may assume that ρ_1 is the identity and $\rho_2 = \rho$. Thus ρ induces a local isomorphism of $SL_2(\mathbf{Z}_p)$ into itself. We look at its effect on the Lie algebra. Let $X = \begin{pmatrix} 0 & 1 \\ 0 & 0 \end{pmatrix}$, $Y = \begin{pmatrix} 0 & 0 \\ 1 & 0 \end{pmatrix}$, and $H = \begin{pmatrix} 1 & 0 \\ 0 & -1 \end{pmatrix}$. Then $[X, H] = 2X$, $[Y, H] = -2Y$ and $[X, Y] = H$. Since ρ maps H on a semisimple element, after a conjugation by an element of $GL_2(\mathbf{Q}_p)$ if necessary we may assume that ρ sends H into a scalar multiple of H. Looking at the above brackets shows that this scalar is ± 1, and another conjugation reduces us to the case when ρ leaves H fixed. Again looking at the effect of ρ on the brackets we conclude that ρ sends X into aX and Y into bY, and then that $b = a^{-1}$. Conjugation by $\begin{pmatrix} 1 & 0 \\ 0 & a^{-1} \end{pmatrix}$ then returns aX to X and bY to Y. Hence the effect of ρ on the Lie algebra is inner. It follows that it is locally given by a conjugation on the group.

Corollary. *Let A, A' be elliptic curves over K, and assume that $K(A^{(p)})$ and $K(A'^{(p)})$ have an intersection which is of infinite degree over K. Assume that the representations of* $\mathrm{Gal}(K(A^{(p)})/K)$ *on $T_p(A)$ and* $\mathrm{Gal}(K(A'^{(p)})/K)$ *on $T_p(A')$ map the Galois groups onto open subgroups of $SL_2(\mathbf{Z}_p)$. Then there is a finite extension E of K such that*

$$E(A^{(p)}) = E(A'^{(p)}),$$

and Theorem 1 applies.

Proof. Since the Lie algebra of $SL_2(\mathbf{Z}_p)$ is simple, there exists an open subgroup W of $\mathrm{Gal}(K(A^{(p)})/K)$ having the following properties:

i) W has no finite subgroup other than 1.

ii) Any closed normal non-trivial subgroup of W is also open, and hence of finite index.

Let K_1 be the fixed field of W. We consider the inclusion of fields:

$$K_1 \subset K_1(A^{(p)}) \cap K_1(A'^{(p)}) \subset K_1(A^{(p)}).$$

The intermediate field is of infinite degree over K_1, and is the fixed field of a closed normal subgroup of W. By the above two properties, it must be equal to $K_1(A^{(p)})$. Arguing the same way with respect to A', i.e. selecting an open subgroup W' in a similar way, we can find a finite extension K_2 of K such that

$$K_2(A^{(p)}) = K_2(A'^{(p)}).$$

This proves our corollary, with $E = K_2$.

The assumptions of the corollary concerning A and K are always satisfied in the following cases.

i) K is obtained from a number field by adjoining all roots of unity, and then making a finite extension. A has no complex multiplication. This is a theorem of Serre, whose proof will be reproduced in the next chapter, in case the invariant of A is not integral over $\mathbf{Z}$.

ii) K is finitely generated over an algebraically closed field of characteristic 0, and A has transcendental invariant over this field. This is clear from Chapter 6.

§2. RESULTS OF KUMMER THEORY

In this section we assume that K has characteristic 0. We let μ_n be the group of p^n-th roots of unity in an algebraic closure K_a. We thus use the p-logarithmic notation, and similarly let A_n denote what we would otherwise write as A_{p^n}, the group of points of order p^n on an elliptic curve A.

We let $G = G_K = \mathrm{Gal}(K_a/K)$ *through the section.*

We suppose that K is the quotient field of a ring R, complete, local Noetherian, integrally closed, and we assume that the prime p lies in the maximal ideal $\mathfrak{m}$.

Let q, q' be elements of $\mathfrak{m}$ and let $A = A(q)$ and $A' = A(q')$ be the elliptic curves as in the Tate parametrization, defined over K. Let $D_q = q^{\mathbf{Z}}$. We know that there is an isomorphism

$$D_q^{1/p^n}/D_q \approx A_n,$$

Actually, the elliptic curve will be irrelevant for this section, and one could phrase all the statements completely in terms of the Kummer extensions $K(D_q^{1/p^n})$, letting the above $\approx$ be an equality.

As in Kummer theory, if $z \in D_q^{1/p^n}$ then z^{p^n} lies in D_q, and there is an integer c such that

$$z^{p^n} = q^c.$$

The association $z \mapsto$ class of $c \bmod p^n\mathbf{Z}$ defines a homomorphism of A_n onto $\mathbf{Z}/p^n\mathbf{Z}$, and hence gives rise to the exact sequence

$$(1) \qquad 0 \to \mu_n \to A_n \to \mathbf{Z}/p^n\mathbf{Z} \to 0$$

of G-modules, the Galois group acting trivially on $\mathbf{Z}/p^n\mathbf{Z}$. Taking the limit, we obtain an exact sequence

$$(2) \qquad 0 \to T_p(\mu) \to T_p(A) \to \mathbf{Z}_p \to 0,$$

where G operates trivially on $\mathbf{Z}_p$. Tensoring with $\mathbf{Q}_p$ yield the exact sequence of G-modules,

$$(3) \qquad 0 \to V_p(\mu) \to V_p(A) \to \mathbf{Q}_p \to 0.$$

Lemma 1. *The above sequence does not split.*

To prove Lemma 1, we introduce an invariant x which belongs to the group

$$\varprojlim H^1(G, \mu_n).$$

Let d be the coboundary homomorphism

$$d\colon H^0(G, \mathbf{Z}/p^n\mathbf{Z}) \to H^1(G, \mu_n)$$

with respect to the exact sequence (1), and let $x_n = d(1)$. We define x to be the element of $\varprojlim H^1(G, \mu_n)$ defined by the family $\{x_n\}$, $n \geqq 1$.

Lemma 2. i) *The isomorphism*

$$\delta\colon K^*/K^{*p^n} \to H^1(G, \mu_n)$$

of Kummer theory transforms the class of q mod K^{*p^n} *into x_n.*

ii) *The element x is of infinite order.*

Proof. Recall that δ is induced by the coboundary map relative to the exact sequence

$$1 \to \mu_n \to K_a^{*p^n} \to K_a^* \to 1.$$

The first assertion of Lemma 2 is immediate from the definitions, because the isomorphism of Kummer theory transforms an element $\alpha \in K^*$ into the class of the cocycle α^σ/α, $\sigma \in G$.

To prove the second assertion, let v be a discrete valuation on K which is admissible, i.e. induces the given topology on R. Then the valuation defines a homomorphism

$$f_n\colon K^*/K^{*p^n} \to \mathbf{Z}/p^n\mathbf{Z},$$

and hence a homomorphism

$$f\colon \varprojlim K^*/K^{*p^n} \to \mathbf{Z}_p.$$

If we identify x with the corresponding element of $\varprojlim K^*/K^{*p^n}$, as in (i), then we have

$$f(x) = v(q),$$

and hence x is of infinite order, proving Lemma 2.

We can now prove Lemma 1. Suppose the sequence (3) splits. There is a G-subspace W of $V_p(A)$ which is mapped isomorphically onto $\mathbf{Q}_p$. Let

$$W_T = W \cap T_p(A).$$

The image of W_T in $\mathbf{Z}_p$ is $p^N\mathbf{Z}_p$ for some $N \geqq 0$. But then it follows immediately that $p^N x = 0$, contradicting the fact that x has infinite order.

Lemma 3. *Let R_∞ be the integral closure of R in*

$$K_\infty = K(\mu^{(p)}, q^{1/p^\infty}) = K(A^{(p)}).$$

Let $\mathfrak{M}$ be the maximal ideal of R_∞ lying above $\mathfrak{m}$. Let I be the inertia group of $\mathfrak{M}$ in $\mathrm{Gal}(K_\infty/K)$. *Then I is of finite index in* $\mathrm{Gal}(K_\infty/K)$.

Proof. Let v be an admissible discrete valuation on K. We denote an extension of this valuation to K_∞ by the same letter. Let I_v be the inertia group for this extended valuation. It will suffice to prove that I_v is of finite index in $\mathrm{Gal}(K_\infty/K)$, because $I \supset I_v$. Without loss of generality, we may therefore assume that R is a discrete valuation ring. Let K_v be the completion of K at v, and let L be the completion of the maximal unramified extension of K_v. Then L again has a discrete valuation v. It will suffice to prove that $\mathrm{Gal}(L(A^{(p)})/L)$, identified in the usual manner with a subgroup of $\mathrm{Gal}(K(A^{(p)})/K)$, is of finite index. The picture of Galois theory is as follows.

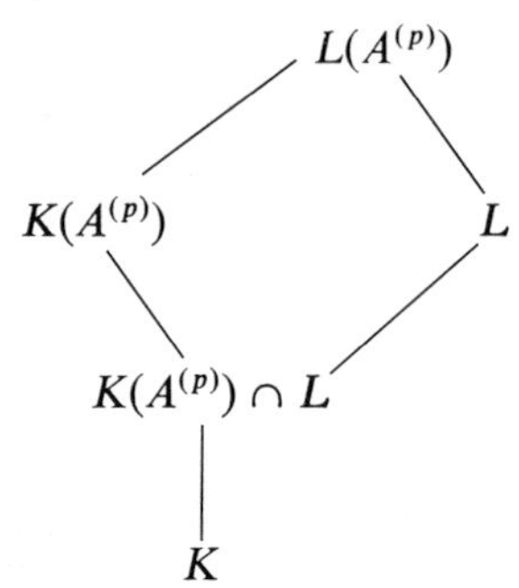

It is known from elementary algebraic number theory that if ζ is a primitive p^n-th root of unity, then $1 - \zeta$ has order $1/\varphi(p^n) = 1/(p-1)p^{n-1}$ at the p-adic valuation giving p order 1. Since v is a discrete valuation, it follows that there is a constant c such that for all n,

$$[L(\mu_n) : L] \geqq cp^n,$$

and in fact the ramification index of $L(\mu_n)$ over L satisfies a similar inequality. The operation of the Galois group on $T_p(A)$ is represented, relative to a basis by matrices

$$\begin{pmatrix} a & b \\ 0 & d \end{pmatrix}$$

with components in $\mathbf{Z}_p$. There exists some positive integer r such that the equation

$$X^{p^r} - q = 0$$

has no root in $L(\mu^{(p)})$. For otherwise, we obtain

$$L(\mu^{(p)}) = L(\mu^{(p)}, q^{1/p^\infty}) = L(A^{(p)}),$$

and the Galois group of $L(A^{(p)})$ over L is abelian, which means that the above matrices must be diagonal, whence the representation is reducible, contradicting Lemma 1, applied to the field L, with its discrete valuation ring. By an elementary irreducibility criterion, or even Kummer Theory, this implies that the degree of

$$L(\mu^{(p)}, q^{1/p^n})$$

over $L(\mu^{(p)})$ satisfies an inequality of the same kind as above, i.e. it is at least equal to cp^n for some constant c. Hence there is a constant c such that for all n,

$$[L(A_n) : L] \geqq cp^{2n}.$$

Since the Galois groups of $K(A_n)$ over K and $L(A_n)$ over L have an order of magnitude at most equal to $c'p^{2n}$ for some constant c', it follows that

$$\mathrm{Gal}(L(A^{(p)})/L)$$

is of finite index in $\mathrm{Gal}(K(A^{(p)})/K)$. Since L is maximal unramified, it follows that $L(A^{(p)})$ is totally ramified over L, thereby proving our lemma.

§3. THE LOCAL ISOGENY THEOREMS

Serre [35] discovered that over a $\mathfrak{p}$-adic field, an elliptic curve whose j-invariant is not integral satisfies the isogeny theorem: If A, B are such elliptic curves, and their p-adic representations on V_p are Galois isomorphic, then the curves are isogenous. It turns out that his proof, with minor modifications, is valid over a more general type of local ring [28a]. Thus we shall prove:

Theorem 2. *Let R be a Noetherian complete local ring, integrally closed, without divisors of* 0, *and of characteristic* 0. *Let K be its quotient field. Assume that the maximal ideal $\mathfrak{m}$ of R contains the prime number p, and that $R/\mathfrak{m}$ is finite. Let A, A' be elliptic curves defined over K, with invariants j, j' such that $1/j$ and $1/j'$ are contained in $\mathfrak{m}$. Suppose that $V_p(A)$ and $V_p(A')$ are G_K-isomorphic. Then A and A' are isogenous.*

Proof. It will suffice to prove that there exist integers i, i' such that $q^i = q'^{i'}$, by Theorem 2 of the preceding chapter. Let

$$\varphi\colon V_p(A) \to V_p(A')$$

be a G_K-isomorphism. By Lemma 1 we know that $V_p(\mu)$ is the only 1-dimensional subspace of $V_p(A)$ (resp. $V_p(A')$) which is stable by G_K. Hence φ maps $V_p(\mu)$ into itself. Moreover, after multiplying φ by some p-adic integer, we may suppose that φ maps $T_p(A)$ into $T_p(A')$. We then have a commutative diagram:

$$\begin{array}{ccccccccc} 0 & \to & T_p(\mu) & \to & T_p(A) & \to & \mathbf{Z}_p & \to & 0 \\ & & {\scriptstyle r}\downarrow & & {\scriptstyle \varphi}\downarrow & & {\scriptstyle s}\downarrow & & \\ 0 & \to & T_p(\mu) & \to & T_p(A') & \to & \mathbf{Z}_p & \to & 0 \end{array} \tag{4}$$

where the vertical arrows on the ends are multiplication by p-adic integers r and s respectively. Let x, x' be the elements of $\lim H^1(G, \mu_n)$ associated to A and A' above, then the commutativity of (4) shows that

$$rx = sx'.$$

Using again our discrete valuation v, we get a homomorphism

$$\varprojlim H^1(G, \mu_n) = \varprojlim K^*/K^{*p^n} \to \mathbf{Z}_p,$$

and we have seen that the image of x is $v(q)$ and the image of x' is $v(q')$. Hence

$$rv(q) = sv(q').$$

It will now suffice to prove that

$$\alpha = q^{v(q')}/q'^{v(q)}$$

is a root of unity.

We look at the image of α in $\lim K^*/K^{*p^n}$. This image is

$$v(q')x - v(q)x',$$

and multiplying by s, we find 0 by using the above relations. Hence the image of α in $\varprojlim K^*/K^{*p^n}$ is 0.

We are thus reduced to proving that the kernel of the canonical map

$$K^* \to \lim K^*/K^{*p^n}$$

is finite. If an element α lies in the kernel, then α must be a p^n-th power in K for all n. If α does not lie in R, then $1/\alpha$ does not generate the unit ideal in $R[1/\alpha]$, for otherwise α would be integral over R, whence in R, a contradiction. A minimal prime over the ideal $(1/\alpha)$ in the integral closure of $R[1/\alpha]$ would give rise to a discrete valuation where α has a pole, and hence could not be a p^n-power for large n. So α lies in R. Similarly, α cannot lie in $\mathfrak{m}$, otherwise $1/\alpha$ does not lie in R. Hence α is a unit in R. Since the residue class field is finite, and R is complete, there is a finite subgroup k^* in R representing the non-zero elements of $R/\mathfrak{m}$, and the group of units U of R is isomorphic to a product

$$U \approx k^* \times U_1,$$

where U_1 consists of the units congruent to 1 mod $\mathfrak{m}$. If $w \in \mathfrak{m}$, then $(1 + w)^{p^n}$ lies in $1 + \mathfrak{m}^n$. From this it is clear that α must lie in k^*. This concludes the proof of Theorem 2.

Remark 1. Having proved that two integral powers of q and q' are equal, it is then also true that the curves are isogenous over K. This follows from the general Tate theory viewing $q^{\mathbf{Z}}$ and $q'^{\mathbf{Z}}$ as "lattices".

Remark 2. In higher dimensions, one can define the analogue of the "multiplicative" parametrization given here for certain abelian varieties. However, Ribet has given an example where the corresponding local isogeny theorem is false in dimension 2, over an ordinary p-adic field. There remains the problem of determining if it is true for "generic" abelian varieties.

Theorem 3. *Let R be a complete Noetherian local ring, without divisors of zero, integrally closed, with maximal ideal* $\mathfrak{m}$, *and quotient field K of characteristic* 0. *Assume that $R/\mathfrak{m}$ is finite. Let A be an elliptic curve defined over K,* *with invariant $j \in R$, and let A' be defined over K, with invariant j' such that $1/j' \in \mathfrak{m}$. Then the representations of G_K on $V_p(A)$ and $V_p(A')$ for any prime p are not isomorphic.*

Proof. Passing to a finite extension of K and the integral closure of R in this extension if necessary, we may assume that A has non-degenerate reduction mod $\mathfrak{m}$. Furthermore, A' becomes isomorphic over a finite extension of K to the curve having the Tate parametrization in terms of q', and hence again without loss of generality, we may assume that A' is the Tate curve. We now distinguish two cases.

The reduction $\bar{A}$ of A mod $\mathfrak{m}$ has a point of order p in the algebraic closure of the residue class field $\bar{R} = R/\mathfrak{m}$. Then $K(A^{(p)})$ contains an infinite unramified part, corresponding to the infinite residue class field extension

$$\bar{R}(\bar{A}^{(p)}).$$

On the other hand, by Lemma 3, we know that $K(A'^{(p)})$ is almost totally ramified, in the sense of that lemma. Hence A and A' cannot be isogenous. (If $p \neq$ characteristic of $R/\mathfrak{m}$, then all of $K(A^{(p)})$ is unramified, and the argument works even more strongly.)

The reduction $\bar{A}$ of A mod $\mathfrak{m}$ is supersingular, i.e. has no point of order p, so that $\bar{A}^{(p)} = 0$. In that case, we use an admissible discrete valuation v. The representation of G_K on $V_p(A')$ is triangular, and has in particular an invariant subspace of dimension 1, corresponding to $V_p(\mu)$. On the other hand, Serre has proved that the representation of G_K on $V_p(A)$ is irreducible, [36], p. 128, Prop. 8. [For the convenience of the reader, we shall reproduce the proof in §4.] Hence these representations cannot be isomorphic, and the curves are not isogenous, as was to be proved.

Remark. The assumption that the residue class field is finite can be weakened to finitely generated over the prime field, since it is known that for such field k, the extension $k(\bar{A}^{(p)})$ of k has an infinite separable part if $\bar{A}$ is not supersingular. However, we shall not use this in the sequel.

§4. SUPERSINGULAR REDUCTION

We now deal with the irreducibility property mentioned above. For the rest of this section, we let A be an elliptic curve defined over a field K of characteristic 0, with a discrete valuation. We let $\mathfrak{o}_K$ be the ring of integers of the valuation, $\mathfrak{m}_K$ its maximal ideal. To prove that $V_p(A)$ is G_K-irreducible, it suffices to do so with respect to any closed subgroup of G_K. Thus we may assume without loss of generality that K is complete. We let $\mathfrak{o}$ be the ring of integers in the algebraic closure of K, and we let $\mathfrak{m}$ be the maximal ideal of $\mathfrak{o}$. We assume that $\mathfrak{o}/\mathfrak{m}$ has characteristic p.

Suppose that A has non-degenerate reduction mod $\mathfrak{m}_K$. We want to find an appropriate parametrization of the points of $A^{(p)}$ which will exhibit their ramification properties. This is done by studying the formal law defined by A over $\mathfrak{o}_K$. (Cf. Serre's *Lie Algebras and Lie Groups*, Chapter 4 and Appendix 1, §3.) Assume for simplicity that the characteristic of the residue class field is $\neq 2, 3$ and that A is in Weierstrass form,

$$y^2 = x^3 + bx + c, \qquad b, c \in \mathfrak{o}_K,$$

with non-degenerate reduction. The origin is represented by the point at infinity. Let (x_1, y_1) be a point in A_K which is in the kernel of the reduction map. Then x_1, y_1 cannot lie in $\mathfrak{o}_K$.

It is clear by comparing poles that

$$x_1 = u_1\pi^{-2m}, \qquad y_1 = v_1\pi^{-3m},$$

with some positive integer m, units u_1, v_1, where π is an element of order 1 at the discrete valuation of K. Let

$$t = \frac{x}{y} \quad \text{and} \quad s = \frac{1}{y}.$$

The correspondence $(x, y) \mapsto (t, s)$ changes the Weierstrass model into the curve defined by

$$s = t^3 + bts^2 + cs^2.$$

The kernel of the reduction map is then represented by points in the (s, t) plane, with coordinates in $\mathfrak{m}$, and the origin of A_K has coordinates $(0, 0)$ in the (s, t) plane. Observe that t is a local uniformizing parameter at the origin of A. [The only use we have made of the assumption that the characteristic of $\mathfrak{o}/\mathfrak{m}$ is $\neq 2, 3$ is to give this explicit parameter. Except for this, all the arguments which follow hold quite generally. The p-adic analytic study of the points on an elliptic curve was originated by E. Lutz, "Sur l'equation $y^2 = x^3 - Ax - B$ sur les corps p-adiques," *J. reine angew. Math.* 177 (1937), p. 204.]

Let z be the point with affine coordinates (x, y) on A, and write $t = t(z)$. It is easily shown by an explicit computation that multiplication by p on A is represented by a power series with coefficients in $\mathfrak{o}_K$. In other words,

$$t(pz) = f(t) = pt + a_2t^2 + a_3t^3 + \cdots,$$

with $a_1 = p$, and $a_n \in \mathfrak{o}_K$ for all n. [In general, the formal group law is defined by a power series in two variables,

$$t(z + z') = F(t(z), t(z')),$$

and f is obtained by iterating F, p times, setting $z = z'$.] See Appendix 1, §3.

Assume that the absolute value on K defined by the discrete valuation is normalized in such a way that $|p| = 1/p$. Let in general

$$f(t) = a_1 t + a_2 t^2 + \cdots$$

be a power series with coefficients in $\mathfrak{o}_K$. Let h be a positive integer such that $|a_i| < 1$ for $1 \leqq i \leqq h - 1$, and suppose that a_h is a unit u. Then the Weierstrass preparation theorem tells us that we can factor f as

$$f(t) = g(t)\psi(t),$$

where $g(t)$ is a polynomial of degree h, and $\psi(t)$ is a unit in the power series ring $\mathfrak{o}_K[[t]]$, i.e. a power series starting with a unit. In particular, a zero of f in $\mathfrak{m}$ is a root of g. For the proof, see A. Fröhlich, *Formal Groups*, Lecture Notes 74, Springer-Verlag, 1968, Chapter I, §3, Theorem 3.

We apply this to the power series obtained from $t(pz)$ on our elliptic curve.

$$t(pz) = f(t) = pt + a_2 t^2 + \cdots + a_{h-1} t^{h-1} + ut^h + \cdots,$$

where u is a unit, and $|a_i| < 1$, $1 \leqq i \leqq h - 1$. We see that a point $Q \in A_{K_a}$ lies in the kernel of the reduction map if and only if $\overline{t(Q)} = 0$, and this occurs if and only if $t(Q) \in \mathfrak{m}$.

Assume now that A has supersingular reduction, i.e. that $\bar{A}^{(p)}$ consists only of the origin. Then all points of $A^{(p)}$ lie in the kernel of the reduction, and in particular, there are p^2 elements in A_p, so that $h \geqq p^2$. Indeed, if $pQ = O$, then $t(Q)$ is a zero of f, because $t(O) = 0$.

Theorem 4. *Assume that A has supersingular reduction*, i.e. *that $\bar{A}^{(p)} = O$. Let $w = (w_1, w_2, \ldots) \in T_p(A)$, so that $pw_{n+1} = w_n$, and suppose $w_1 \neq O$. There exists a number $C > 0$ such that the ramification index of $K(w_n)$ over K is $\geqq Cp^{2n}$.*

Proof. Let $t_n = t(w_n)$. Then $|t_n| < 1$, and we have the relation

$$t_n = pt_{n+1} + a_2 t_{n+1}^2 + \cdots + a_{h-1} t_{n+1}^{h-1} + ut_{n+1}^h + \cdots.$$

First let us prove that

$$\lim_{n \to \infty} |t_n| = 1.$$

We cannot have $|t_{n+1}| \leqq |t_n|$, because the right-hand side would then have an absolute value $< |t_n|$. Furthermore, the absolute value of the right-hand side is at most

$$\max\{|p|\,|t_{n+1}|, |t_{n+1}|^2\},$$

and its absolute value must be the same as $|t_n|$. This shows that

$$|t_{n+1}| \geqq p|t_n| \qquad \text{or} \qquad |t_{n+1}| \geqq |t_n|^{\frac{1}{2}}.$$

From this we conclude that $|t_n| \to 1$ as $n \to \infty$.

If $|t_n|$ is sufficiently close to 1, then the term ut_{n+1}^h on the right-hand side has absolute value strictly greater than any other term, because $|a_i| < 1$ for $1 \leqq i \leqq h - 1$. We must therefore have

$$|t_n| = |ut_{n+1}^h| = |t_{n+1}|^h.$$

Thus from a certain n_0 on, the ramification index at the n-th step increases at least by a factor of $h \geqq p^2$. This proves our theorem.

Theorem 5. *Let A be an elliptic curve defined over a field K of characteristic 0, complete with respect to a discrete valuation, and with non-degenerate reduction $\bar{A}$, which we assume supersingular. Then $V_p(A)$ is G_K-irreducible.*

Proof. Let $w = (w_1, w_2, \ldots) \in T_p(A)$ and suppose that $w_1 \neq 0$. It suffices to prove that there exists $\sigma \in G_K$ such that σw does not lie in the 1-dimensional module over $\mathbf{Z}_p$ generated by w, because then w and σw form a basis of $V_p(A)$ over $\mathbf{Q}_p$, whence $V_p(A)$ is G_K-irreducible. We use Theorem 4, and need only that

$$[K(w_n) : K] \geqq Cp^{2n}.$$

Suppose that σw is a p-adic multiple of w for all $\sigma \in G_K$. Take n large. Then for all $\sigma \in G_K$, the point σw_n is an integral multiple of w_n, and there are at most p^n such multiples. This contradicts the degree inequality above, and proves that $V_p(A)$ is irreducible, as desired.

§5. THE GLOBAL ISOGENY THEOREMS

We shall now see that the isogeny theorem holds globally, over a number field, for an elliptic curve having non-integral invariant; and over a function field for an elliptic curve having transcendental invariant, both when the function field has a constant field which is a number field, and when it is over the complex numbers. The first case, over number fields, is due to Serre.

Theorem 6. *Let A, A' be elliptic curves over a number field K, with invariants j, j'. Assume that j is not $\mathfrak{p}$-integral for some prime $\mathfrak{p}$ of K, dividing p. Assume that $V_p(A)$ and $V_p(A')$ are G_K-isomorphic. Then the curves are isogenous.*

Proof. We have seen in §3 that j' is necessarily not $\mathfrak{p}$-integral. The Galois representations being isomorphic on G_K, they are isomorphic on any closed subgroup, in particular the subgroup which is the Galois group over the $\mathfrak{p}$-adic field $K_{\mathfrak{p}}$. This reduces our problem to the local case, and concludes the proof by Theorem 2.

At the time this book is written, the isogeny theorem in general over number fields is not known.

Next we deal with the generic case. Deligne [2] proved it over the complex numbers by using Hodge structures. I showed [28a] that the Serre arguments for p-adic fields hold also in this case by working over $\mathbf{Z}[1/j]$ as follows.

Theorem 7. *Let A, A' be elliptic curves over a field K, finitely generated over the rationals. Assume that they have transcendental j-invariants. Let p be a prime number, and assume that $V_p(A)$ and $V_p(A')$ are G_K-isomorphic. Then the curves are isogenous.*

Proof. It is trivial that j, j' must be algebraically dependent. Hence K can be selected to be a finite extension of $\mathbf{Q}(j, j')$, of transcendence degree 1 over $\mathbf{Q}$.

Next we prove that j' is integral over $\mathbf{Z}[j]$ and vice versa. Suppose this is not the case. There exists a homomorphism of $\mathbf{Z}[j]$ which extends to $\mathbf{Z}[j, 1/j']$ sending $1/j'$ to 0. Let R be the integral closure of $\mathbf{Z}[j, 1/j']$ in K. Extend the homomorphism to R. By composing our homomorphism with another one if necessary, we may assume that our homomorphism takes on its values in a finite field. Let $\mathfrak{m}$ be the kernel in R. The completion $\hat{R}_{\mathfrak{m}}$ has no divisors of 0 by *EGA*, Chapter IV, 7.8.3 and 7.8.6. The Galois representations being isomorphic on G_K, they are isomorphic with respect to any closed subgroup, in particular the. subgroup arising from the extension $K_{\mathfrak{m}}(A^{(p)}) = K_{\mathfrak{m}}(A'^{(p)})$, where $K_{\mathfrak{m}}$ is the quotient field of $\hat{R}_{\mathfrak{m}}$. This is a contradiction in view of Theorem 3. Hence j' is integral over $\mathbf{Z}[j]$.

(For the reference to commutative algebra, the reader can also look up Matsumura's book on the subject, W. A. Benjamin, Reading, Mass., 1970, Chapter XIII.)

Consider the ring $\mathbf{Z}[1/j, 1/j']$. We contend that the ideal generated by p, $1/j$, $1/j'$ is not the unit ideal. Let $\mathfrak{o}$ be the local ring in $\mathbf{Q}(j)$ of the homomorphism of $\mathbf{Z}[1/j]$ which sends p and $1/j$ to 0. Any place of $\mathbf{Q}(j)$ over this homomorphism must send $1/j'$ to 0. Otherwise, suppose $1/j'$ goes to a finite element $c \neq 0$. Then j' goes to $1/c$, and j goes to infinity, which we have already seen is impossible. Similarly, $1/j'$ cannot go to infinity. This proves our contention.

Let R be the integral closure of $\mathbf{Z}[1/j, 1/j']$ in K and let $\mathfrak{m}$ be a maximal ideal of R containing p, $1/j$, $1/j'$. We now argue as in the first part of the proof, with $\hat{R}_{\mathfrak{m}}$, reducing our problem to the local case, and cite Theorem 2 to conclude the proof.

Theorem 8. *Let K be a finitely generated field over an algebraically closed field k of characteristic 0. Let A, A' be elliptic curves defined over K, with invariants j, j' which are transcendental over k. Assume that $V_p(A)$ and $V_p(A')$ are G_K-isomorphic. Then the elliptic curves are isogenous.*

Proof. As in Theorem 7, the invariants j, j' must be algebraically dependent over k, and we can assume K finite over $k(j, j')$. Without loss of generality, we can assume that A is defined by a Weierstrass equation

$$y^2 = 4x^3 - gx - g,$$

and that A' is defined by

$$y^2 = 4x^3 - g'x - g',$$

after replacing K with a finite extension if necessary. Then $k(g) = k(j)$ and $k(j') = k(g')$. There exists a function field K_0 with constant field k_0, such that k_0 is contained in k, is finitely generated over $\mathbf{Q}$, K_0 is a finite extension of $k_0(j, j')$, and K is obtained from K_0 by extending the constants from k_0 to k. The picture is as follows.

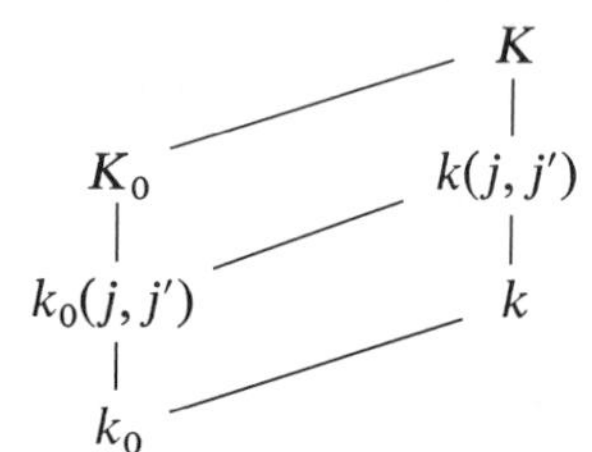

The only new constants introduced over $\mathbf{Q}$ by the points $A^{(p)}$ are the p-power roots of unity (Chapter 6, §3). Let k_1 be the constant field of $K_0(A^{(p)})$, i.e. the algebraic closure of k_0 in $K_0(A^{(p)})$. Let $K_1 = k_1K_0$ be the corresponding constant field extension. We must then have

$$K_1(A^{(p)}) = K_1(A'^{(p)}).$$

Indeed, if we make the constant field extension to k, the two fields

$$K_1(A^{(p)}) \quad \text{and} \quad K_1(A'^{(p)})$$

become equal. Let

$$E = K_1(A^{(p)}) \cap K_1(A'^{(p)}).$$

If E is a proper subfield of $K_1(A^{(p)})$, then there is an element $\neq 1$ of the Galois group of $K_1(A^{(p)})$ over E which extends to an element of the Galois group of $K_1(A^{(p)}, A'^{(p)})$ over $K_1(A'^{(p)})$, thus acting trivially on $A'^{(p)}$, contradicting the hypothesis that the Galois representations over K on $V_p(A)$ and $V_p(A')$ are isomorphic.

We now conclude that

$$K_0(A^{(p)}) = K_0(A'^{(p)}).$$

Let G be the Galois group of the extension $K_0(A^{(p)})$ over K_0. We have two representations

$$\rho: G \to \operatorname{Aut} T_p(A) \quad \text{and} \quad \rho': G \to \operatorname{Aut} T_p(A')$$

onto open subgroups of these automorphism groups, each one of which is isomorphic to $GL_2(\mathbf{Z}_p)$ after a choice of basis over $\mathbf{Z}_p$. Let S be the Galois group of $K_1(A^{(p)})$ over K_1. Then the image of S under both ρ and ρ' is an open subgroup of the special linear subgroup of $\operatorname{Aut} T_p(A)$ and $\operatorname{Aut} T_p(A')$, respectively, that is this image is open in $SL_2(\mathbf{Z}_p)$ under both representations. The center of S maps onto open subgroups of the diagonal groups, formed with units in $\mathbf{Z}_p$. An open subgroup W of the center, not containing -1, is then such that $W \cap S = 1$, whence $S \times W$ is open in G.

The representations of W on $V_p(A)$ and $V_p(A')$ give rise to two characters ψ, ψ' of W into the group of p-adic units, such that if $\sigma \in W$, then the matrix representation of σ on $T_p(A)$ is a diagonal matrix

$$\begin{pmatrix} \psi(\sigma) & 0 \\ 0 & \psi(\sigma) \end{pmatrix},$$

and similarly for σ'. The effect of σ on a p-power root of unity ζ has been shown to be

$$\sigma(\zeta) = \zeta^{\det \rho(\sigma)}.$$

This implies that $\psi(\sigma^2) = \psi'(\sigma^2)$ for all $\sigma \in W$. Since $(W : W^2)$ is finite, passing to an open subgroup of W is necessary, we may assume without loss of generality that $\psi = \psi'$ on W.

By hypothesis, we know that there is a $\mathbf{Q}_p$-isomorphism

$$h: V_p(A) \to V_p(A')$$

which is also an S-isomorphism. Since W acts on $V_p(A)$ and $V_p(A')$ as the same group of p-adic multiplications, it follows that h is also a W-isomorphism, in other words, h is a G-isomorphism for the group $G = S \times W$. The fixed field of G is a finite extension of K_0, finitely generated over the rationals, and we are therefore reduced to the situation of Theorem 7, thus concluding the proof of Theorem 8.

The argument given at the end also shows:

Theorem 9. *Let A, A' be elliptic curves defined over a field K. Assume that the representations*

$$\rho: G_K \to \operatorname{Aut} V_p(A) \qquad \text{and} \qquad \rho': G_K \to \operatorname{Aut} V_p(A')$$

map G_K onto open subgroups of Aut $V_p(A)$ *and* Aut $V_p(A')$ *respectively. Let $L = K(\mu^{(p)})$ be the field obtained by adjoining all p-power roots of unity to K. If the restrictions of ρ and ρ' to G_L are isomorphic, then ρ and ρ' are isomorphic on an open subgroup of G_K.*

A result of Serre states that the hypotheses of Theorem 9 are satisfied in the case of number fields, for elliptic curves without complex multiplication.

In each one of the cases of Theorems 7 and 8, the curves are actually isogenous over the given field K. This comes from an easy additional argument, as in Serre, namely:

Theorem 10. *Let A, A' be elliptic curves defined over a field K of characteristic* 0. *Assume that $V_p(A)$ and $V_p(A')$ are G_K-isomorphic, and that the images* *of* Gal(K_a/K) *in* $\mathrm{End}_{\mathbf{Z}_p}(V_p(A))$ *and* $\mathrm{End}_{\mathbf{Z}_p}(V_p(A'))$ *contain an open subgroup of* $SL_2(\mathbf{Z}_p)$. *Let* $\lambda: A \to A'$ *be an isogeny. Then* λ *is defined over* K.

Proof. First λ must be defined over a finite extension of K (otherwise λ would have infinitely many distinct conjugates, corresponding to distinct images of a smallest field of definition under isomorphisms in a sufficiently large algebraically closed field). Say λ is defined over a Galois extension L of K. Let G_L and G_K be the Galois groups of K_a over L and K respectively, and let $G = \mathrm{Gal}(L/K)$. It suffices to prove that $\lambda^\sigma = \lambda$ for all $\sigma \in G$, and since

$$\lambda^\sigma(a^\sigma) = \lambda(a)^\sigma$$

for any point a of A rational over the algebraic closure of K, it suffices to prove that the endomorphism

$$V_p(\lambda): V_p(A) \to V_p(A')$$

commutes with all $\sigma \in G$. As noted already in §1, we know that $V_p(\lambda)$ lies in $\mathrm{Hom}_{G_L}(V, V')$ (writing $V = V_p(A)$ and similarly for V'). It will suffice to prove that

$$\mathrm{Hom}_{G_L}(V, V') = \mathrm{Hom}_{G_K}(V, V').$$

We know that V and V' are G_K-isomorphic. Hence it will suffice to prove that

$$\mathrm{End}_{G_L}(V) = \mathrm{End}_{G_K}(V).$$

Having assumed that the image of G_K, and hence G_L, in $\mathrm{End}(V)$ contains an open subgroup of $SL_2(\mathbf{Z}_p)$, it follows that the only G_L-endomorphisms of V must be scalar multiples of the identity, i.e. are the endomorphisms αI with $\alpha \in \mathbf{Q}_p$. These are also G_K-endomorphisms, and our assertion is proved.

In the generic case, we know from function theory that the image of the Galois group G_K (when K is finitely generated over $\mathbf{Q}$) in $\mathrm{End}(V)$ contains an open subgroup of $SL_2(\mathbf{Z}_p)$. In the next chapter, this will be proved over number fields for curves with invariant which is not p-integral, so that our remark applies to this case too. As mentioned before, the proof when j is integral over $\mathbf{Z}$ (and A has no complex multiplication) is harder and won't be given in this book.

17 *Division Points over Number Fields*

We know from Chapter 2, §1 that over any field K, the Galois group of the field obtained by adjoining to K all coordinates of points of finite order on an elliptic curve A defined over K is representable as a closed subgroup of the product

$$\prod_{\ell} GL_2(\mathbf{Z}_\ell),$$

taken over primes ℓ. In this chapter, we reproduce Serre's fundamental work that over a number field this Galois group is always open in the product, in the case that the elliptic curve has a non-integral invariant at some prime p. Serre also proved the theorem in general, when the curve does not have complex multiplication, but the proof involves different, and in many respects deeper, techniques. The special case to be given here is sufficiently important, and fits in well enough with the preceding chapters to be included, since the proof is quite short.

§1. A THEOREM OF SHAFAREVIČ

Let K be a number field. Let $\mathfrak{o}$ be the ring of algebraic integers $\mathfrak{o}_K$, and let S be a finite set of primes of K. We let $\mathfrak{o}_S$ be the ring of S-integers, i.e. elements of K which are integral for all $\mathfrak{p} \notin S$. The group of units of $\mathfrak{o}_S$ is denoted by $\mathfrak{o}_S^*$.

Let A be an elliptic curve defined over K. We shall say that A has **good reduction** at a prime $\mathfrak{p}$ of K (or at one of the discrete valuations v of K) if A is isomorphic over K to an elliptic curve defined by an equation having non-degenerate reduction at the local ring $\mathfrak{o}_\mathfrak{p}$ (resp. $\mathfrak{o}_v$). If $\mathfrak{p}$ does not divide 2 or 3, we know that this equation can then be chosen to be a Weierstrass equation whose discriminant is a unit in the local ring.

Theorem 1. (***Shafarevič***) *There is only a finite number of K-isomorphism classes of elliptic curves over K having good reduction at all primes of K outside S.*

Proof. Shafarevič deduced his theorem from a theorem of Siegel on integral points on curves of genus 1. The particular exposition given here is due to Tate—it is also the one in Serre's book [B11]. Suppose that A is defined by the equation

$$y^2 = 4x^3 - g_2x - g_3,$$

with $g_2, g_3 \in K$, and has good reduction outside S. Without loss of generality we can assume that S contains all primes dividing 2 and 3. For each $v \notin S$ there exists an elliptic curve isomorphic to A over K, defined by an equation

$$y^2 = 4x^3 - g_{2,v}x - g_{3,v}$$

with $g_{2,v}$ and $g_{3,v} \in \mathfrak{o}_v$, and discriminant $\Delta_v \in \mathfrak{o}_v^*$, so that there exists $c_v \in k$ such that

$$g_2 = c_v^4 g_{2,v}, \qquad g_3 = c_v^6 g_{3,v}, \qquad \Delta = c_v^{12}\Delta_v.$$

We may also enlarge S so that $\mathfrak{o}_S$ is principal, because making S bigger only strengthens the theorem. For almost all $v \notin S$ we can take $c_v = 1$, i.e. wherever Δ is a unit. Write

$$c_v = p_v^{r_v} \cdot u_v,$$

where u_v is a unit in $\mathfrak{o}_v$. Let

$$c = \prod_v p_v^{r_v}.$$

Then let

$$g_2' = c^{-4}g_2 \qquad \text{and} \qquad g_3' = c^{-6}g_3,$$

so that $\Delta' = c^{-12}\Delta$. It follows that the curve A' defined by

$$y^2 = 4x^3 - g_2'x - g_3'$$

is K-isomorphic to A, and has non-degenerate reduction at all v outside S. We can still change A' by changing the coefficients with a factor $b \in \mathfrak{o}_S^*$ so that $\Delta' \mapsto b^{12}\Delta' = \Delta''$. Therefore A is K-isomorphic to an elliptic curve A'' with coefficients in $\mathfrak{o}_S$ and discriminant defined in $\mathfrak{o}_S^*/\mathfrak{o}_S^{*12}$. Thus we can insure that Δ'' lies among a finite set of representatives F of this factor group of S-units.

But according to the theorem of Siegel (extended by Mahler and Lang, cf. my *Diophantine Geometry*) for $r \in F$, the equation

$$U^3 - 27V^2 = r$$

has only a finite number of solutions in $\mathfrak{o}_S$. This proves Shafarevič's theorem.

Remark. The theorem of Shafarevič extends as follows. Let R be a finitely generated ring over $\mathbf{Z}$, without divisors of zero, and integrally closed. A minimal prime of R is called a prime divisor. It gives rise to a discrete valuation of the quotient field K. We can form as usual the group of divisor classes, which is known to be finitely generated (cf. *DG* again). Thus by localizing, e.g. considering $R[1/x]$ for some $x \in R$, we can kill the finite number of generators of this group, and end up with a factorial ring. One calls $X = \text{spec}(R)$ an absolute affine model of K. Theorem 1 extends to the following statement.

Let S be a finite set of prime divisors of an absolute affine model of a field K, finitely generated over $\mathbf{Q}$. The set of isomorphism classes of elliptic curves over K, with good reduction at all prime divisors of the model not in S, is finite.

The proof is the same as the above, because we only used the unique factorization in R, the finite generation of the group of units (also known, cf. *DG*), and the finiteness of the number of points in R, of a curve $U^3 - 27V^2 = r$, an extension of Siegel's theorem which is also known (loc. cit.).

The importance of Theorem 1 for what follows lies in the fact that we can combine it with a known result:

If A is an elliptic curve over a field K with nondegenerate reduction at a discrete valuation ring $\mathfrak{o}$ of K, and if B is an elliptic curve over K isogenous to A over K, then B has good reduction at $\mathfrak{o}$.

This theorem was proved by Koizumi-Shimura [27], and Serre-Tate [37] for abelian varieties. I don't know a convenient (perhaps computational) proof for elliptic curves, although it is quite plausible. For instance, it is obvious that B has good reduction at the valuation ring in a finite extension, which could easily be taken of degree 4 or 6.

We shall now give an alternate proof for the Shafarevič theorem.

Lemma 1. *Let K be a number field, let S be a finite set of primes in K, and let d be a positive integer. There is only a finite number of extensions of K of degree $\leqq d$, unramified outside S.*

Proof. By taking a sufficiently large set of prime numbers, including all those divisible by primes in S, and those which ramify in K, we see that any extension of K satisfying the hypotheses as stated in the lemma will give rise to an extension of the rationals satisfying similar hypotheses. Thus we may assume that $K = \mathbf{Q}$. It will therefore suffice to prove that the Galois extensions of $\mathbf{Q}$ of bounded degree, unramified outside a finite set of primes S, are finite in number. For each prime $p \in S$, let E_p be the smallest Galois extension of $\mathbf{Q}_p$

containing all the extensions of $\mathbf{Q}_p$ of degree $\leqq d$. (There is only a finite number of these, see for instance [B7], II, §5, Proposition 14.) Let E be a Galois extension of $\mathbf{Q}$ whose completions at all primes dividing those of S contain E_p. If F is a Galois extension of $\mathbf{Q}$ of degree $\leqq d$, unramified outside S, then the completion F_p for any prime p dividing $p \in S$ has degree $\leqq d$ over $\mathbf{Q}_p$, and hence is contained in E_p. This implies that FE over E is unramified (in fact splits completely) at any prime lying above a prime p in S. If F over $\mathbf{Q}$ is also assumed to be unramified outside S, then it follows that FE over E is everywhere unramified. The different of E over $\mathbf{Q}$ is fixed, and is equal to the different of FE over $\mathbf{Q}$ ([B7], III, §1, Proposition 5). Its norm down to $\mathbf{Q}$ from FE is the discriminant of FE over $\mathbf{Q}$, and is therefore bounded. But a classical elementary theorem of Minkowski says that there is only a finite number of extensions of $\mathbf{Q}$ with bounded degree and bounded discriminant ([B7], V, §4, Theorem 5). This proves our lemma.

Lemma 2. *Let K be a number field and $j_0 \in K$. Let S be a finite set of primes of K. There exists only a finite number of K-isomorphism classes of elliptic curves over K with good reduction outside S, having invariant j_0.*

Proof. Let A, B be such curves. There is an isomorphism

$$\alpha: A \to B$$

defined over an extension of K of degree $\leqq 6$. We contend that α is defined over an extension which is unramified outside S. To prove this, we may replace K by its completion $K_\mathfrak{p}$ for $\mathfrak{p} \notin S$. Let $\sigma_1\alpha, \ldots, \sigma_n\alpha$ be the distinct conjugates of α over $K_\mathfrak{p}$, where $\sigma_1, \ldots, \sigma_n$ are automorphisms of the algebraic closure of $K_\mathfrak{p}$ over $K_\mathfrak{p}$. Then

$$\overline{\sigma_1\alpha}, \ldots, \overline{\sigma_n\alpha}$$

are distinct, and are equal to $\bar{\sigma}_1\bar{\alpha}, \ldots, \bar{\sigma}_n\bar{\alpha}$ respectively, where $\bar{\sigma}_\nu$ is the automorphism on the residue class field extension determined by σ_ν. Hence the embeddings $\bar{\sigma}_1, \ldots, \bar{\sigma}_n$ are distinct. This implies that the smallest field of definition for α containing $K_\mathfrak{p}$ is unramified over $K_\mathfrak{p}$.

Using Lemma 1, we conclude that there is a finite extension E of K, which we may assume Galois, such that any two elliptic curves A, B over K, with good reduction outside S, having the same invariant j_0, become isomorphic over E. If $\alpha: A \to B$ is an isomorphism over E, then

$$\sigma \mapsto \alpha^{-1} \circ \alpha^\sigma, \qquad \sigma \in \mathrm{Gal}(E/K)$$

is a function of $\mathrm{Gal}(E/K)$ into $\mathrm{Aut}(A)$, and the set of such functions is finite. If we fix A, and consider elliptic curves B_1, B_2 having the same associated function as above, say by isomorphisms

$$\alpha: A \to B_1 \quad \text{and} \quad \beta: A \to B_2,$$

then B_1, B_2 are isomorphic over K. Indeed, let $\lambda = \beta\alpha^{-1}$. From $\alpha^{-1}\alpha^\sigma = \beta^{-1}\beta^\sigma$ we see that $\lambda^\sigma = \lambda$, so λ is an isomorphism defined over K. This proves our lemma.

To prove the theorem of Shafarevič from the lemmas, let N be an integer so that the genus of the modular function field F_N is $\geqq 1$. Let R_N be the integral closure of $\mathbf{Z}[j]$ in F_N. Enlarge the set S to contain all prime divisors of N. Let A be an elliptic curve defined over K with good reduction outside S. Then the extension $K(A_N)$ of K is unramified outside S, and has degree bounded by N^4. Hence there is a finite extension E of K, which we may assume Galois, such that for all elliptic curves A over K, with invariant $j_0, \in \mathfrak{o}_{K,S}$, and good reduction outside S, we have

$$K(A_N) \subset E.$$

Let $\mathfrak{o}_{E,S}$ be the integral closure of $\mathfrak{o}_{K,S}$ in E. Then any specialization $j \mapsto j_0$ in $\mathfrak{o}_{K,S}$ extends to a point of $\operatorname{spec}(R_N)$ in $\mathfrak{o}_{E,S}$. By the Siegel–Mahler–Lang result, we conclude that there is only a finite number of possible values of such j_0 in $\mathfrak{o}_{K,S}$. The proof of Shafarevič's theorem is finished by using Lemma 2.

The advantage of the above proof over the previous one is that it exhibits better the connection of the theorem with the moduli scheme, which in our case is $\operatorname{spec}(R_N)$. A similar proof could be given for higher dimensional abelian varieties if one knew the finiteness of integral points on the higher dimensional moduli schemes.

§2. THE IRREDUCIBILITY THEOREM

***Theorem* 2.** *Let A be an elliptic curve without complex multiplication, defined over a number field K. Let $G = \operatorname{Gal}(K_a/K)$. Then:*

i) *For almost all primes p, A_p is G-irreducible.*

ii) *For all primes p, $V_p(A)$ is G-irreducible.*

Proof. Suppose that A_p is not irreducible for infinitely many p, and let W_p be an irreducible subspace, necessarily of dimension 1, over $\mathbf{F}_p$. Then W_p is cyclic of order p, and A/W_p is an elliptic curve which can be defined over K, and is isogenous to A over K. If W, W' are cyclic subgroups of A of different prime orders, then A/W and A/W' cannot be isomorphic, otherwise we get a non-trivial endomorphism of A from the following diagram,

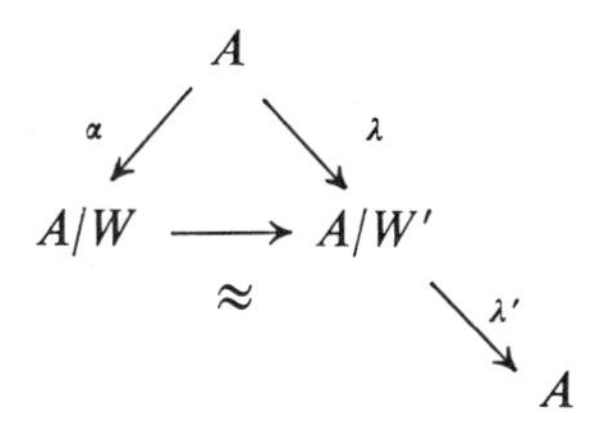

where $\lambda\lambda' = \nu(\lambda)$, and the endomorphism is $\lambda' \circ \approx \circ \alpha$. This contradicts the hypothesis that A has no complex multiplication. Theorem 1 (the theorem of Shafarevič), together with the remarks at the end of §1, now show that there can only be a finite number of W_p as above, in other words, only a finite number of primes p such that A_p is reducible.

The proof of (ii) is similar, except that we work vertically. Suppose that $V_p(A)$ is not irreducible. Then there is a G-irreducible 1-dimensional subspace over $\mathbf{Q}_p$, and therefore after multiplying a generator for this subspace with a suitable p-adic integer, we get a G-invariant 1-dimensional $\mathbf{Z}_p$-subspace Z of $T_p(A)$. Let Z_n be the projection of Z in A_{p^n}. Then the order of Z_n goes to infinity with n, and each Z_n is cyclic, invariant under G. We form $A/Z_n = B_n$ as before, defined over K, and with good reduction by the assumed result mentioned above. The curves B_n cannot be isomorphic, for if $B_m \approx B_n$, say $Z_m \subset Z_n$, then we have a sequence of isogenies

$$A/Z_n \overset{\approx}{\to} A/Z_m \overset{\text{can}}{\to} A/Z_n,$$

whose composite has cyclic kernel, whence is a complex multiplication, contrary to hypothesis. This proves Theorem 2, again in view of the theorem of Shafarevič.

§3. THE HORIZONTAL GALOIS GROUP

Let A be an elliptic curve defined over a number field K. For each prime ℓ, let $A^{(\ell)}$ be the group of points of order a power of ℓ on A, in a fixed algebraic closure. (When we consider A with invariant j_A which is not $\mathfrak{p}$-integral for some prime $\mathfrak{p}$ of K, it is convenient to take this algebraic closure to be in an algebraic closure of the completion $K_{\mathfrak{p}}$.) Let A_{tor} denote the group of torsion points of A, and $K(A_{\text{tor}})$ be the field generated over K by all the coordinates of the torsion points of A.

Let $G = \text{Gal}(K(A_{\text{tor}})/K)$ be the Galois group of the torsion points of A. By the representation on the product

$$\prod T_\ell(A),$$

taken over all primes ℓ, we get an embedding

$$\rho: G \to \prod_\ell GL_2(\mathbf{Z}_\ell)$$

of G as a closed subgroup of the product of the linear groups $GL_2(\mathbf{Z}_\ell)$. At each ℓ, we get a similar embedding

$$\rho_\ell: G_\ell \to GL_2(\mathbf{Z}_\ell),$$

where $G_\ell = \text{Gal}(K(A^{(\ell)})/K)$. We shall often identify G and G_ℓ with their images under this representation.

Serre has proved:

Theorem 3. *Let A be an elliptic curve over a number field K, without complex multiplication. Then the Galois group of $K(A_{\mathrm{tor}})$ over K is open in the product, taken over all primes ℓ,*

$$\prod_{\ell} GL_2(\mathbf{Z}_\ell).$$

We shall prove Serre's theorem here only when A has an invariant $j = j_A$ which is not integral at some prime $\mathfrak{p}$ of K. The proof in general uses quite different techniques.

In this section, we prove one portion of the theorem, namely:

Step 1. *The Galois group of $K(A_\ell)$ over K is $GL_2(\mathbf{Z}/\ell\mathbf{Z})$ for almost all ℓ.*

We note that the local extension $K_{\mathfrak{p}}(A_\ell)$ has a local group which acts on A_ℓ, which we know is Galois-isomorphic to

$$D_q^{1/\ell}/D_q,$$

under the Tate parametrization, as in Chapter 15, §2. Here, D_q is the cyclic group generated by q, and $q = \pi^e u$ has order $e > 0$ at $\mathfrak{p}$, and can be expressed as the above product with some unit u in $K_{\mathfrak{p}}$. For all ℓ not dividing e, the field

$$K_{\mathfrak{p}}(\zeta_\ell, q^{1/\ell})$$

admits an automorphism σ over $K_{\mathfrak{p}}$ which leaves ζ_ℓ fixed and such that

$$\sigma q^{1/\ell} = \zeta_\ell q^{1/\ell}.$$

Thus in a suitable basis of A, the matrix of σ is

$$\begin{pmatrix} 1 & 1 \\ 0 & 1 \end{pmatrix}.$$

On the other hand, A_ℓ is a vector space of dimension 2 over $\mathbf{F}_\ell$, and is irreducible for almost all ℓ by Theorem 2. Since σ leaves a 1-dimensional subspace of A_ℓ fixed (corresponding to ζ_ℓ), there exists some $\tau \in \mathrm{Gal}(K(A_\ell)/K)$ which moves that subspace to another. Then $\sigma' = \tau\sigma\tau^{-1}$ leaves the other subspace fixed. If we select for basis eigenvectors of σ and σ' respectively, then σ and σ' have matrices of the form

$$\begin{pmatrix} 1 & b \\ 0 & 1 \end{pmatrix} \quad \text{and} \quad \begin{pmatrix} 1 & 0 \\ c & 1 \end{pmatrix}$$

with $b, c \neq 0$. These matrices generate $SL_2(\mathbf{Z}/\ell\mathbf{Z})$, thus proving that

$$\mathrm{Gal}(K(A_\ell)/K)$$

at least contains $SL_2(\mathbf{Z}/\ell\mathbf{Z})$.

We also know that the roots of unity lie in $K(A_\ell)$, and hence for almost all ℓ, we get a subgroup $(\mathbf{Z}/\ell\mathbf{Z})^*$ as a factor group of the Galois group. This implies that $\mathrm{Gal}(K(A_\ell)/K)$ must be the whole group

$$GL_2(\mathbf{Z}/\ell\mathbf{Z}),$$

as desired.

Step 2. *For all ℓ, the Galois group of $K(A^{(\ell)})$ over K contains an open subgroup of $GL_2(\mathbf{Z}_\ell)$.*

Proof. We first consider the matter locally over $K_\mathfrak{p}$. As r goes to infinity, q^{1/ℓ^r} generates an extension of arbitrarily high degree over the field generated by all ℓ^ν-th roots of unity over $K_\mathfrak{p}$ (notation as in Chapter 16, and justification by Lemma 1 of Chapter 16, §2 concerning the non-split exact sequence

$$0 \to V_p(\mu) \to V_p(A) \to \mathbf{Q}_p \to 0$$

locally, for $\ell = p$.) For $\ell \neq p$, the extension by ℓ^r-th roots of unity is unramified, and so our assertion is even more trivial.

Hence there is an automorphism σ of the algebraic closure of $K_\mathfrak{p}$ leaving $K_\mathfrak{p}$ and all ℓ^r-th roots of unity fixed, such that the matrix of σ has the form

$$\begin{pmatrix} 1 & a \\ 0 & 1 \end{pmatrix}$$

with some $a \neq 0$ in $\mathbf{Z}_\ell$. By the irreducibility Theorem 2 (ii), there exists globally an element τ in $\mathrm{Gal}(K(A^{(\ell)})/K)$ which moves the 1-dimensional subspace of V_p left invariant by σ, and $\tau\sigma\tau^{-1}$ leaves another subspace invariant. In a suitable basis, we conclude that there exist automorphisms in the global Galois group $\mathrm{Gal}(K(A^{(\ell)})/K)$ represented by the matrices

$$\begin{pmatrix} 1 & a \\ 0 & 1 \end{pmatrix} \quad \text{and} \quad \begin{pmatrix} 1 & 0 \\ b & 1 \end{pmatrix}.$$

Hence the closure of the subgroup generated by these matrices contains the analytic subgroups

$$\begin{pmatrix} 1 & a\mathbf{Z}_\ell \\ 0 & 1 \end{pmatrix} \quad \text{and} \quad \begin{pmatrix} 1 & 0 \\ b\mathbf{Z}_\ell & 1 \end{pmatrix}$$

as well as their product. It is therefore locally a 3-dimensional analytic subgroup of $SL_2(\mathbf{Z}_\ell)$, whence is open in $SL_2(\mathbf{Z}_\ell)$. (Again for the elementary theory of Lie subgroups, cf. Serre's Notes, *Lie Algebras and Lie Groups*.)

To get an open subgroup of $GL_2(\mathbf{Z}_\ell)$, we merely consider the exact sequence

$$0 \to SL_2(\mathbf{Z}_\ell) \to GL_2(\mathbf{Z}_\ell) \xrightarrow{\det} \mathbf{Z}_\ell^* \to 0,$$

and observe that since the field of all ℓ^r-th roots of unity (for all r) over the rationals has a Galois group isomorphic to $\mathbf{Z}_\ell^*$, the translation of this field to a number field has a Galois group open in $\mathbf{Z}_\ell^*$. From this it follows at once that $\mathrm{Gal}(K(A^{(\ell)})/K)$ is open in $GL_2(\mathbf{Z}_\ell)$.

Step 3. *Given a positive integer N, let $A^{(N)}$ be the group of points of order divisible by prime powers only for primes dividing N. Then the Galois group of $K(A^{(N)})$ over K contains an open subgroup of*

$$\prod_{\ell|N} GL_2(\mathbf{Z}_\ell).$$

Proof. Again, we do the SL_2 part first. For each $\ell|N$, a suitably small open subgroup W_ℓ of $SL_2(\mathbf{Z}_\ell)$ is a pro-ℓ-group (it is a subgroup of those elements $\equiv 1 \pmod{\ell}$). The field $K(A^{(N)})$ is the composite of the fields $K(A^{(\ell)})$ for $\ell|N$, and passing to a finite extension E of K (corresponding to an open subgroup of the Galois group) we know that $E(A^{(N)})$ is the composite of the fields $E(A^{(\ell)})$ for $\ell|N$. Taking E sufficiently large, we see that $E(A^{(\ell)})$ is a union of Galois extensions over E, finite, of degree a power of ℓ. Hence for different ℓ, these extensions are linearly disjoint, thus proving our assertion.

Again, using the roots of unity takes care of the GL_2 part.

§4. THE VERTICAL GALOIS GROUP

Now let us prove that for almost all ℓ, we get all of $SL_2(\mathbf{Z}_\ell)$ in the Galois group. The proof is based on the following lemma.

Lemma. *Let H be a closed subgroup of $GL_2(\mathbf{Z}_\ell)$ whose projection* mod ℓ *contains $SL_2(\mathbf{Z}/\ell\mathbf{Z})$. Then H contains $SL_2(\mathbf{Z}_\ell)$ if $\ell \geqq 5$.*

Proof. Let $s \in SL_2(\mathbf{Z}_\ell)$. We must show that $s \in H$. There exists $x_1 \in H$ such that

$$x_1 \equiv s \pmod{\ell},$$

so

$$x_1^{-1}s \equiv 1 \pmod{\ell}.$$

Without loss of generality, we may thus assume that $s \equiv 1 \pmod{\ell}$, and write

$$s = 1 + \ell u,$$

with

$$u = \begin{pmatrix} a & b \\ c & d \end{pmatrix} \in M_2(\mathbf{Z}_\ell).$$

Then

$$\det s \equiv 1 + \ell(a + d) \pmod{\ell^2},$$

and therefore $a + d = \mathrm{tr}(u) \equiv 0 \pmod{\ell}$.

We can write u as a sum

$$u \equiv u_1 + \cdots + u_n \pmod{\ell},$$

where $u_i \in M_2(\mathbf{Z}_\ell)$ and $u_i^2 = 0$, $\operatorname{tr}(u_i) = 0$ for all i. For instance,

$$\begin{pmatrix} a & b \\ c & d \end{pmatrix} = \begin{pmatrix} 0 & b \\ 0 & 0 \end{pmatrix} + \begin{pmatrix} 0 & 0 \\ c & 0 \end{pmatrix} + \begin{pmatrix} a & 0 \\ 0 & -a \end{pmatrix},$$

and the last matrix in the sum can be written as a scalar times

$$\begin{pmatrix} 1 & 0 \\ 0 & -1 \end{pmatrix} = \begin{pmatrix} 1 & 1 \\ -1 & -1 \end{pmatrix} + \begin{pmatrix} 0 & -1 \\ 0 & 0 \end{pmatrix} + \begin{pmatrix} 0 & 0 \\ 1 & 0 \end{pmatrix}.$$

Then we have

$$(1 + \ell u_1) \cdots (1 + \ell u_n) \equiv s \pmod{\ell^2}.$$

Let $s_i = 1 + \ell u_i$. Then $\det s_i = 1 + \ell \operatorname{tr}(u_i) + \ell^2 \det(u_i) = 1$. We see that our s is a product of the s_i, and we are reduced to studying each s_i separately.

Suppose therefore that

$$s = 1 + \ell u,$$

with $u^2 = 0$, $\operatorname{tr}(u) = 0$. We want to show that there exists $x_2 \in H$ such that

$$x_2 \equiv s \pmod{\ell^2}.$$

By hypothesis, there exists $y \in H$ such that $y \equiv 1 + u \pmod{\ell}$, so

$$y = 1 + u + \ell v, \qquad \text{with } v \in M_2(\mathbf{Z}_\ell).$$

Then H contains y^ℓ, and

$$\begin{aligned} y^\ell &= 1 + \ell(u + \ell v) + \sum_{\nu=2}^{\ell-1} \binom{\ell}{\nu}(u + \ell v)^\nu + (u + \ell v)^\ell \\ &\equiv 1 + \ell u \pmod{\ell^2}. \end{aligned}$$

The binomial coefficients $\binom{\ell}{\nu}$ contain ℓ for $\nu = 2, \ldots, \ell - 1$, and the terms in the sum contain either $u^2 = 0$, or ℓ for $\ell \geqq 5$, so these terms contain ℓ^2.

The last term $(u + \ell v)^\ell$ contains $(u + \ell v)^3$ because $\ell \geqq 5$, and

$$(u + \ell v)^2 = 0 + \ell(uv + vu) + \ell^2 v^2,$$

so

$$(u + \ell v)^4 \text{ contains } \ell^2.$$

This proves that H contains $y^\ell \equiv s \pmod{\ell^2}$.

We can now proceed inductively, writing $s = 1 + \ell^n u$, and take

$$y = 1 + \ell^{n-1} u.$$

This proves the lemma.

We can combine the lemma with the result of the preceding section, and find:

Step 4. *Let $A^{(\ell)}$ be the group of points of ℓ-power order on A, defined over a number field K, and with a non-integral invariant at some prime $\mathfrak{p}$. Then $\mathrm{Gal}(K(A^{(\ell)})/K)$ contains $SL_2(\mathbf{Z}_\ell)$ for almost all ℓ.*

Let $G = \mathrm{Gal}(K(A_{\mathrm{tor}})/K)$. Then we have a closed embedding of G in the product of all $GL_2(\mathbf{Z}_\ell)$. For each ℓ we have the determinant $GL_2(\mathbf{Z}_\ell) \to \mathbf{Z}_\ell^*$, which extends to the product over all ℓ componentwise, and induces a homomorphism of G onto a subgroup of $\prod \mathbf{Z}_\ell^*$, denoted by Z, with kernel W. Note that Z is open in the product because all the roots of unity lie in $K(A_{\mathrm{tor}})$. Thus we have a pair of exact sequences

$$\begin{array}{ccccccccc} 0 \to & W & \to & G & \to & Z & \to 0 \\ & \downarrow & & \downarrow & & \downarrow & \\ 0 \to & \prod_\ell SL_2(\mathbf{Z}_\ell) & \to & \prod_\ell GL_2(\mathbf{Z}_\ell) & \to & \prod_\ell \mathbf{Z}_\ell^* & \to 0. \end{array}$$

Furthermore we know from our above results that there is a finite set F of primes such that the projection of G on

$$\prod_{\ell \in F} GL_2(\mathbf{Z}_\ell)$$

is an open subgroup of this product by Step 3. Also, the projection of G on the ℓ-th factor contains $SL_2(\mathbf{Z}_\ell)$ for almost all ℓ.

In the next section we conclude the proof, using only group theory.

§5. END OF THE PROOF

The end of the proof depends on a formal juggling with groups and factor groups, and prime factorizations, and we don't use elliptic curves any more, just group theory. Again we let

$$G = \mathrm{Gal}(K(A_{\mathrm{tor}})/K).$$

Step 5. *The group G contains*

$$\Gamma_p = (\ldots, 1, 1, SL_2(\mathbf{Z}_p), 1, 1, \ldots)$$

for almost all p.

Proof. A group X is called **profinite** if it is a projective limit of finite groups. Galois groups of infinite Galois extensions are of this type. If X is profinite and S is a finite simple group, we shall say that S **occurs** in X if there exist subgroups $X_1 \subset X_2 \subset X$ such that X_1 is normal in X_2 and $X_1/X_1 \approx S$.

Using elementary isomorphism theorems, one sees that if X is a closed normal subgroup of the profinite group Y, then S occurs in X or S occurs in Y/X.

Let $S_p = SL_2(\mathbf{Z}/p\mathbf{Z})/\pm 1$ for a prime p. It is well known that S_p is simple for $p \geqq 5$.

We use the exact sequences of the last section. We know that S_p occurs in G for almost all p, by projecting on the p-factors. We want to conclude that G contains the factor

$$\Gamma_p = (\ldots, 1, 1, SL_2(\mathbf{Z}_p), 1, 1, \ldots)$$

for almost all p. We show first that S_p occurs in $G \cap \Gamma_p$. Let

$$U_p = (\ldots, 1, 1, GL_2(\mathbf{Z}_p), 1, 1, \ldots).$$

We have an injection

$$G/(G \cap U_p) \to (\prod_{\ell} U_\ell)/U_p.$$

But S_p does not occur in any $GL_2(\mathbf{Z}_\ell)$, for $\ell \neq p$ and $p > 5$. Hence S_p does not occur in $G/(G \cap U_p)$, so S_p occurs in $G \cap U_p$, whence it occurs in $G \cap \Gamma_p$, which is closed in Γ_p, and projects into $PSL_2(\mathbf{Z}/p\mathbf{Z}) = SL_2(\mathbf{Z}/p\mathbf{Z})/\pm 1$. Let H_p be its image. We contend that $H_p = PSL_2(\mathbf{Z}/p\mathbf{Z})$. If not, H_p is a proper subgroup, so S_p occurs in the kernel of the projection, i.e. in

$$\{u \in SL_2(\mathbf{Z}_p), \quad u \equiv 1 \pmod{p}\}.$$

This is impossible because this group is solvable, while S_p is simple.

We have therefore shown that $G \cap \Gamma_p$ projects onto $SL_2(\mathbf{Z}/p\mathbf{Z})$, whence $G \cap \Gamma_p = SL_2(\mathbf{Z}_p)$ for p large by the lemma of §4, combined with our preceding results. This finishes Step 5.

We now conclude that G contains finite products

$$(\ldots, 1, 1, SL_2(\mathbf{Z}_{\ell_1}), SL_2(\mathbf{Z}_{\ell_2}), \ldots, SL_2(\mathbf{Z}_{\ell_m}), 1, 1, \ldots)$$

for ℓ_i sufficiently large. Since G is closed in $\prod GL_2(\mathbf{Z}_\ell)$, it follows that there is a finite set S of primes such that G contains

$$\prod_{\ell \notin S} SL_2(\mathbf{Z}_\ell).$$

Step 6. *The group G contains an open subgroup of $\prod SL_2(\mathbf{Z}_\ell)$.*

Proof. Let S be as above. Let G_S be the projection of G into

$$\prod_{\ell \in S} GL_2(\mathbf{Z}_\ell),$$

and G'_S the projection into the complementary product

$$\prod_{\ell \notin S} GL_2(\mathbf{Z}_\ell).$$

Let

$$H_S = G \cap \prod_{\ell \in S} GL_2(\mathbf{Z}_\ell) \qquad \text{and} \qquad H'_S = G \cap \prod_{\ell \notin S} GL_2(\mathbf{Z}_\ell),$$

so that $H_S \subset G_S$ and $H'_S \subset G'_S$. We have canonical isomorphisms

$$G_S/H_S \approx G/(H_S \times H'_S) \approx G'_S/H'_S.$$

Step 5 shows that H'_S contains $\prod_{\ell \notin S} SL_2(\mathbf{Z}_\ell)$, so that G'_S/H'_S is abelian. Hence

G_S/H_S is abelian, and H_S contains the closure of the commutator group of G_S. But the Lie algebra of SL_2 is equal to its own derived Lie algebra. Hence the closure of the commutator subgroup of an open subgroup of $SL_2(\mathbf{Z}_\ell)$ contains an open subgroup of $SL_2(\mathbf{Z}_\ell)$. This implies that H_S contains an open subgroup W of

$$\prod_{\ell \in S} SL_2(\mathbf{Z}_\ell),$$

Combined with Step 5, this yields Step 6.

Final Step. *We now consider the determinant map*

$$G \to \prod_\ell \mathbf{Z}_\ell^*,$$

induced from the product of the determinant maps

$$\prod_\ell GL_2(\mathbf{Z}_\ell) \to \prod_\ell \mathbf{Z}_\ell^*.$$

Since $K(A_{\text{tor}})$ contains all roots of unity, it follows that the determinant map sends G onto an open subgroup of $\prod_\ell \mathbf{Z}_\ell^*$, necessarily of finite index. By Step 6, we know that the kernel contains an open subgroup of $\prod_\ell SL_2(\mathbf{Z}_\ell)$, also of finite index. From the commutative exact sequences at the end of §4, it follows that G is of finite index in

$$\prod_\ell GL_2(\mathbf{Z}_\ell),$$

and must therefore be open because G is closed in this product. This concludes the proof.

Part Four

Theta Functions and Kronecker Limit Formulas

This last part enters into the multiplicative theory of the elliptic functions, and its connection with L-series. Chapters 18 and 19 are immediate continuations of Chapters 1 and 4, and could have been treated much earlier, with the obvious exception of the arithmetic application of Shimura's reciprocity law to the special values of the Siegel function. We deal first with the analytic construction of modular functions by "multiplicative" means, and then study the special values at imaginary quadratic numbers.

The reader can read the first Kronecker limit formula independently of the other chapters, and immediately in connection with Chapter 18. He can then read the chapter on the fundamental theta function in connection with the second Kronecker limit formula. The treatment of these limit formulas follows Siegel's exposition [B15]. A complete account, including relations to L-series, and real quadratic fields, is also given in Meyer's book [B8].

18 Product Expansions

§1. THE SIGMA AND ZETA FUNCTIONS

Both in number theory and analysis one factorizes elements into prime powers. In analysis, this means that a function gets factored into an infinite product corresponding to its zeros and poles. Taking the values at special points, such an analytic expression reflects itself into special properties of the values, for which it becomes possible to determine the prime factorization in number fields.

In this chapter, we are concerned with the analytic expressions.

Our first task is to give a universal gadget allowing us to factorize an elliptic function, with a numerator and denominator which are entire functions, and are as periodic as possible.

One defines a **theta function** (on **C**) with respect to a lattice L, to be an entire function θ satisfying the condition

$$\theta(z + u) = \theta(z)e^{2\pi i[l(z,u) + c(u)]}, \qquad z \in \mathbf{C}, u \in L,$$

where l is **C**-linear in z, **R**-linear in u, and $c(u)$ is some function depending only on u. We shall construct a theta function.

We write down the **Weierstrass sigma function,** which has zeros of order 1 at all lattice points, by the Weierstrass product

$$\sigma(z) = z \prod_{\omega \in L'} \left(1 - \frac{z}{\omega}\right) e^{z/\omega + \frac{1}{2}(z/\omega)^2}.$$

Here L' means the lattice from which 0 is deleted, i.e. we are taking the product over the non-zero periods. We note that σ also depends on L, and so we write $\sigma(z, L)$, which is homogeneous of degree 1, namely

$$\boxed{\sigma(\lambda z, \lambda L) = \lambda \sigma(z, L)} \qquad \lambda \in \mathbf{C}.$$

Taking the logarithmic derivative formally yields the **Weierstrass zeta function**

$$\zeta(z, L) = \zeta(z) = \frac{\sigma'(z)}{\sigma(z)} = \frac{1}{z} + \sum_{\omega \in L'} \left[\frac{1}{z - \omega} + \frac{1}{\omega} + \frac{z}{\omega^2} \right].$$

It is clear that the sum on the right converges absolutely and uniformly for z in a compact set not containing any lattice point, and hence integrating and exponentiating shows that the infinite product for $\sigma(z)$ also converges absolutely and uniformly in such a region. Differentiating $\zeta(z)$ term by term shows that

$$\zeta'(z) = -\wp(z) = -\frac{1}{z^2} - \sum_{\omega \in L'} \left[\frac{1}{(z - \omega)^2} - \frac{1}{\omega^2} \right].$$

Also from the product and sum expressions, we see at once that *both σ and ζ are odd functions*, i.e.

$$\sigma(-z) = -\sigma(z) \qquad \text{and} \qquad \zeta(-z) = -\zeta(z).$$

The series defining $\zeta(z, L)$ shows that it is homogenous of degree -1, that is

$$\boxed{\zeta(\lambda z, \lambda L) = \frac{1}{\lambda} \zeta(z, L).}$$

Differentiating the function $\zeta(z + \omega) - \zeta(z)$ for any $\omega \in L$ yields 0 because the $\wp$-function is periodic. Hence there is a constant $\eta(\omega)$ (sometimes written η_ω) such that

$$\zeta(z + \omega) = \zeta(z) + \eta(\omega).$$

It is clear that $\eta(\omega)$ is $\mathbf{Z}$-linear in ω. If $L = [\omega_1, \omega_2]$, then one uses the notation

$$\eta(\omega_1) = \eta_1 \qquad \text{and} \qquad \eta(\omega_2) = \eta_2.$$

As with ζ, the form $\eta(\omega)$ satisfies the homogeneity relation

$$\eta(\lambda\omega) = \frac{1}{\lambda} \eta(\omega),$$

as one verifies directly from the similar relation for ζ. Observe that the lattice should strictly be in the notation, so that in full, the above relations should read

$$\boxed{\begin{aligned} \zeta(z + \omega, L) &= \zeta(z, L) + \eta(\omega, L) \\ \eta(\lambda\omega, \lambda L) &= \frac{1}{\lambda} \eta(\omega, L). \end{aligned}}$$

Remark. For those who like to connect with other ideas, the map

$$(z, t) \mapsto (1, \wp(z), \quad \wp'(z), \quad t - \wp(z))$$

sends $\mathbf{C}^2$ onto a 2-dimensional group variety, which projects on the elliptic curve parametrized by the $\wp$ and $\wp'$-functions. We observe that the above map is genuinely periodic, with periods (ω_1, η_1) and (ω_2, η_2). The group variety is that associated with integrals of the second kind on the elliptic curve, and is a group extension of the elliptic curve by an additive group.

Theorem 1. *The function σ is a theta function, and in fact*

$$\frac{\sigma(z + \omega)}{\sigma(z)} = \psi(\omega)e^{\eta(\omega)(z+\omega/2)}$$

where

$$\psi(\omega) = 1 \quad \text{if} \quad \omega/2 \in L$$

$$\psi(\omega) = -1 \quad \text{if} \quad \omega/2 \notin L.$$

Proof. We have

$$\frac{d}{dz}\log\frac{\sigma(z + \omega)}{\sigma(z)} = \eta(\omega).$$

Hence

$$\log\frac{\sigma(z + \omega)}{\sigma(z)} = \eta(\omega)z + c(\omega),$$

whence exponentiating yields

$$\sigma(z + \omega) = \sigma(z)e^{\eta(\omega)z+c(\omega)},$$

which shows that σ is a theta function. We write the quotient as in the statement of the theorem, thereby defining $\psi(\omega)$, and it is then easy to determine $\psi(\omega)$ as follows.

Suppose that $\omega/2$ is not a period. Set $z = -\omega/2$ in the above relation. We see at once that $\psi(\omega) = -1$ because σ is odd. On the other hand, consider

$$\frac{\sigma(z + 2\omega)}{\sigma(z)} = \frac{\sigma(z + 2\omega)}{\sigma(z + \omega)}\frac{\sigma(z + \omega)}{\sigma(z)}.$$

Using the functional equation twice and comparing the two sides, we see that $\psi(2\omega) = \psi(\omega)^2$. In particular, if $\omega/2 \in L$, then

$$\psi(\omega) = \psi(\omega/2)^2.$$

Dividing by 2 until we get some element of the lattice which is not equal to twice a period, we conclude at once that $\psi(\omega) = (-1)^{2n} = 1$.

The numbers η_1 and η_2 are called **basic quasi periods of** ζ.

Legendre Relation. *We have*

$$\eta_2\omega_1 - \eta_1\omega_2 = 2\pi i.$$

Proof. We integrate around a fundamental parallelogram P, just as we did for the $\wp$-function:

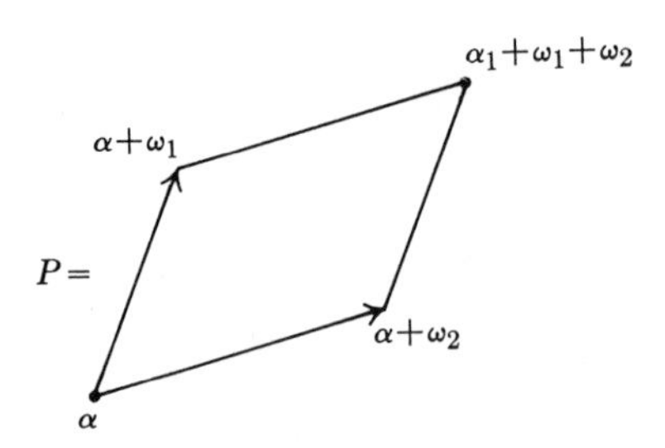

Fig. 18-1

The integral is equal to

$$\int_{\partial P} \zeta(z)\,dz = 2\pi i \sum \text{residues of } \zeta$$
$$= 2\pi i$$

because ζ has residue 1 at 0 and no other pole in a fundamental parallelogram containing 0. On the other hand, using the quasi periodicity, the integrals over opposite sides combine to give

$$\eta_2\omega_1 - \eta_1\omega_2,$$

as desired.

Next, we show how the sigma function can be used to factorize elliptic functions. We know that the sum of the zeros and poles of an elliptic function must be congruent to zero modulo the lattice. Selecting suitable representatives of these zeros and poles, we can always make the sum equal to 0.

For any $a \in \mathbf{C}$ we have

$$\frac{\sigma(z + a + \omega)}{\sigma(z + a)} = \psi(\omega)e^{\eta(\omega)(z+\omega/2)}\, e^{\eta(\omega)a}.$$

Observe how the term $\eta(\omega)a$ occurs linearly in the exponent. It follows that if $\{a_i\}$, $\{b_i\}$ $(i = 1, \ldots, n)$ are families of complex numbers such that

$$\sum a_i = \sum b_i,$$

then the function

$$\frac{\prod \sigma(z - a_i)}{\prod \sigma(z - b_i)}$$

is periodic with respect to our lattice, and is therefore an elliptic function. Conversely, any elliptic function can be so factored into a numerator and denominator involving the sigma function. We write down explicitly the special case with the $\wp$-function.

Theorem 2. *For any $a \in \mathbf{C}$ not in L, we have*

$$\wp(z) - \wp(a) = -\frac{\sigma(z + a)\sigma(z - a)}{\sigma^2(z)\sigma^2(a)}.$$

Proof. The function $\wp(z) - \wp(a)$ has zeros at a and $-a$, and has a double pole at 0. Hence

$$\wp(z) - \wp(a) = C\frac{\sigma(z + a)\sigma(z - a)}{\sigma^2(z)}$$

for some constant C. Multiply by z^2 and let $z \to 0$. Then $\sigma^2(z)/z^2$ tends to 1 and $z^2\wp(z)$ tends to 1. Hence we get the value $C = -1/\sigma^2(a)$, thus proving our theorem.

APPENDIX. THE SKEW SYMMETRIC PAIRING

As an application of the sigma function, we shall carry out the details of the skew-symmetric pairing between points of order N on an elliptic curve mentioned in Chapter 6, §3.

Recall that a **divisor** (or a **0-cycle**) on the elliptic curve (torus) A is an element of the free abelian group generated by the points, and can therefore be written in the form

$$\mathfrak{a} = \sum m_i(a_i),$$

with integer coefficients m_i. We take a_i to be a point in $\mathbf{C}$ representing a point on $A_{\mathbf{C}} = \mathbf{C}/L$. We say that $\mathfrak{a}$ has **degree** 0 if $\sum m_i = 0$. We write $\mathfrak{a} \sim 0$ if $\mathfrak{a}$ is the divisor of a function, and we say then that $\mathfrak{a}$ is **linearly equivalent** to 0. We let

$$S(\mathfrak{a}) = \sum m_i a_i \pmod{L}$$

be the point on the torus obtained by summing the a_i in $\mathbf{C}$ (as distinguished from the formal sum giving the divisor). Then the representation of a function as a product of sigma factors shows that $\mathfrak{a} \sim 0$ if and only if $S(\mathfrak{a}) = 0$.

Let g be a non-zero function on A such that none of the components (a_i) of $\mathfrak{a}$ are zeros or poles of g. Then we define

$$g(\mathfrak{a}) = \prod g(a_i)^{m_i}.$$

If f, g are non-zero rational functions on A, then we have the reciprocity law

$$\boxed{f((g)) = g((f))}$$

provided of course that the expressions are defined, i.e. the divisors of f and g have no point in common. This fact holds on arbitrary curves (Weil, 1940), and a suitably formulated generalization holds on arbitrary varieties (Lang, 1958). In the case of elliptic curves, the relation is obvious if we make use of the sigma function. Indeed, if

$$(f) = \sum m_i(a_i) \qquad \text{and} \qquad (g) = \sum n_j(b_j),$$

where a_i, b_j are complex numbers corresponding to points on the torus, such that $\sum m_i a_i = \sum n_j b_j = 0$, then

$$f(z) = c \prod_i \sigma(z - a_i)^{m_i},$$

with some constant c. Consequently

$$f((g)) = \prod_{i,j} \sigma(b_j - a_i)^{m_i n_j} = g((f)),$$

because σ is an even function and $\sum m_i = 0$.

Now let $\mathfrak{a}$, $\mathfrak{b}$ be divisors such that $N\mathfrak{a}$ and $N\mathfrak{b} \sim 0$. Say

$$N\mathfrak{a} = (f) \qquad \text{and} \qquad N\mathfrak{b} = (g).$$

Assume that $\mathfrak{a}$ and $\mathfrak{b}$ have no point in common. We define

$$\langle \mathfrak{a}, \mathfrak{b} \rangle = \frac{f(\mathfrak{b})}{g(\mathfrak{a})}.$$

Theorem. *The symbol $\langle \mathfrak{a}, \mathfrak{b} \rangle$ depends only on the linear equivalence classes of $\mathfrak{a}$ and $\mathfrak{b}$. It induces a skew-symmetric non-degenerate pairing*

$$A_N \times A_N \to \mu_N,$$

where μ_N is the group of N-th roots of unity.

Proof. If $\mathfrak{b}' \sim \mathfrak{b}$ and $\mathfrak{a}$, $\mathfrak{b}'$ have no point in common, it is immediately verified from the reciprocity law that $\langle \mathfrak{a}, \mathfrak{b} \rangle = \langle \mathfrak{a}, \mathfrak{b}' \rangle$. Thus our pairing depends only on the linear equivalence classes of $\mathfrak{a}$ and $\mathfrak{b}$ respectively. In particular, if a, b are points of order N on A, we may let $\mathfrak{a} = (a) - (0)$ and $\mathfrak{b} = (b) - (0)$, and define

$$\langle a, b \rangle = \langle \mathfrak{a}, \mathfrak{b} \rangle = \langle \mathfrak{a}', \mathfrak{b}' \rangle,$$

where $\mathfrak{a}' \sim \mathfrak{a}$, $\mathfrak{b}' \sim \mathfrak{b}$, and $\mathfrak{a}'$, $\mathfrak{b}'$ have no point in common. (We can always find such $\mathfrak{a}'$, $\mathfrak{b}'$ by making appropriate translations.)

It is also an immediate consequence of the reciprocity law that the pairing is skew-symmetric, and takes its value in the N-th roots of unity. We shall obtain an analytic expression for this root of unity, which will automatically show that the pairing is non-degenerate.

The symbol $\langle \mathfrak{a}, \mathfrak{b} \rangle$ is independent of the linear equivalence classes of $\mathfrak{a}$ and $\mathfrak{b}$ respectively. We let a, b be complex numbers representing the points $S(\mathfrak{a})$ and $S(\mathfrak{b})$ respectively, so that

$$Na = \omega \quad \text{and} \quad Nb = \omega'$$

are periods. To compute $\langle \mathfrak{a}, \mathfrak{b} \rangle$ we may then take $\mathfrak{a}$ and $\mathfrak{b}$ to be the divisors

$$\mathfrak{a} = (u + a) - (u) \quad \text{and} \quad \mathfrak{b} = (v + b) - (v),$$

where u, v are sufficiently general. Again letting

$$N\mathfrak{a} = (f) \quad \text{and} \quad N\mathfrak{b} = (g),$$

we see that the factorization of f and g in terms of the sigma functions is given by:

$$f(z) = \frac{\sigma(z - (u + a))^N}{\sigma(z - u)^{N-1}\,\sigma(z - u - \omega)}$$

$$g(z) = \frac{\sigma(z - (v + b))^N}{\sigma(z - v)^{N-1}\,\sigma(z - v - \omega')}.$$

If we now make the appropriate substitutions for $f(\mathfrak{a})/g(\mathfrak{b})$, and use the functional equation for the sigma function, together with the fact that the sigma function is odd, we find the value

$$\frac{f(\mathfrak{a})}{g(\mathfrak{b})} = \frac{e^{\eta(\omega)\omega'/N}}{e^{\eta(\omega')\omega/N}}$$

Let us select $\omega = \omega_1$ and $\omega' = \omega_2$, and use the Legendre relation. We find

$$\boxed{\langle \mathfrak{a}_1, \mathfrak{a}_2 \rangle = e^{-2\pi i/N}}$$

for the special divisors $\mathfrak{a}_1, \mathfrak{a}_2$ such that $S(\mathfrak{a}_1)$ is represented by the complex number ω_1/N and $S(\mathfrak{a}_2)$ is represented by the complex number ω_2/N. Expressing ω and ω' as linear combinations of ω_1, ω_2 with integer coefficients, we see at once that our pairing $\langle a, b \rangle$ is non-degenerate. This proves everything we wanted.

Remark. The symbol $\langle a, b \rangle$ can also be given in terms of Kummer theory. Cf. my book *Abelian Varieties* for the general statement in higher dimensions. We leave it as an exercise to the reader to give the proofs in terms of sigma function on elliptic curves. Shimura [B12] treats the pairing directly from the Kummer point of view.

§2. A NORMALIZATION AND THE q-PRODUCT FOR THE σ-FUNCTION

We normalize our lattice to be $L_\tau = [\tau, 1]$, so that the corresponding sigma function is $\sigma(z; \tau)$. We wish to multiply σ by a trivial theta function, of the form

$$e^{az^2+bz},$$

i.e. let

$$\varphi(z) = e^{az^2+bz}\sigma(z),$$

such that φ has period τ (we shall see afterwards how φ behaves under translation by 1). This is a trivial problem in solving for a and b. We let

$$a = -\tfrac{1}{2}\eta(1) \qquad \text{and} \qquad b = i\pi.$$

Computing $\varphi(z + \tau)/\varphi(z)$, and using the functional equation for σ, yields the first part of the next theorem.

Theorem 3. *Let*

$$\varphi(z; \tau, 1) = \varphi(z) = e^{-\frac{1}{2}\eta z^2} q_z^{\frac{1}{2}} \sigma(z; \tau),$$

where $\eta = \eta(1)$ ($= \eta_2$ *for the lattice* $[\tau, 1]$), *and* $q_z = e^{2\pi i z}$. *Then*

$$\varphi(z + 1) = \varphi(z) \qquad \text{and} \qquad \varphi(z + \tau) = -\frac{1}{q_z}\varphi(z).$$

Proof. The first relation was achieved by construction. The second part of the theorem comes by expanding

$$\begin{aligned}\varphi(z + \tau) &= e^{a(z+\tau)^2+b(z+\tau)}\psi(\tau)\, e^{\eta(\tau)(z+\tau/2)}\sigma(z)\\ &= \varphi(z) \text{ times an obvious exponential factor.}\end{aligned}$$

Write down the exponential factor explicitly, and use the Legendre relation, which reads

$$\eta(1)\tau - \eta(\tau)\cdot 1 = 2\pi i.$$

You get at once

$$\varphi(z + \tau) = \varphi(z)(-1)\, e^{-2\pi i z}.$$

This proves the second part, as described.

One also wants the formulation of Theorem 3 in its homogeneous form as follows.

Theorem 3′. *Let* $L = [\omega_1, \omega_2]$ *and*

$$\varphi(z; \omega_1, \omega_2) = e^{-\frac{1}{2}\eta_2\omega_2(z/\omega_2)^2} q_{z/\omega_2}^{\frac{1}{2}} \sigma(z; L).$$

Then

$$\varphi(z + \omega_2; \omega_1, \omega_2) = \varphi(z; \omega_1, \omega_2)$$

and

$$\varphi(z + \omega_1; \omega_1, \omega_2) = -\frac{1}{q_{z/\omega_2}} \varphi(z; \omega_1, \omega_2).$$

Remark. In the relation between φ and σ, observe that σ is homogeneous of degree 1, and that the exponential factors in front are homogeneous of degree 0 (that is the products $\eta_2\omega_2$ and z/ω_2). In particular, φ is homogeneous of degree 1, that is:

$$\boxed{\varphi(\lambda z; \lambda\omega_1, \lambda\omega_2) = \lambda\varphi(z; \omega_1, \omega_2).}$$

We want product expansions for $\sigma(z)$ and $\varphi(z)$, which are entire, with zeros of order 1 at the lattice points of $[\tau, 1]$. Let $q_\tau = e^{2\pi i\tau}$ and $q_z = e^{2\pi iz}$.

Theorem 4. *Let $\varphi(z)$ be as in Theorem 3. Then*

$$\varphi(z; \tau) = (2\pi i)^{-1}(q_z - 1) \prod_{n=1}^{\infty} \frac{(1 - q_\tau^n q_z)(1 - q_\tau^n/q_z)}{(1 - q_\tau^n)^2}$$

and

$$\sigma(z; \tau) = (2\pi i)^{-1} e^{\frac{1}{2}\eta z^2}(q_z^{\frac{1}{2}} - q_z^{-\frac{1}{2}}) \prod_{n=1}^{\infty} \frac{(1 - q_\tau^n q_z)(1 - q_\tau^n/q_z)}{(1 - q_\tau^n)^2}.$$

(Again we put $\eta = \eta(1) = \eta_2$ with respect to the lattice $[\tau, 1]$.)

Proof. Let $g(z)$ be the expression on the right-hand side, which we want to be equal to $\varphi(z)$. It is clear that g has period 1, just like φ, that is

$$g(z + 1) = g(z).$$

Let us compute $g(z + \tau)$. Substituting $z + \tau$ for z in the terms of the product, we essentially get all these terms back, except that the product of terms involving $q_\tau^n q_z$ starts with $n = 2$, and the product of terms involving q_τ^n/q_z starts with $n = 0$. Taking these into account, together with the transformation of $q_z - 1$ into $q_z q_\tau - 1$ arising from the term in front of the product, we find that g satisfies the same functional equation as φ, namely

$$g(z + \tau) = -\frac{1}{q_z} g(z).$$

Therefore φ/g has a period lattice $[\tau, 1]$. On the other hand, our product expansion for g shows that g has exactly the same zeros, of order 1, as σ (and hence φ). Therefore φ/g is constant. Letting $z \to 0$ immediately shows that the constant is 1, thus proving our theorem.

Again for the record, we give the homogeneous form of Theorem 4.

Theorem 4′. *Let* $L = [\omega_1, \omega_2]$, *and let* $\varphi(z\colon \omega_1, \omega_2)$ *be as in Theorem* 3′. *Then*

$$\varphi(z; \omega_1, \omega_2) = \frac{\omega_2}{2\pi i}(q_{z/\omega_2} - 1) \prod_{n=1}^{\infty} \frac{(1 - q_\tau^n q_{z/\omega_2})(1 - q_\tau^n / q_{z/\omega_2})}{(1 - q_\tau^n)^2}$$

and

$$\sigma(z; L) = \frac{\omega_2}{2\pi i} e^{\frac{1}{2}\eta_2 \omega_2 (z/\omega_2)^2} (q_{z/\omega_2}^{\frac{1}{2}} - q_{z/\omega_2}^{-\frac{1}{2}}) \prod_{n=1}^{\infty} \frac{(1 - q_\tau^n q_{z/\omega_2})(1 - q_\tau^n / q_{z/\omega_2})}{(1 - q_\tau^n)^2}$$

Remark. In all our q-expansions, we emphasize that the power of $2\pi i$ occurs with precisely minus the homogeneity degree of the function involved. Thus we have $(2\pi i)^{-1}$ in the q-product for σ, while we have for instance $(2\pi i)^2$ in the q-expansion for $\wp$, in Chapter 4, and say $(2\pi i)^4$ in the q-expansion for g_2.

§3. q-EXPANSIONS AGAIN

This section may be omitted. For the most part we recover q-expansions already obtained in Chapter 4, by using the q-product for σ, and then getting the corresponding q-expansions for $\zeta, \eta_2, \wp$ by differentiation. *In particular, the product expression for* Δ *in the next section is independent of the present section.*

Taking the logarithmic derivative of the product for σ term by term, which we can do by absolute convergence, we obtain:

$$(1) \qquad \zeta(z) = \eta_2 z + \pi i \frac{q_z + 1}{q_z - 1} + 2\pi i \sum_{n=1}^{\infty} \left[\frac{q_\tau^n / q_z}{1 - q_\tau / q_z} - \frac{q_\tau^n q_z}{1 - q_\tau^n q_z} \right]$$

where $\eta_2 = \eta_2(\tau, 1)$. On the other hand, going back to the additive expression for ζ obtained from the logarithmic derivative of the Weierstrass product for σ, we get the power series expansion of ζ at the origin,

$$(2) \qquad \zeta(z) = \frac{1}{z} - s_4 z^3 - s_6 z^5 - \cdots$$

where

$$s_m = \sum_{\omega \in L'} \frac{1}{\omega^m}, \qquad L' = L - \{0\}.$$

Furthermore, we have trivially

$$\frac{q_z + 1}{q_z - 1} = \frac{e^{\pi i z} + e^{-\pi i z}}{e^{\pi i z} - e^{-\pi i z}} = -i \frac{\cos \pi z}{\sin \pi z},$$

whose power series expansion at the origin is immediate from Taylor's formula, and yields

$$\pi i \frac{q_z + 1}{q_z - 1} = \pi\left[\frac{1}{\pi z} - \frac{\pi z}{3} - \frac{(\pi z)^3}{45} - \frac{2(\pi z)^5}{45 \cdot 21} - \cdots\right].$$

To get a power series in z for the sum in (1), let $q = q_\tau$ and $w = q_z$ for simplicity. Then for $|q| < |w| < |q|^{-1}$ we have

$$\sum_{n=1}^{\infty}\left[\frac{q^n/w}{1 - q^n/w} - \frac{q^n w}{1 - q^n w}\right] = \sum_{n=1}^{\infty}\sum_{m=1}^{\infty}\left[\left(\frac{q^n}{w}\right)^m - (q^n w)^m\right],$$

which by interchanging the two sums is equal to

$$\sum_{m=1}^{\infty} \frac{q^m}{1 - q^m}(w^{-m} - w^m).$$

Substituting back $w = e^{2\pi i z}$, we obtain another power series in z for ζ. Comparing the coefficient of z yields the q-expansion

$$\eta_2(\tau, 1) = \frac{(2\pi i)^2}{12}\left[-1 + 24 \sum_{n=1}^{\infty} \frac{n q_\tau^n}{1 - q_\tau^n}\right]. \tag{2}$$

Similarly, comparing the coefficients of z^3 and z^5 would yield the same expansions for g_2 and g_3 that we found in Chapter 4.

Differentiating (1) with respect to z also gives us another derivation of the q-expansion for $\wp(z; \tau)$ found in Chapter 4. Observe that one needs here the intermediate step giving us η_2 in (2). There is no need to write these expansions again, as they have been tabulated previously.

§4. THE q-PRODUCT FOR Δ

We shall obtain the product expansion for $\Delta = \Delta(\tau, 1)$.

Theorem 5.

$$\Delta = (2\pi i)^{12} q_\tau \prod_{n=1}^{\infty} (1 - q_\tau^n)^{24}.$$

By definition, the discriminant of our cubic polynomial is given in terms of the roots by

$$\Delta = 16[(e_2 - e_1)(e_3 - e_2)(e_3 - e_1)]^2,$$

where

$$e_k = \wp\left(\frac{\omega_k}{2}\right).$$

We shall actually find q-products for the differences $e_i - e_k$, and even their square roots.

We continue to work with our normalized lattice $[\tau, 1]$. Then by Theorem 2,

$$e_2 - e_1 = \wp\left(\frac{1}{2}\right) - \wp\left(\frac{\tau}{2}\right) = -\frac{\sigma\left(\frac{\tau+1}{2}\right)\sigma\left(\frac{1-\tau}{2}\right)}{\sigma^2\left(\frac{1}{2}\right)\sigma^2\left(\frac{\tau}{2}\right)}$$

$$e_2 - e_3 = \wp\left(\frac{1}{2}\right) - \wp\left(\frac{\tau+1}{2}\right) = -\frac{\sigma\left(\frac{\tau}{2}+1\right)\sigma\left(-\frac{\tau}{2}\right)}{\sigma^2\left(\frac{1}{2}\right)\sigma^2\left(\frac{\tau+1}{2}\right)}$$

$$e_3 - e_1 = \wp\left(\frac{\tau+1}{2}\right) - \wp\left(\frac{\tau}{2}\right) = -\frac{\sigma\left(\tau+\frac{1}{2}\right)\sigma\left(\frac{1}{2}\right)}{\sigma^2\left(\frac{\tau+1}{2}\right)\sigma^2\left(\frac{\tau}{2}\right)}$$

We use the functional equation of the sigma function in Theorem 1 on each one of the numerators of the expressions on the right hand side. For instance,

$$\sigma\left(\frac{\tau-1}{2}\right) = \sigma\left(\frac{\tau+1}{2} - 1\right) = -e^{-\eta(1)\frac{1}{2}\tau}\sigma\left(\frac{\tau+1}{2}\right),$$

and similarly for the other cases. We also use the fact that σ is an odd function. Then our expressions for the differences of the e_k become:

E$_{21}$.
$$e_2 - e_1 = -e^{-\eta(1)\frac{1}{2}\tau}\frac{\sigma^2\left(\frac{\tau+1}{2}\right)}{\sigma^2\left(\frac{1}{2}\right)\sigma^2\left(\frac{\tau}{2}\right)}$$

E$_{23}$.
$$e_2 - e_3 = -e^{\eta(1)\frac{1}{2}(\tau+1)}\frac{\sigma^2\left(\frac{\tau}{2}\right)}{\sigma^2\left(\frac{1}{2}\right)\sigma^2\left(\frac{\tau+1}{2}\right)}$$

E$_{31}$.
$$e_3 - e_1 = e^{\eta(\tau)\frac{1}{2}(\tau+1)}\frac{\sigma^2\left(\frac{1}{2}\right)}{\sigma^2\left(\frac{\tau+1}{2}\right)\sigma^2\left(\frac{\tau}{2}\right)}$$

Remark 1. Each expression on the right is a perfect square, which shows that the square roots $\sqrt{e_k - e_i}$ are holomorphic on $\mathfrak{H}$.

Remark 2. Using the q-product expression for σ found in Theorem 4 and substituting the special values for z yields the corresponding q-product expressions for the differences of the e_k. We can tabulate these, although we won't need them in what follows. Let $q = q_\tau$ and let as in Fricke,

$$P_0 = \prod_{n=1}^{\infty} (1 - q^n) \qquad P_1 = \prod_{n=1}^{\infty} (1 - q^{n-\frac{1}{2}})$$

$$P_2 = \prod_{n=1}^{\infty} (1 + q^n) \qquad P_3 = \prod_{n=1}^{\infty} (1 + q^{n-\frac{1}{2}})$$

Then:

$\mathbf{E}_{21}^{\frac{1}{4}}$ $$(e_2 - e_1)^{\frac{1}{4}} = \sqrt{\pi}\, P_0 P_3^2$$

$\mathbf{E}_{23}^{\frac{1}{4}}$ $$(e_2 - e_3)^{\frac{1}{4}} = \sqrt{\pi}\, P_0 P_1^2$$

$\mathbf{E}_{31}^{\frac{1}{4}}$ $$(e_3 - e_1)^{\frac{1}{4}} = \sqrt{\pi}\, 2q^{\frac{1}{8}} P_0 P_2^2.$$

Since $P_0 P_1 P_2 P_3 = P_0$ trivially, we see that

$$P_1 P_2 P_3 = 1.$$

It then follows that the product expansion for Δ is the desired one. However, we shall do it directly again below.

Remark 3. Having given the differences of the e_k in terms of the $\wp$-function, we see that these differences are modular forms of appropriate weight. The classical literature went overboard on this. To read Weber, just to find the q-product expansion of Δ, one has to plow through all the formalism of these differences and the names given to the numerators and denominators occurring on the right in $\mathbf{E}_{21}$, $\mathbf{E}_{23}$, $\mathbf{E}_{31}$ (they are theta functions with various indices). Of course, these modular forms of low level are very useful in other applications, and provide computational data which should not be disregarded, but should be tabulated in its proper place.

Let us now multiply together all the expressions $\mathbf{E}_{ik}$. We get cancellations, giving us

$$\sqrt{\Delta} = 4 \frac{e^{\eta(1)/2 + \eta(\tau)((\tau+1)/2)}}{\sigma^2\left(\frac{1}{2}\right)\sigma^2\left(\frac{\tau}{2}\right)\sigma^2\left(\frac{\tau+1}{2}\right)}.$$

We use the q-product for σ found in Theorem 4. To figure out Δ, we must therefore keep track of the exponential term, a rational function in q, and three types of infinite products.

The exponential term is dealt with by using the Legendre relation

$$\eta(1)\tau - \eta(\tau) = 2\pi i,$$

which will cause all the transcendental terms in the exponent to cancel. It comes out neatly, and we won't clutter up the page with it.

For the product, let

$$P(z) = \prod_{n=1}^{\infty} (1 - q_\tau^n q_z)(1 - q_\tau^n / q_z).$$

We have to study the product

$$P = P\left(\frac{1}{2}\right)P\left(\frac{\tau}{2}\right)P\left(\frac{\tau + 1}{2}\right)$$

$$= \prod_{n=1}^{\infty} (1 + q^n)(1 - q^{2n+1})(1 + q^n)(1 - q^{2n-1}).$$

Let $P_0 = \prod_{n=1}^{\infty} (1 - q^n)$. Then we get the efficient relation

$$PP_0^2 = \frac{P_0^2}{1 - q},$$

so that (miracle)

$$P = \frac{1}{1 - q} \;!!$$

This contribution from the infinite product therefore reduces to a contribution of a rational function of q, which we can combine with the other rational functions of q arising from the expression for the σ-function in Theorem 4. We are therefore left only with the product

$$\prod_{n=1}^{\infty} (1 - q^n)^{24}.$$

You can work out the rational function in q which must appear in front, and you will find that all the terms cancel out except the desired $q = q_\tau$. The power of $2\pi i$ must be 12, and is 12 (corresponding to the homogeneity degree of Δ). This gives us our desired q-product for Δ.

§5. THE ETA FUNCTION OF DEDEKIND

We now use the symbol η for a new function, and not for the quasi periods of ζ.

We define the **Dedekind eta function** by

$$\eta(\tau) = q_\tau^{1/24} \prod_{n=1}^{\infty} (1 - q_\tau^n),$$

where $q = q_\tau = e^{2\pi i \tau}$. It is holomorphic on the upper half plane $\mathfrak{H}$.

Theorem 6. *The eta function satisfies*

$$\eta(\tau + 1) = e^{2\pi i/24}\,\eta(\tau)$$

$$\eta\left(-\frac{1}{\tau}\right) = \sqrt{-i\tau}\,\eta(\tau),$$

where the square root is the obviously normalized one for $\tau \in \mathfrak{H}$, taking positive values on the positive real axis.

Proof. The first relation is trivial from the q-product. As for the second, we know that Δ, viewed as a function of two variables, i.e.

$$\Delta\begin{pmatrix}\omega_1\\ \omega_2\end{pmatrix}$$

is homogeneous of degree -12, so that

$$\Delta(-1/\tau) = \Delta\begin{pmatrix}-1/\tau\\ 1\end{pmatrix} = \Delta\left(\frac{1}{\tau}\begin{pmatrix}0 & -1\\ 1 & 0\end{pmatrix}\begin{pmatrix}\tau\\ 1\end{pmatrix}\right) = \tau^{12}\Delta(\tau).$$

Taking the 24-th root shows that

$$|\eta(-1/\tau)| = |\sqrt{\tau}|\,|\eta(\tau)|.$$

Note that $\sqrt{\tau}$ is holomorphic on $\mathfrak{H}$. Hence the function

$$\frac{\eta(-1/\tau)}{\sqrt{\tau}\,\eta(\tau)}$$

is holomorphic on $\mathfrak{H}$ and has absolute value 1. By the maximum modulus principle, it must be constant. Putting $\tau = i$ shows that

$$1 = C\sqrt{i},$$

whence $C = 1/\sqrt{i} = \sqrt{-i}$. This proves our theorem.

We can now recover a fact used in our analysis of ramification in the modular function field. We have the definition of J,

$$J = g_2^3/\Delta.$$

We want to see that its cube root exists as a modular function of level 3. Since g_2 is homogeneous of degree -4 as a function of two variables, we find that

$$g_2(-1/\tau) = \tau^4 g_2(\tau).$$

On the other hand, from Theorem 6 we get

$$\eta^8(-1/\tau) = \tau^4\eta^8(\tau).$$

Furthermore,

$$\eta^8(\tau + 1) = e^{2\pi i/3}\,\eta^8(\tau) \qquad \text{and} \qquad g_2(\tau + 1) = g_2(\tau).$$

Hence we obtain the transformation rule for $J^{\frac{1}{3}}$ under the modular group.

Theorem 7. *Let* $J^{\frac{1}{3}} = g_2/\eta^8$. *Then*

$$J^{\frac{1}{3}}(\tau + 1) = e^{2\pi i/3} J^{\frac{1}{3}}(\tau) \qquad \text{and} \qquad J^{\frac{1}{3}}(-1/\tau) = J^{\frac{1}{3}}(\tau).$$

Similarly,

Theorem 8. *Let* $f = \sqrt{J - 1} = 27 g_3/\eta^{12}$. *Then*

$$f(\tau + 1) = -f(\tau) \qquad \text{and} \qquad f(-1/\tau) = f(\tau).$$

Corollary. *The functions* $J^{\frac{1}{3}}$ *and* $\sqrt{J - 1}$ *are modular functions of level* 3 *and* 2 *respectively.*

Proof. Let $\Gamma = SL_2(\mathbf{Z})$ as usual, and let $g = J^{\frac{1}{3}}, f = \sqrt{J - 1}$. We have a representation of Γ on the space generated by f, g over $\mathbf{C}$, which is abelian, with characters of order 3 and 2 respectively. However, letting S, T be the mappings

$$S(\tau) = -1/\tau \qquad \text{and} \qquad T(\tau) = \tau + 1,$$

we know that S, T generate the modular group, and so do S, ST which have order 2, 3 respectively. The abelianized modular group can therefore have order at most 6, and has order 6 since we just found the appropriate representation for it. Let Γ_3 and Γ_2 be the congruence subgroups of level 3 and 2 respectively. Then $\Gamma/\pm\Gamma_3$ has order 12 and Γ/Γ_2 has order 6. Also, $\Gamma/\pm\Gamma_3$ has a normal subgroup whose factor group is cyclic of order 3, and Γ/Γ_2 has a normal subgroup whose factor group has order 2. In this way we obtain another representation of Γ into a cyclic group of order 6, whose kernel must be the same as that of the previous one, because the abelianized modular group has order at most 6. This proves that Γ_3 and Γ_2 leave f and g fixed, as was to be shown.

§6. MODULAR FUNCTIONS OF LEVEL 2

This section will not be used anywhere else, and is included as an example, for the sake of completeness, and because it fits with the computations involving e_1, e_2, e_3.

We consider the congruence subgroup $\Gamma(2)$ consisting of all elements α of $SL_2(\mathbf{Z})$ satisfying the condition

$$\alpha \equiv I \pmod{2}.$$

Such α can be written in the form

$$\alpha = \begin{pmatrix} a & b \\ c & d \end{pmatrix}$$

with a, d odd and b, c even. Using arguments similar to those involved in determining a fundamental domain for the modular group, one sees that the elements

$$T_2 = \begin{pmatrix} 1 & 2 \\ 0 & 1 \end{pmatrix} \quad \text{and} \quad S_2 = \begin{pmatrix} 1 & 0 \\ 2 & 1 \end{pmatrix}$$

generate $\Gamma(2)$, and that a fundamental domain for $\Gamma(2)$ consists of the shaded region in the next figure. The mapping S_2 carries the semicircle on the left onto the semicircle on the right.

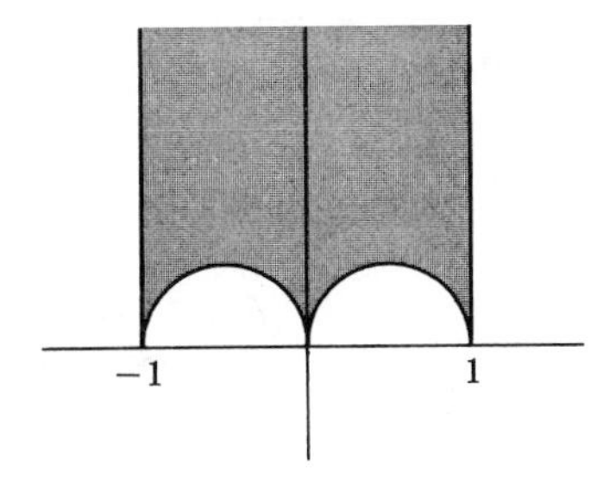

Fig. 18-2

Let $G_6 = G = \Gamma/\Gamma(2)$ be the factor group, which is of order 6. It is represented by the matrices:

$$\begin{pmatrix} 1 & 0 \\ 0 & 1 \end{pmatrix}, \begin{pmatrix} 0 & 1 \\ -1 & 0 \end{pmatrix}, \begin{pmatrix} 1 & 1 \\ -1 & 0 \end{pmatrix}, \begin{pmatrix} 1 & 1 \\ 0 & 1 \end{pmatrix}, \begin{pmatrix} 1 & 0 \\ 1 & 1 \end{pmatrix}, \begin{pmatrix} 0 & 1 \\ -1 & 1 \end{pmatrix}.$$

Define the function

$$\lambda(\tau) = \frac{e_2 - e_3}{e_1 - e_3}.$$

The homogeneity properties of the quasi periods of the Weierstrass zeta function, and of the σ-function, show that the above ratio is homogeneous of degree 0, and that our notation as a function of τ is legitimate. Indeed, in the relations $\mathbf{E}_{jk}$, the exponential factor is homogeneous of degree 0, and each factor involving σ is homogeneous of degree 2, so that we get homogeneity of degree 0 when taking the quotient.

It is now verified by direct computation that the six transformations of G_6 transform the function λ into the following six functions.

$$\lambda = \frac{e_2 - e_3}{e_1 - e_3} \qquad \frac{1}{1 - \lambda} = \frac{e_3 - e_1}{e_2 - e_1} \qquad \frac{\lambda - 1}{\lambda} = \frac{e_1 - e_2}{e_3 - e_2}$$

$$\frac{1}{\lambda} = \frac{e_1 - e_3}{e_2 - e_3} \qquad \frac{\lambda}{\lambda - 1} = \frac{e_3 - e_2}{e_1 - e_2} \qquad 1 - \lambda = \frac{e_2 - e_1}{e_3 - e_1}$$

This yields a faithful representation of G_6 on those six functions, and the fixed field consists of rational functions in j, with rational coefficients. The function λ generates the modular function field of level 2, which we denoted by F_2. We shall express $J(\tau)$ as a rational function of $\lambda(\tau)$, namely we shall prove:

$$\boxed{j(\tau) = 2^8 \frac{(\lambda^2 - \lambda + 1)^3}{\lambda^2(\lambda - 1)^2}.}$$

(Recall that $j(\tau) = 12^3 J(\tau)$ is the normalization whose q-expansion starts with $1/q$.)

To derive the above rational expression as in Ford's *Automorphic Functions*, consider the rational function

$$\begin{aligned} Q &= (\lambda + 1)\left(\frac{1}{1 - \lambda} + 1\right)\left(\frac{\lambda - 1}{\lambda} + 1\right)\left(\frac{1}{\lambda} + 1\right)\left(\frac{\lambda}{\lambda - 1} + 1\right)(1 - \lambda + 1) \\ &= -\frac{(\lambda + 1)^2(\lambda - 2)^2(2\lambda - 1)^2}{\lambda^2(\lambda - 1)^2}. \end{aligned}$$

In terms of e_1, e_2, e_3, it becomes

$$Q = -\frac{(e_2 + e_1 - 2e_3)^2(e_2 + e_3 - 2e_1)^2(e_1 + e_3 - 2e_2)^2}{(e_1 - e_2)^2(e_2 - e_3)^2(e_1 - e_3)^2}.$$

But

$$e_1 + e_2 + e_3 = 0 \qquad \text{and} \qquad e_1 e_2 e_3 = \tfrac{1}{4} g_3.$$

The numerator of Q is then equal to

$$(-3e_3)^2(-3e_1)^2(-3e_2)^2 = \frac{3^6}{2^4} g_3^2.$$

The denominator is equal to

$$\tfrac{1}{16}(g_2^3 - 27g_3^2),$$

which is Δ, up to the factor $1/16$. Therefore

$$Q = 27(1 - J).$$

Since we had the original expression of Q as a rational function of λ, it is then trivial to get the rational expression of j in terms of λ, and it is the stated one.

The function λ is used by Deuring [8]. It is also taken by Igusa as one of the fundamental parameters in his theory of abstract elliptic functions [25].

It is advantageous because it can be used instead of j to parametrize elliptic curves in a non-degenerate way, by means of the equation

$$y^2 = x(x-1)(x-\lambda), \qquad \lambda \neq 0, 1, \infty.$$

The point $\lambda = 0$ lies above $j = \infty$, and is ramified of order 2. One can see directly (and thus confirm the general fact) that $\mathbf{Q}(\lambda)$ is ramified over $\mathbf{Q}(J) = \mathbf{Q}(j)$ of order 3 over $j = 0$ and of order 2 over $j = 12^3$, i.e. $J = 1$. A direct computation shows that the j-invariant of the above curve is precisely j, for $\lambda \neq 0, 1$. As Igusa points out, the same parametrization is valid for all characteristics $\neq 2$.

One can look at the function λ from another point of view, namely as the analogue of a "Minkowski" unit in the function field. It can be generalized as follows. For an integer $N > 1$, let

$$\lambda_N(\tau) = \frac{\wp(\omega_2/N) - \wp(\omega_3/N)}{\wp(\omega_1/N) - \wp(\omega_3/N)}.$$

The expression on the right is homogeneous of degree 0, and hence gives rise to a function of $\tau \in \mathfrak{H}$, modular of level N. The function λ_N obviously has no zero or pole on $\mathfrak{H}$. It would be interesting to determine the part of the unit group it generates in the integral closure of $\mathbf{Z}[j]$ in the modular function field of level N, and to investigate its special values at imaginary quadratic points, to see if they generate the ray class fields.

[*Added in the Second Edition*: The study of the unit group of such functions was carried out in Kubert-Lang. See the series of papers in *Math. Annalen*, and "Modular Units", Springer-Verlag, 1981.]

19 *The Siegel Functions and Klein Forms*

§1. THE KLEIN FORMS

This chapter is entirely rewritten for the second edition, and follows Kubert–Lang (*Units in the Modular Function Field* I, Math. Ann. 218 (1975), pp. 67–96; see also "Modular Units", Springer-Verlag, 1981).

In line with the terminology which has become standard, we speak of modular forms instead of automorphic forms. We let:

$\Gamma(1) = SL_2(\mathbf{Z})$,
$\Gamma(N) =$ subgroup of elements

$$\alpha = \begin{pmatrix} a & b \\ c & d \end{pmatrix}$$

such that $\alpha \equiv 1 \bmod N$. By 1 we mean the unit 2×2 matrix, so this congruence condition is equivalent with

$$a \equiv d \equiv 1 \bmod N \qquad \text{and} \qquad c \equiv b \equiv 0 \bmod N.$$

Let k be an integer. By a **form of degree** k **(weight** $-k$**)** we mean a function

$$h\begin{pmatrix} \omega_1 \\ \omega_2 \end{pmatrix} = h(W) \qquad \text{where} \quad W = \begin{pmatrix} \omega_1 \\ \omega_2 \end{pmatrix},$$

of two complex variables, with $\mathrm{Im}(\omega_1/\omega_2) > 0$, satisfying the homogeneity property

MF 1. $$h\left(\lambda\begin{pmatrix} \omega_1 \\ \omega_2 \end{pmatrix}\right) = \lambda^k h\begin{pmatrix} \omega_1 \\ \omega_2 \end{pmatrix}, \qquad \lambda \in \mathbf{C}^*.$$

Let Γ be a subgroup of $SL_2(\mathbf{Z})$, of finite index. We say that a form h as above is **modular on** Γ, or **with respect to** Γ, if it satisfies the additional properties:

MF 2. $$h\left(\alpha\begin{pmatrix} \omega_1 \\ \omega_2 \end{pmatrix}\right) = h\begin{pmatrix} \omega_1 \\ \omega_2 \end{pmatrix} \qquad \textit{for all} \quad \alpha \in \Gamma.$$

MF 3. *For τ in the upper half plane, i.e.* $\operatorname{Im} \tau > 0$, *the function* $h(\tau, 1)$ *is meromorphic at infinity, meaning that it has a convergent Laurent series expansion for some* N:

$$h(\tau, 1) = \sum_{n=n_0}^{\infty} a_n q^{n/N} \qquad \textit{for some} \quad n_0 \in \mathbf{Z},$$

where $q = e^{2\pi i \tau}$. *A similar expansion should hold for the function* $h \circ \alpha$ *with any* $\alpha \in SL_2(\mathbf{Z})$.

Given a modular form h, the composite functions $h \circ \alpha$ with $\alpha \in SL_2(\mathbf{Z})$ are called the **conjugates** of h. We often write $h(\tau)$ instead of $h(\tau, 1)$.

A form of weight 0 is called a **modular function**.

If a form is modular on $\Gamma(N)$, then we also say that the form has level N. We shall use the dot product notation

$$z = a_1\omega_1 + a_2\omega_2 = a \cdot W, \qquad \text{where} \quad W = \begin{pmatrix} \omega_1 \\ \omega_2 \end{pmatrix}.$$

We also write

$$W_\tau = \begin{pmatrix} \tau \\ 1 \end{pmatrix}.$$

Let $\eta(z, L)$ be the Weierstrass eta function as in Chapter 18, §1. Define the **Klein forms**

$$\mathfrak{k}(z, L) = e^{-\eta(z, L)z/2}\, \sigma(z, L).$$

We also write

$$\mathfrak{k}(z, L) = \mathfrak{k}_a(W).$$

Then the Klein forms satisfy the following properties. First, they are homogeneous of degree 1:

K 0. $\quad \mathfrak{k}(\lambda z, \lambda L) = \lambda \mathfrak{k}(z, L) \quad$ *and* $\quad \mathfrak{k}_a(\lambda W) = \lambda \mathfrak{k}_a(W)$.

This is obvious from the homogeneity property of the sigma function and the Weierstrass zeta function. The next property is also clear.

K 1. *If* $\alpha = \begin{pmatrix} a & b \\ c & d \end{pmatrix}$ *is in* $SL_2(\mathbf{Z})$, *then*

$$\mathfrak{k}_a(\alpha W) = \mathfrak{k}_{a\alpha}(W).$$

We recall the standard transformation property of the sigma function with respect to the periods. Let $\omega = b_1\omega_1 + b_2\omega_2$ with integers b_1, b_2. Then

$$\sigma(z + \omega, L) = (-1)^{b_1 b_2 + b_1 + b_2}\, e^{\eta(\omega, L)(z + \omega/2)}\, \sigma(z, L).$$

Then we find:

K 2. $$\mathfrak{k}_{a+b}(W) = \varepsilon(a, b)\mathfrak{k}_a(W),$$

where $\varepsilon(a, b)$ *has absolute value* 1, *and is given explicitly by*

$$\varepsilon(a, b) = (-1)^{b_1 b_2 + b_1 + b_2} e^{-2\pi i(b_1 a_2 - b_2 a_1)/2}.$$

This follows easily from the Legendre relation

$$\eta_2 \omega_1 - \eta_1 \omega_2 = 2\pi i.$$

We leave the computation to the reader.

So far, we needed no further assumption on a_1, a_2. Assume now that they are rational numbers, with denominators dividing an integer $N > 1$, say

$$a_1 = r/N \qquad \text{and} \qquad a_2 = s/N.$$

Let

$$\alpha = \begin{pmatrix} a & b \\ c & d \end{pmatrix} \equiv \begin{pmatrix} 1 & 0 \\ 0 & 1 \end{pmatrix} \bmod N$$

be in $\Gamma(N)$, and write

$$ar + cs = r + \left(\frac{a-1}{N} r + \frac{c}{N} s\right)N, \qquad br + ds = s + \left(\frac{b}{N} r + \frac{d-1}{N} s\right)N.$$

Then we find from **K 2**:

K 3. $$\mathfrak{k}_a\left(\alpha\begin{pmatrix} \omega_1 \\ \omega_2 \end{pmatrix}\right) = \mathfrak{k}_{a\alpha}\begin{pmatrix} \omega_1 \\ \omega_2 \end{pmatrix} = \varepsilon_a(\alpha)\mathfrak{k}_a\begin{pmatrix} \omega_1 \\ \omega_2 \end{pmatrix},$$

where $\varepsilon_a(\alpha)$ *is a* $(2N)$*th root of unity, given precisely by*

$$\varepsilon_a(\alpha) = \varepsilon(\alpha) = -(-1)^{\left(\frac{a-1}{N} r + \frac{c}{N} s + 1\right)\left(\frac{b}{N} r + \frac{d-1}{N} s + 1\right)} e^{2\pi i(br^2 + (d-a)rs - cs^2)\, 2N^2}.$$

From this transformation law, we get:

Theorem 1. *Let* $a \in (1/N)\mathbf{Z}^2$ *but* $a \notin \mathbf{Z}^2$. *Then* $\varepsilon_a(\alpha)^{2N} = 1$. *Hence* $\mathfrak{k}_a$ *is a modular form on* $\Gamma(2N^2)$, *and* $\mathfrak{k}_a^{2N}$ *is on* $\Gamma(N)$. *If* N *is odd, then* $\mathfrak{k}_a^N$ *is on* $\Gamma(N)$.

The proof is immediate, by considering the cases when r, s are both even, or one of them is odd, or both are odd, and using $ad - bc = 1$, so that for instance, not both c, d are even.

§2. THE SIEGEL FUNCTIONS

We now take $\omega_1 = \tau$ and $\omega_2 = 1$, so that

$$z = a_1\tau + a_2.$$

Using the q-expansion for the sigma function, we can easily derive the q-product for the Klein forms. We let the **Siegel functions** be defined by

$$g_a(\tau) = \mathfrak{k}_a(\tau)\Delta(\tau)^{1/12},$$

where $\Delta(\tau)^{1/12}$ is the square of the Dedekind eta function, namely the natural q-product for the 12-th root of Δ, which is

$$\text{Dedekind } \eta(\tau)^2 = 2\pi i \cdot q^{1/12} \prod_{n=1}^{\infty} (1 - q^n)^2,$$

where $q^{1/12} = e^{2\pi i\tau/12}$. We shall use the notation

$$q = q_\tau = e^{2\pi i\tau} \qquad \text{and} \qquad q_z = e^{2\pi iz}.$$

Then from the q-product for the sigma function we obtain the q-product for the Siegel functions:

S 1. $$g_a(\tau) = -q_\tau^{(1/2)\mathbf{B}_2(a_1)}\, e^{2\pi i a_2(a_1-1)/2}(1 - q_z)\prod_{n=1}^{\infty}(1 - q_\tau^n q_z)(1 - q_\tau^n/q_z),$$

where $\mathbf{B}_2(X) = X^2 - X + \frac{1}{6}$ is the second Bernoulli polynomial.

Remarks. If we change a by an integral vector in $\mathbf{Z}^2$, then **K 2** shows that g_a changes by a root of unity. We can always make such a change so that a representative in the class mod $\mathbf{Z}^2$ has coordinates $a = (a_1, a_2)$ such that

$$0 \leqq a_1 < 1 \qquad \text{and} \qquad 0 \leqq a_2 < 1.$$

These are the standard representatives $\langle a_1 \rangle$ and $\langle a_2 \rangle$.

The next theorem is immediate from the formalism of the Klein forms.

Theorem 2. *Assume that a has denominator dividing N. Then the Siegel functions g_a are modular functions, and g_a^{12N} is on $\Gamma(N)$. Furthermore, g_a has no zeros or poles on the upper half plane.*

Remark. Because the Dedekind function $\eta(\tau)$ is not a function of lattices, but depends on the choice of a basis, we could not define the Siegel functions in terms of a lattice. However, if we take the 12-th powers, then we may define for any complex number z, the functions

$$g^{12}(z, L) = \mathfrak{k}^{12}(z, L)\Delta(L).$$

These give rise to modular forms when

$$z \in \frac{1}{N} L \qquad \text{and} \qquad z \notin L,$$

for some positive integer N.

Also from the formalism of the Klein forms and the fact that the Siegel functions have weight 0, we find:

S 2. *If α is in $SL_2(\mathbf{Z})$, then*

$$g_{a\alpha}(\tau) = g_a(\alpha\tau).$$

As in the study of the Fricke functions, and the automorphisms of the modular function field, let σ_d (for d prime to N) be the automorphism of the modular function field F_N induced by the action

$$\zeta_N \mapsto \zeta_N^d$$

on the roots of unity, and leaving the local uniformizing parameter $q_\tau^{1/N}$ fixed. Suppose a_1, a_2 have denominator N. Let

$$f_a(\tau) = g_a(\tau)^{12N}.$$

Then

S 3.

$$\sigma_d f_{a_1, a_2} = f_{a_1, da_2}.$$

This is immediate from the q-expansion for the Siegel functions.

Recall also that we had defined automorphisms of the modular function field, namely $\sigma(\alpha)$ for $\alpha \in GL_2^+(\mathbf{Q})$ and $\sigma(u)$ for

$$u \in U = \prod_p GL_2(\mathbf{Z}_p).$$

The automorphism $\sigma(u)$ was defined relative to the coordinatization obtained by the Fricke functions. Their effect on the Siegel functions is, however, easily determined.

S 4. *If $\alpha \in GL_2^+(\mathbf{Q})$, then $f_a^{\sigma(\alpha)}(\tau) = f_a(\alpha\tau)$.*

This is merely the definition of how $\sigma(\alpha)$ operates on modular functions.

S 5. *Let $u \in U$ and write*

$$u \equiv \begin{pmatrix} 1 & 0 \\ 0 & d \end{pmatrix} \alpha \pmod{N}$$

with some positive integer d satisfying $d \equiv \det g_p \pmod{N}$ for all $p | N$, and $\alpha \in SL_2(\mathbf{Z})$. Then on F_N we have

$$\sigma(u) = \sigma_d \sigma(\alpha),$$

and for any rational a_1, a_2 *with denominator* N, *we have*

$$f_{a_1,a_2}^{\sigma(u)}(\tau) = f_{a_1,da_2}(\alpha\tau).$$

Proof. The first assertion is merely a repetition of the fact that we have a homomorphism of U into the group of automorphisms of F_N, and of the matrix representation of the automorphism σ_d on the Fricke functions as described in Chapter 6, §3. The second assertion follows by definition, and **S 3**.

§3. SPECIAL VALUES OF THE SIEGEL FUNCTIONS

Let $\mathfrak{f} \neq (1)$ be an ideal of the imaginary quadratic field k. We let $\mathrm{Cl}(\mathfrak{f})$ denote the generalized ideal class group of conductor $\mathfrak{f}$. Let N be the smallest positive integer in $\mathfrak{f}$, and let (a_1, a_2) be rational numbers with denominator N. If $L = [\omega_1, \omega_2]$ is a lattice and $\tau = \omega_1/\omega_2$, we write

$$f_{a_1,a_2}(\tau) = f(a_1\omega_1 + a_2\omega_2, L).$$

This is the same notation already used for the Klein forms. Here

$$f_a = g_a^{12N}.$$

Note that in what follows, we use only certain properties of the Siegel functions, which can easily be axiomatized, cf. "Fricke Families" in Kubert–Lang. We let

$K(\mathfrak{f})$ = ray class field of conductor $\mathfrak{f}$ over k.

Let C be a ray class in $\mathrm{Cl}(\mathfrak{f})$. We define the **Siegel–Ramachandra invariant**

$$\boxed{g_{\mathfrak{f}}(C) = g^{12N}(C) = f(C) = f(1, \mathfrak{f}\mathfrak{c}^{-1})}$$

where $\mathfrak{c}$ is any ideal in C (and in particular, $\mathfrak{c}$ is prime to $\mathfrak{f}$). This value is independent of the choice of $\mathfrak{c}$. Indeed, if $\mathfrak{c}_1$ is another such ideal, there exists $\alpha \in k^*$ such that $\mathfrak{c}_1 = \alpha\mathfrak{c}$ and $\alpha \equiv 1 \bmod^* \mathfrak{f}$. Since $\mathfrak{c}_1 \subset \mathfrak{o}$ (by definition of an ideal), it follows that $\alpha \in \mathfrak{c}^{-1}$. It is then immediate that

$$\alpha - 1 \in \mathfrak{f}\mathfrak{c}^{-1},$$

whence $f(1, \mathfrak{f}\mathfrak{c}^{-1}) = f(\alpha, \mathfrak{f}\mathfrak{c}^{-1})$, thus proving our assertion, since $f(\lambda t, \lambda L) = f(t, L)$ because f is of weight 0.

Theorem 3. *Let* $\{f_a\}$ *be the Siegel functions of level* N, *where* N *is the smallest positive integer in* $\mathfrak{f}$. *Then*

$$f(C) \in K(\mathfrak{f}), \qquad \text{and} \qquad f(C)^{\sigma(C')} = f(CC').$$

Proof. We shall use Shimura's reciprocity law, cf. [Sh], and Chapter 11, §1, Theorem 1.

Let $\mathfrak{c} \in C$ and let $\mathfrak{a} \in C'$ be such that $\mathbf{N}\mathfrak{a}$, $\mathbf{N}\mathfrak{c}$ are relatively prime to $\mathbf{N}\mathfrak{f}$. Then by definition,

$$f(C) = f(1, \mathfrak{f}\mathfrak{c}^{-1}) \qquad \text{and} \qquad f(CC') = f(1, \mathfrak{f}\mathfrak{c}^{-1}\mathfrak{a}^{-1}).$$

Let $\mathfrak{f}\mathfrak{c}^{-1} = [z_1, z_2]$ with $z = z_1/z_2$ in the upper half plane as usual. Let $\alpha \in \mathrm{Mat}_2^+(\mathbf{Z})$ be an integral matrix with positive determinant such that

$$\alpha^{-1}\begin{pmatrix} z_1 \\ z_2 \end{pmatrix} \qquad \text{is a basis of} \quad \mathfrak{f}\mathfrak{c}^{-1}\mathfrak{a}^{-1}.$$

Then $\det \alpha = \mathbf{N}\mathfrak{a}$. Let s be an idele of K such that:

$$s_p = 1 \qquad \text{if} \quad p \mid \mathbf{N}\mathfrak{f},$$

$$s_p\mathfrak{o}_p = \mathfrak{a}_p \qquad \text{if} \quad p \nmid \mathbf{N}\mathfrak{f}.$$

Then we also have $s_p^{-1}\mathfrak{o}_p = \mathfrak{a}_p^{-1}$. Furthermore $(s, k) = \sigma(C')$ on the ray class field $K(\mathfrak{f})$ because for all $\mathfrak{p} \nmid \mathbf{N}\mathfrak{f}$, $\mathrm{ord}_\mathfrak{p}\, s_\mathfrak{p} = \mathrm{ord}_\mathfrak{p}\, \mathfrak{a}$. Write

$$1 = a_1 z_1 + a_2 z_2 \qquad \text{with} \quad a_1, a_2 \in \frac{1}{N}\mathbf{Z}.$$

Then $a = (a_1, a_2)$ is primitive of order N mod $\mathbf{Z}^2$. By the Shimura Reciprocity Law, we get

$$f(C)^{(s,k)} = f_a^\sigma(z) \qquad \text{where} \quad \sigma = \sigma(q_z(s^{-1})).$$

Since for all primes p,

$$q_{z,p}(s_p^{-1})\begin{pmatrix} z_1 \\ z_2 \end{pmatrix} \qquad \text{and} \qquad \alpha^{-1}\begin{pmatrix} z_1 \\ z_2 \end{pmatrix}$$

are bases of $(\mathfrak{f}\mathfrak{c}^{-1}\mathfrak{a}^{-1})_p$, it follows that there exists $u_p \in GL_2(\mathbf{Z}_p)$ such that

$$q_{z,p}(s_p^{-1}) = u_p\alpha^{-1}.$$

We let $u = (u_p) \in \prod GL_2(\mathbf{Z}_p)$, so that

$$q_z(s^{-1}) = u\alpha.$$

Then

$$\begin{aligned} f_a^\sigma(z) &= f_a^{\sigma(u\alpha^{-1})}(z) \\ &= f_{au}(\alpha^{-1}(z)) \\ &= f_{a\alpha}(\alpha^{-1}(z)) \end{aligned}$$

[because for $p \mid \mathbf{N}\mathfrak{f}$, we have $1 = u_p\alpha^{-1}$, so $u_p = \alpha$ and $au = a\alpha$]

$$\begin{aligned} &= f\left(a\alpha\alpha^{-1}\begin{pmatrix} z_1 \\ z_2 \end{pmatrix}, \alpha^{-1}\begin{pmatrix} z_1 \\ z_2 \end{pmatrix}\right) \\ &= f(1, \mathfrak{f}\mathfrak{c}^{-1}\mathfrak{a}^{-1}) \\ &= f(CC'). \end{aligned}$$

This shows that $f(C)^{(s,k)} = f(CC')$, and hence that $f(C)^{(s,K)}$ depends only on $\sigma(C')$. Hence $f(C)$ lies in $K(\mathfrak{f})$, and we know that $(s, k) = \sigma(C')$ on $K(\mathfrak{f})$. This concludes the proof of the theorem.

In Ramachandra's paper, there is a twist which is explained by the next theorem.

Theorem 4. *Let $\{f_a\}$ be the Siegel functions of level N as above. Let $\mathfrak{c}$ be an ideal in a ray class C mod $\mathfrak{f}$. Let $\mathfrak{d}$ be the different of $K/\mathbf{Q}$, and let*

$$\mathfrak{c}\mathfrak{d}^{-1}\mathfrak{f}^{-1} = \mathfrak{a} = [z_1, z_2], \qquad z_1/z_1 \in \mathfrak{H}.$$

Then

$$f((\mathrm{tr}\ z_2)z_1 - (\mathrm{tr}\ z_1)z_2, \mathfrak{a}) = f(\bar{C}),$$

where $\bar{C}$ is the complex conjugate of C.

Proof. We need a lemma.

Lemma. *Let $\mathfrak{a} = [z_1, z_2]$ be a fractional ideal with $\mathrm{Im}(z_1/z_2) > 0$. Let $D = D(\mathfrak{o}_K)$ be the discriminant. Then*

$$(\mathrm{tr}\ z_2)z_1 - (\mathrm{tr}\ z_1)z_2 = \sqrt{D}\mathbf{N}\mathfrak{a}.$$

Proof. If we replace (z_1, z_2) by $(\lambda z_1, \lambda z_2)$ with $\lambda \in K^*$, then both sides change by $\lambda\bar{\lambda} = \mathbf{N}\lambda$. Hence it suffices to prove the lemma for $\mathfrak{a} = [z, 1]$, i.e. $z_2 = 1$, and $z = x + y\sqrt{D}$, $y > 0$. The left-hand side of our formula is equal to

$$2z - 2x = 2y\sqrt{D}.$$

Hence we have only to show that $2y = \mathbf{N}\mathfrak{a}$. But

$$D(\mathfrak{a}) = \mathbf{N}\mathfrak{a}^2 D(\mathfrak{o}) = \mathbf{N}\mathfrak{a}^2 \cdot D,$$

and

$$D(\mathfrak{a}) = \begin{vmatrix} 1 & z \\ 1 & \bar{z} \end{vmatrix}^2 = (2y)^2 D.$$

Since $y > 0$, our lemma is proved.

For the theorem, observe that $\mathfrak{d} = \mathfrak{o}\sqrt{D}$. Then we just substitute the expression of the lemma in the left-hand side of the formula to be proved, and we find the value

$$f(-1, \bar{\mathfrak{f}}\bar{\mathfrak{c}}^{-1}) = f(1, \bar{\mathfrak{f}}\bar{\mathfrak{c}}^{-1}) = f(\bar{C}),$$

as was to be shown.

20 The Kronecker Limit Formulas

§1. THE POISSON SUMMATION FORMULA

Let f be a function on $\mathbf{R}$. We shall say that f **tends to 0 rapidly at infinity** if for each positive integer m the function

$$x \mapsto |x|^m f(x)$$

is bounded. We define the **Schwartz space** S to be the set of functions on $\mathbf{R}$ which are infinitely differentiable and which tend to 0 rapidly at infinity, as well as their derivatives of all orders.

Example. The function e^{-x^2} is in the Schwartz space. Any C^∞ function with compact support is in the Schwartz space.

We define the **Fourier transform** of a function f in S by the integral

$$\hat{f}(y) = \int_{-\infty}^{\infty} f(x)\, e^{-2\pi i x y}\, dx.$$

Differentiating under the integral sign shows that $\hat{f}$ is C^∞ and tends rapidly to zero at infinity (it is in fact in the Schwartz space but we won't need this).

Poisson Summation Formula. *Let f be in the Schwartz space. Then*

$$\sum_{n \in \mathbf{Z}} f(n) = \sum_{n \in \mathbf{Z}} \hat{f}(n).$$

Proof. Let

$$g(x) = \sum_{k \in \mathbf{Z}} f(x + k).$$

The convergence is obviously absolute and uniform on compact sets, and we see that g is periodic with period 1, and C^∞. Its Fourier coefficients are defined by

$$c_m = \int_0^1 g(x)\, e^{-2\pi imx}\, dx.$$

Integrating by parts, one sees that $|c_m| \leqq C/|m|^2$ for some constant C (essentially the sup norm of the first two derivatives of g). Hence the Fourier series converges to g. We have

$$\sum_{m\in\mathbf{Z}} c_m = g(0) = \sum_{m\in\mathbf{Z}} f(m).$$

On the other hand, interchanging a sum and an integral, we get

$$\begin{aligned} c_m = \int_0^1 g(x)\, e^{-2\pi imx}\, dx &= \sum_n \int_0^1 f(x+n)\, e^{-2\pi imx}\, dx \\ &= \sum_n \int_0^1 f(x+n)\, e^{-2\pi im(x+n)}\, dx \\ &= \int_{-\infty}^{\infty} f(x)\, e^{-2\pi imx}\, dx = \hat{f}(m). \end{aligned}$$

This proves the formula.

§2. EXAMPLES

The function $h(x) = e^{-\pi x^2}$ is self dual, i.e. $\hat{h} = h$. One merely has to differentiate under the integral sign and integrate by parts to see that

$$\hat{h}'(y) = -2\pi y\hat{h}(y).$$

It follows that

$$\hat{h}(y) = C\, e^{-\pi y^2}$$

for some construct C. Using the standard integral $\hat{h}(0)$ shows that $C = 1$.

Let f be in the Schwartz space, and let $g(x) = f(x + c)$ for some constant c. Then

$$\hat{g}(y) = e^{2\pi icy}\hat{f}(y).$$

This just comes from changing variables in the integral defining g.

Similarly, let $g(x) = f(bx)$ where $b > 0$. Then

$$\hat{g}(y) = \frac{1}{b}\hat{f}\left(\frac{y}{b}\right).$$

Again this comes from a trivial change of variables in the integral.

If we let

$$\theta(t) = \sum_{n\in\mathbf{Z}} e^{-\pi n^2 t}$$

with $t > 0$, then we obtain the relation $\theta(t^{-1}) = t^{\frac{1}{2}}\theta(t)$, or

$$\sum_{n\in\mathbf{Z}} e^{-\pi n^2 t} = \sum_{n\in\mathbf{Z}} \frac{1}{\sqrt{t}}\, e^{-\pi n^2/t}$$

known as the **functional equation of the theta function.**

From it we shall obtain the functional equation of the zeta function defined for $\operatorname{Re}(s) > 1$ by the series

$$\zeta(s) = \sum \frac{1}{n^s}.$$

Recall that

$$\Gamma(s) = \int_0^\infty e^{-t}\, t^s \frac{dt}{t},$$

and also recall the invariance of the integral with respect to multiplicative translations, that is

$$\int_0^\infty f(at)\frac{dt}{t} = \int_0^\infty f(t)\frac{dt}{t}$$

if $a > 0$ and f is absolutely integrable. Select $a = \pi n$. Then let

$$F(s) = \pi^{-s/2}\Gamma\left(\frac{s}{2}\right)\zeta(s) = \int_0^\infty \sum_{n=1}^\infty e^{-\pi n^2 t}\, t^{s/2}\frac{dt}{t}.$$

Under the integral on the right we have essentially the theta function, except for its term with $n = 0$.

Let

$$\varphi(t) = \sum_{n=1}^\infty e^{-\pi n^2 t},$$

so that $2\varphi(t) = \theta(t) - 1$. Then we obtain

$$\begin{aligned} F(s) &= \int_0^\infty t^{s/2}\,\varphi(t)\frac{dt}{t} \\ &= \int_1^\infty t^{s/2}\,\varphi(t)\frac{dt}{t} + \int_1^\infty t^{-s/2}\,\varphi(1/t)\frac{dt}{t}. \end{aligned}$$

The functional equation of the theta function immediately implies that

$$\pi^{-s/2}\Gamma\left(\frac{s}{2}\right)\zeta(s) = \frac{1}{s-1} - \frac{1}{s} + \int_1^\infty \varphi(t)[t^{\frac{s}{2}} + t^{\frac{1-s}{2}}]\frac{dt}{t}.$$

The right-hand integral is absolutely convergent for all s, and this whole expression is invariant under $s \mapsto 1 - s$, so we obtain the analytic continuation and functional equation of the zeta function, following Riemann's original proof.

The argument is typical of all proofs of functional equations, and of expressions for functions of zeta type, especially those which we shall give for the Kronecker limit formula in a moment.

§3. THE FUNCTION $K_s(x)$

Let a, b be real numbers > 0. Define

K1.
$$K_s(a, b) = \int_0^\infty e^{-(a^2t + b^2/t)}\, t^s \frac{dt}{t}.$$

This is like an integral for the gamma function, but is much better, because first, it is more symmetric, involving both t and $1/t$, and second, it converges absolutely for all complex s, because the presence of $1/t$ cures the blow up which occurs for the gamma integral near 0.

Let us use the invariance of the integral under multiplicative translations, and let $t \mapsto \frac{b}{a}t$. We find that

K2.
$$K_s(a, b) = \left(\frac{b}{a}\right)^s K_s(ab)$$

where for $c > 0$ we define

K3.
$$K_s(c) = \int_0^\infty e^{-c(t+1/t)}\, t^s \frac{dt}{t}.$$

In general, this integral cannot be changed any further, and we note that

K4.
$$K_s(c) = K_{-s}(c),$$

proved by letting $t \mapsto t^{-1}$ and using the invariance of the integral on $\mathbf{R}^+$ by this transformation.

However, for $s = \frac{1}{2}$, the integral collapses to

K5.
$$K_{\frac{1}{2}}(c) = \sqrt{\frac{\pi}{c}}\, e^{-2c}$$

whence

K6.
$$K_{\frac{1}{2}}(a, b) = \frac{\sqrt{\pi}}{a} e^{-2ab}.$$

The proof of **K5** is easy, and runs as follows. Let

$$g(x) = K_{\frac{1}{2}}(x) = \int_0^\infty e^{-x(t+1/t)} t^{\frac{1}{2}} \frac{dt}{t}.$$

Let $t \mapsto t/x$. Then

$$g(x) = \frac{1}{\sqrt{x}} \int_0^\infty e^{-(t+x^2/t)} t^{\frac{1}{2}} \frac{dt}{t}.$$

Let $h(x) = \sqrt{x}\, g(x)$. We can differentiate $h(x)$ under the integral sign to get

$$h'(x) = -2x \int_2^\infty e^{-(t+x^2/t)} t^{-\frac{1}{2}} \frac{dt}{t}.$$

Let $t \mapsto t^{-1}$, use the invariance of the integral under this transformation, and then let $t \mapsto t/x$. We then find that

$$h'(x) = -2h(x),$$

whence

$$h(x) = Ce^{-2x}$$

for some constant C. We can let $x = 0$ in the integral for $h(x)$ (but not in the integral for $g(x)$!) to evaluate C, which comes out as $\Gamma(\frac{1}{2}) = \sqrt{\pi}$. This proves **K5**.

It is also useful to have an estimate for $K_s(x)$, namely:

K7. *Let $x_0 > 0$ and $\sigma_0 \leqq \sigma \leqq \sigma_1$. There is a number $C(x_0, \sigma_0, \sigma_1) = C$ such that if $x \geqq x_0$, then*

$$K_\sigma(x) \leqq C\, e^{-2x}.$$

Proof. First note that $t + 1/t \geqq 2$, if $t > 0$. Split up the integral as

$$\int_0^\infty = \int_0^{1/8} + \int_{1/8}^8 + \int_8^\infty.$$

The middle integral obviously gives an estimate of the type Ce^{-2x}. To estimate the first integral, note that if $t \leqq 1/8$, then

$$\frac{1}{t} \geqq 4 + \frac{1}{2t}.$$

Hence

$$\int_0^{1/8} e^{-x(t+1/t-2)} t^\sigma \frac{dt}{t} \leqq e^{-2x} \int_0^{1/8} e^{-x_0(t+1/2t)} t^\sigma \frac{dt}{t}$$

which is of the desired type. The integral to infinity is estimated in the same way, to conclude the proof.

The preceding formulas provide the basic formalism of the K-function. We suggest to the reader that he skip the following properties until he needs to use them, or to give alternate proofs for identities which he knows already.

K8. $$\int_{-\infty}^{\infty} \frac{1}{(u^2+1)^s}\, du = \sqrt{\pi}\,\frac{\Gamma(s-\frac{1}{2})}{\Gamma(s)} \qquad \text{for} \quad \mathrm{Re}(s) > \tfrac{1}{2}.$$

Proof. Consider

$$\Gamma(s)\int_{-\infty}^{\infty} \frac{1}{(u^2+1)^s}\, du = \int_{-\infty}^{\infty}\int_{0}^{\infty} e^{-t}\,\frac{1}{(u^2+1)^s}\, t^s\,\frac{dt}{t}\, du.$$

Let $t \mapsto (u^2+1)t$ and use the invariance of the integral with respect to dt/t, relative to multiplicative translations. The formula **K8** drops out.

The above formula allows us to find the first term of the expansion of the right-hand side at $s = 1$, which will be needed. There are of course alternate proofs for this (using the functional equation of the gamma function), but it does no harm to get it in the spirit of the present section. Putting $s = 1$ in the integral of **K8** yields the value π because $1/(u^2+1)$ integrates to the arctangent. To get the coefficient of $s - 1$ in the expansion, we differentiate under the integral sign with respect to s, and we must evaluate the integral

$$\int_{-\infty}^{\infty} \frac{\log(u^2+1)}{u^2+1}\, du.$$

To do this, I use a trick shown to me by Seeley. Let

$$g(x) = \int_{0}^{\infty} \frac{\log(u^2x^2+1)}{u^2+1}\, du$$

so that $g(0) = 0$. Differentiating under the integral sign and using a trivial partial fraction decomposition yields

$$g'(x) = \frac{\pi}{1+x}, \qquad x > 0.$$

Hence $g(1) = \pi \log 2$. This gives us

$$\frac{\Gamma(s-\frac{1}{2})}{\Gamma(s)} = \sqrt{\pi}(1 - (s-1)\log 4 + \cdots).$$

Finally in questions related to the second Kronecker limit formula it is sometimes useful to know the next identity.

K9. $$\Gamma(s)\int_{-\infty}^{\infty} \frac{e^{ixu}}{(u^2+1)^s}\, du = 2\sqrt{\pi}\left(\frac{x}{2}\right)^{s-\frac{1}{2}} K_{s-\frac{1}{2}}(x) \qquad \text{for} \quad \mathrm{Re}(s) > \tfrac{1}{2}.$$

Proof. Again as in **K8**, write down the integral for $\Gamma(s)$, interchange the order of integration, let $t \mapsto (u^2 + 1)t$ and use the fact that $e^{-x^2/2}$ is self-dual for the Fourier transform normalized as

$$\int_{-\infty}^{\infty} f(x)\, e^{-ixy}\, dx.$$

Then let $t \mapsto xt$. The desired formula drops out.

§4. THE KRONECKER FIRST LIMIT FORMULA

Let $\tau = x + iy$ be in the upper half plane, $y > 0$. We are interested in the function $E(\tau, s)$ defined by the series

$$E(\tau, s) = \sum_{m,n}' \frac{y^s}{|m\tau + n|^{2s}}, \qquad \text{Re}(s) > 1,$$

the sum being taken for all integers $(m, n) \neq (0, 0)$.

We want to get its constant term in the expansion at $s = 1$. We shall derive an analytic expression for $E(\tau, s)$ which will exhibit a simple pole at $s = 1$ with residue π, and will show that otherwise it is holomorphic in the complex plane. From this expression, we shall be able to read off the first two terms.

Kronecker first limit formula. *Let* $q_\tau = e^{2\pi i\tau}$, *and let*

$$\eta(\tau) = q_\tau^{1/24} \prod_{n=1}^{\infty} (1 - q_\tau^n).$$

Let γ *be the Euler constant. Then*

$$E(\tau, s) = \frac{\pi}{s - 1} + 2\pi(\gamma - \log 2 - \log(\sqrt{y}\, |\eta(\tau)|)^2) + O(s - 1).$$

Proof. Let $\tau = x + iy$, so that

$$|m\tau + n|^2 = (n + mx)^2 + m^2y^2.$$

As in the functional equational equation for the zeta function, we start with

$$\frac{\pi^{-s}\Gamma(s)}{a^s} = \int_0^\infty e^{-\pi a t}\, t^s \frac{dt}{t}.$$

Therefore summing $E(\tau, s)$ first for $m = 0$ and then $m \neq 0$, we find:

$$\begin{aligned}
(4.1) \qquad & \pi^{-s}\Gamma(s)y^{-s}E(\tau, s) \\
&= 2\pi^{-s}\Gamma(s)\zeta(2s) + 2\sum_{m=1}^{\infty} \int_0^\infty \sum_n e^{-\pi|m\tau+n|^2 t}\, t^s \frac{dt}{t}. \\
&= 2\pi^{-s}\Gamma(s)\zeta(2s) + 2\sum_{m=1}^{\infty} \int_0^\infty \sum_n e^{-\pi(n+xm)^2 t}\, e^{-\pi y^2 m^2 t}\, t^s \frac{dt}{t}. \\
&= \text{I} + \text{II}.
\end{aligned}$$

We now apply the Poisson summation formula on the sum over all $n \in \mathbf{Z}$ under the integral, yielding

$$\sum_n e^{-\pi(n+xm)^2 t} = \frac{1}{\sqrt{t}} \sum_n e^{2\pi i x m n} e^{-\pi n^2/t}.$$

The square root of t in the denominators will combine with t^s to give $t^{s-\frac{1}{2}}$. We now split the sum over n into two parts, with $n = 0$ and $n \neq 0$. When $n = 0$, we essentially get a zeta expression, so that the corresponding term in the right-hand side of (4.1) is

$$\text{(4.2)} \qquad \begin{aligned} \text{II}_{n=0} &= 2 \sum_{m=1}^{\infty} \int_0^{\infty} e^{-\pi y^2 m^2 t} t^{s-\frac{1}{2}} \frac{dt}{t} \\ &= 2\pi^{-(s-\frac{1}{2})} y^{-2(s-\frac{1}{2})} \Gamma(s - \tfrac{1}{2})\zeta(2s-1). \end{aligned}$$

Next we deal with the term $\text{II}_{n\neq 0}$, which is

$$\text{(4.3)} \qquad \begin{aligned} \text{II}_{n\neq 0} &= 2 \sum_{m=1}^{\infty} \sum_{n\neq 0} e^{2\pi i x m n} \int_0^{\infty} e^{-\pi(y^2 m^2 t + n^2/t)} t^{s-\frac{1}{2}} \frac{dt}{t} \\ &= 2 \sum_{m=1}^{\infty} \sum_{n\neq 0} e^{2\pi i x m n} K_{s-\frac{1}{2}}(\sqrt{\pi} y m, \sqrt{\pi}|n|). \end{aligned}$$

Therefore the expression in (4.3) is an entire function of s, as one sees by an easy estimate for K, but what concerns us is that this expression is holomorphic at $s = 1$. Using (4.1), (4.2), and (4.3) now gives the analytic continuation of $E(\tau, s)$, and it would be easy to get the functional equation, having a form similar to that of the zeta function. We concentrate our attention at $s = 1$, in which case

$$\text{(4.4)} \qquad \text{II}_{n\neq 0} \quad \text{at} \quad s = 1 \text{ is equal to } 2 \sum_{m=1}^{\infty} \sum_{n\neq 0} e^{2\pi i x m n} \frac{1}{my} e^{-2ym|n|}.$$

Recall that

$$q_\tau = e^{2\pi i(x+iy)}.$$

Looking at $q_\tau^{mn} + \bar{q}_\tau^{mn}$ (arising from positive and negative values of n), and using formula **K6**, we find that

$$\text{(4.5)} \qquad \begin{aligned} \text{II}_{n\neq 0} \quad \text{at} \quad s = 1 \quad \text{is} \quad &= 4 \sum_{m=1}^{\infty} \frac{1}{my} \operatorname{Re} \sum_{n=1}^{\infty} q_\tau^{mn} \\ &= -\frac{4}{y} \sum_{n=1}^{\infty} \log|1 - q_\tau^n| \\ &= -\frac{4}{y}\left(\log|\eta(\tau)| + \frac{\pi y}{12}\right) \\ &= -\frac{4}{y} \log|\eta(\tau)| - \frac{\pi}{3}. \end{aligned}$$

Putting all our terms together, we obtain

$$\text{(4.6)}\quad \pi^{-s}\Gamma(s)y^{-s}E(\tau, s) = 2\pi^{-s}\Gamma(s)\zeta(2s) + 2\pi^{-(s-\frac{1}{2})}y^{-2(s-\frac{1}{2})}\Gamma(s - \tfrac{1}{2})\zeta(2s - 1) - \frac{4}{y}\log|\eta(\tau)| - \frac{\pi}{3} + O(s - 1).$$

Since $\zeta(2) = \pi^2/6$, we see that the term arising from $\zeta(2s)$ will cancel $-\pi/3$. Divide by $\pi^{-s}\Gamma(s)$, setting $s = 1$. From simple identities with the gamma function, or from **K8**, one knows that

$$\frac{\Gamma(s - \frac{1}{2})}{\Gamma(s)} = \sqrt{\pi}\,(1 - (s - 1)\log 4 + \cdots)$$

while

$$\zeta(2s - 1) = \frac{1}{2(s - 1)} + \gamma + O(s - 1).$$

Multiplying by y^s we still have to expand

$$y^s y^{-2(s-\frac{1}{2})} = y^{1-s} = 1 - (s - 1)\log y + O(s - 1)^2.$$

Putting all this together shows that

$$E(\tau, s) = \frac{\pi}{s - 1} - \pi\log y + 2\pi(\gamma - \log 2) - 4\pi\log|\eta(\tau)| + O(s - 1),$$

which is another way of writing Kronecker's formula.

Remark. The formula can be generalized to arbitrary number fields, the case treated above corresponding to the rational numbers. One uses the sum over pairs of integers of that field. For each real absolute value, one takes a copy of the upper half plane. It was unknown until very recently what to do for the complex absolute values, but as shown in Asai [1] one merely has to take the quaternion upper half plane in this case. The quaternion upper half plane can be represented as the set of matrices

$$\begin{pmatrix} z & -u \\ u & z' \end{pmatrix}$$

where z is a complex number, z' its conjugate, and $u > 0$. You then end up with a multiple integral of K-functions, which does not collapse to an exponential function, and yields a function analogous to $\log|\eta(\tau)|$. Asai discusses precisely several aspects of the analogy. However, the connection with abelian functions and moduli remains to be worked out.

§5. THE KRONECKER SECOND LIMIT FORMULA

Let u, v be real numbers which are not both integers. We define

$$E_{u,v}(\tau, s) = \sum_{(m,n)\neq(0,0)} e^{2\pi i(mu+nv)} \frac{y^s}{|m\tau + n|^{2s}}$$

for $\tau = x + iy$ in the upper half plane. The series converges for $\operatorname{Re}(s) > 1$.

Second limit formula. *The function $E_{u,v}(\tau, s)$ can be continued to an entire function of s, and one has*

$$E_{u,v}(\tau, 1) = -2\pi \log |g_{-v,u}(\tau)|,$$

where $g_{u,v}$ is the Siegel function,

$$g_{u,v}(\tau) = -q^{(1/2)\mathbf{B}_2(u)} e^{2\pi i v(u-1)/2} \prod_{n=1}^{\infty} (1 - q^n q_z)(1 - q^n/q_z),$$

and $\mathbf{B}_2(u) = u^2 - u + \frac{1}{6}$.

Proof. We follow Siegel [B14]. We shall not need any property of $g_{u,v}$ other than its definition as the above product. We carry out the proof first for the values

$$0 \leqq u < 1 \quad \text{and} \quad 0 < v < 1.$$

The extension to the general case will be done afterwards.

As in the first formula, we split off the sum taken for $m = 0$, so that, abbreviating $E_{u,v}(\tau, s)$ by $E(\tau, s)$, we get

$$y^{-s}E(\tau, s) = \sum_{n\neq 0} \frac{e^{2\pi i n v}}{|n|^{2s}} + \sum_{m\neq 0} e^{2\pi i m u} \sum_n e^{2\pi i n v} \frac{1}{|m\tau + n|^{2s}}.$$

At $s = 1$, the first sum is a standard Fourier series,

$$\sum_{n\neq 0} \frac{e^{2\pi i n v}}{|n|^2} = 2\pi^2\left(v^2 - v + \frac{1}{6}\right).$$

The second term is dealt with by using the Gamma integral, and is equal to

$$\frac{\pi^s}{\Gamma(s)} \sum_{m\neq 0} e^{2\pi i m u} \int_0^{\infty} \sum_n e^{2\pi i n v} e^{-\pi t|m\tau+n|^2} t^s \frac{dt}{t}.$$

We write $\tau = x + iy$, so that $|m\tau + n|^2 = (n + mx)^2 + m^2y^2$. The second term is equal to

$$\frac{\pi^s}{\Gamma(s)} \sum_{m\neq 0} e^{2\pi i m(u-vx)} \int_0^{\infty} \sum_n e^{-\pi t(n+mx-iv/t)^2} e^{-\pi(ty^2m^2+v^2/t)} t^s \frac{dt}{t}.$$

We apply the Poisson summation formula to the inner sum over n. This sum is then equal to

$$\sum_n = \frac{1}{\sqrt{t}} \sum_n e^{-\pi n^2/t} e^{2\pi i n(mx - iv/t)}.$$

Therefore the second term is equal to an expression which involves a $K_{s-\frac{1}{2}}$ integral, which is seen to be entire in s. At $s = 1$, the second term is equal to

$$\pi \sum_{m \neq 0} e^{2\pi i m(u - vx)} \sum_n e^{2\pi i n m x} \int_0^\infty e^{-\pi[tm^2y^2 + (n-v)^2/t]} t^{\frac{1}{2}} \frac{dt}{t}.$$

The inner integral is of the form $K_{\frac{1}{2}}(a, b) = \dfrac{\sqrt{\pi}}{a} e^{-2ab}$, and therefore our second term is equal to

$$\pi \sum_{m \neq 0} e^{2\pi i m(u - vx)} \sum_n e^{2\pi i n m x} \frac{1}{|m|y} e^{-2\pi y|n-v|\,|m|}.$$

The sums converge exponentially, and can be reversed, so that we sum over n first, and then over $m \neq 0$. This is the only point where we use $v \neq 0$. For $v = 0$ one has to take the term with $n = 0$ into account separately, and then interchange the summations. The arguments are similar. We obtain for $v \neq 0$,

$$(*) \quad E(\tau, 1) = 2\pi^2\left(v^2 - v + \frac{1}{6}\right)y + \pi \sum_n \sum_{m \neq 0} \frac{1}{|m|} e^{2\pi i[m(u - vx) + nmx + iy|n-v|\,|m|]}.$$

For $|r| < 1$ we have

$$-\log(1 - r) = \sum_{m=1}^{\infty} \frac{r^m}{m}.$$

We evaluate the double sum over n and $m \neq 0$ by distinguishing cases, dealing first with $n = 0$, and then with the four cases corresponding to $n \neq 0$, $m \neq 0$.

Take first $n = 0$. Then we have the double sum

$$\sum_{m \neq 0} = \sum_{m=1}^{\infty} + \sum_{m=-1}^{-\infty}$$

which therefore yields

$$\sum_{m=1}^{\infty} \frac{1}{m} e^{2\pi i m[(u - vx) + iyv]} + \sum_{m=1}^{\infty} \frac{1}{m} e^{2\pi i[-(u - vx) + iyv]m}$$
$$= -\log(1 - e^{2\pi i(u - v\bar{\tau})})(1 - e^{-2\pi i(u - v\tau)}).$$

Let $z = u - v\tau$. The term corresponding to $n = 0$ can be rewritten

$$-\log(1 - e^{-2\pi i z})(1 - e^{2\pi i \bar{z}}) = -2\log|1 - q_z^{-1}|$$
$$= -2\log|1 - q_z| + 2 \cdot 2\pi v y.$$

So far, we have obtained

$$E(\tau, 1) = 2\pi^2 \mathbf{B}_2(-v)y - 2\pi \log|1 - q_z| + \pi \sum_{n \neq 0} \sum_{m \neq 0} \frac{1}{|m|} \quad \text{(term as in } (*)).$$

Next, we consider the four sums separately, arising from the cases

$$n > 0, \quad n < 0, \quad m > 0, \quad m < 0.$$

Consider first the case with $n > 0$ and $m > 0$. Summing over m yields

$$\sum_{m=1}^{\infty} e^{2\pi i[u - vx + nx + iy(n-v)]m} = \sum_{m=1} \frac{1}{m} e^{2\pi i(u - v\tau + n\tau)m} = -\log(1 - q_\tau^n q_z).$$

One of the other cases will contribute the complex conjugate of the above expression, and the two other cases will contribute the factors of type $(1 - q_\tau^n/q_z)$ and their complex conjugates. This accounts for the big double product in the q-product, and concludes the proof.

We still have to make the appropriate remarks when u, v do not lie between 0 and 1. In that case, we note that the series defining $E_{u,v}(\tau, s)$ is obviously periodic in u and v. On the other hand, from the definition of $g_{u,v}(\tau)$ as a product, we see that the right-hand side of the formula is obviously periodic in u. A short computation again using the product definition shows that it is also periodic in v. This takes care of the general case.

21 The First Limit Formula and L-series

§1. RELATION WITH L-SERIES

Let k be an imaginary quadratic field, with discriminant $-d_k < 0$ so that d_k is the absolute value of the discriminant. Let $\mathfrak{o}$ be the ring of integers in k, and let $\mathfrak{A}$ be an ideal class. We define the **zeta function**

$$\zeta(s, \mathfrak{A}) = \sum \frac{1}{\mathbf{N}\mathfrak{a}^s}$$

taking the sum over all ideals $\mathfrak{a}$ in the class. We can define $\mathbf{N}\mathfrak{a}$ to be the unique positive integer which generates $\mathfrak{a}\mathfrak{a}'$, where $\mathfrak{a}'$ is the conjugate ideal to $\mathfrak{a}$. (Refer back to Chapter 8, §1, to see that this makes sense.) Fix some ideal $\mathfrak{b}$ in the inverse class $\mathfrak{A}^{-1}$. Then $\mathfrak{a}\mathfrak{b} = (\xi_\mathfrak{a})$ is principal, and the association

$$\mathfrak{a} \mapsto \xi_\mathfrak{a}$$

gives a bijection between the ideals in $\mathfrak{A}$ and equivalence classes of elements of $\mathfrak{b}$. (Two elements of k are called $\mathfrak{o}$-equivalent if their quotient is a unit in $\mathfrak{o}$.) In what we do later, $\mathfrak{b}$ will only enter homogeneously of degree 0, so we assume right away for convenience that $\mathfrak{b} = [\tau, 1]$. Any ideal is always equivalent to an ideal of this type. Then we can rewrite the zeta function in the form

$$\zeta(s, \mathfrak{A}) = \frac{\mathbf{N}\mathfrak{b}^s}{w} \sum_{\xi \in \mathfrak{b}} \frac{1}{\mathbf{N}\xi^s},$$

where w is the number of roots of unity in $\mathfrak{o}$ (the only units in an imaginary quadratic field are roots of unity). This also can be written as

$$\zeta(s, \mathfrak{A}) = \frac{\mathbf{N}\mathfrak{b}^s}{w} \sum_{m,n}' \frac{1}{|m\tau + n|^{2s}},$$

where the sum is taken over all pairs of integers $(m, n) \neq 0, 0)$.

Note that $\mathbf{N}\mathfrak{a}\mathbf{N}\mathfrak{b} = \mathbf{N}\xi_{\mathfrak{a}}$, so that the $\mathfrak{b}$ really appears only as the usual convenient means of making ideals principal.

The discriminant of $\mathfrak{b}$ is given by

$$D(\mathfrak{b}) = \begin{vmatrix} 1 & \tau \\ 1 & \tau' \end{vmatrix}^2 = (\tau' - \tau)^2 = -(2y)^2$$

if $\tau = x + iy$ and $y > 0$. On the other hand we have

$$D(\mathfrak{b}) = \mathbf{N}\mathfrak{b}^2 D(\mathfrak{o}),$$

where $D(\mathfrak{o})$ is the discriminant of $\mathfrak{o}$. Hence we have a third expression for the zeta function, namely

$$(1) \qquad \boxed{\zeta(s, \mathfrak{A}) = \frac{1}{w}\left(\frac{2}{\sqrt{d}}\right)^s \sum_{m,n}{}' \frac{y^s}{|m\tau + n|^{2s}}}$$

where we see appearing the Eisenstein series for which we know the Kronecker limit formula. Here d is the absolute value of the discriminant of $\mathfrak{o}$.

It will be slightly more convenient to deal with Δ than with η, and we note that the absolute value signs in the Kronecker limit formula anyhow eliminate the ambiguity of the possible roots of unity. As in Chapter 12, define

$$g(\mathfrak{b}) = (2\pi)^{-12}\mathbf{N}\mathfrak{b}^6|\Delta(\mathfrak{b})| = (2\pi)^{-12}\mathbf{N}([\tau, 1])^6|\Delta(\tau)|.$$

This function is an invariant of the equivalence class of $\mathfrak{b}$, because considering $\lambda\mathfrak{b}$ instead of $\mathfrak{b}$, we see that $|\lambda|^{12}$ comes out of the norm sign, and $|\lambda|^{-12}$ comes out of the $|\Delta|$. So we can write

$$g(\mathfrak{b}) = g(\mathfrak{B}),$$

where $\mathfrak{B}$ is the ideal class of $\mathfrak{b}$.

We use the beginning of the exponential series

$$\left(\frac{2}{\sqrt{d}}\right)^{s-1} = 1 + (s-1)\log\frac{2}{\sqrt{d}} + \cdots,$$

and we find the expression for $\zeta(s, \mathfrak{A})$ which we wanted, suitably normalized, namely

$$(2) \qquad \zeta(s, \mathfrak{A}) = \frac{1}{w}\frac{2\pi}{\sqrt{d}}\left(\frac{1}{s-1} + 2\gamma - \log d + \frac{1}{6}\log g(\mathfrak{A}^{-1})\right) + O(s-1).$$

Let χ be a character of the ideal class group G. We define the **L-series**

$$L(s, \chi) = \sum_{\mathfrak{A}} \chi(\mathfrak{A})\zeta(s, \mathfrak{A}) = \prod\left(1 - \frac{\chi(\mathfrak{p})}{\mathbf{N}\mathfrak{p}^s}\right)^{-1}$$

with the product taken over all prime ideals. Let 1 be the trivial character. If $\chi \neq 1$, then taking the sum of $\chi(\mathfrak{A})$ over all $\mathfrak{A}$ yields 0. Hence the terms independent of the class (i.e. the polar term involving the residue, and the universal constants) will disappear after taking the sum, leaving us with

Theorem 1. *Let χ be a non-trivial character of the proper ideal class group G of $\mathfrak{o}$ in k. Then*

$$L(1, \chi) = -\frac{\pi}{3w\sqrt{d}} \sum_{\mathfrak{A}} \chi(\mathfrak{A}) \log g(\mathfrak{A}^{-1}).$$

On the other hand, if h is the order of G, then

$$L(s, 1) = \zeta(s) = \frac{2\pi h}{w\sqrt{d}} \frac{1}{s-1} + \cdots.$$

Let $K = k(j(\mathfrak{o}))$ be the Hilbert class field. Then we have a formal relation

$$\prod_{\mathfrak{P}|\mathfrak{p}} \left(1 - \frac{1}{\mathbf{N}\mathfrak{P}^s}\right) = \prod_{\text{all } \chi} \left(1 - \frac{\chi(\mathfrak{p})}{\mathbf{N}\mathfrak{p}^s}\right),$$

whose proof we reproduce for the convenience of the reader. Let

$$u = \mathbf{N}\mathfrak{p}^s.$$

Then $\mathbf{N}\mathfrak{P} = (\mathbf{N}\mathfrak{p})^f$ and so our relation amounts to

$$(1 - u^f)^r = \prod_{\chi} (1 - \chi(\mathfrak{p})u),$$

if $\mathfrak{p}\mathfrak{o}_K = \mathfrak{P}_1 \cdots \mathfrak{P}_r$. The cyclic group generated by $\mathfrak{p}$ in G has order f by definition of the Frobenius automorphism. Let $\psi_1, \ldots, \psi_f$ be the distinct characters of this cyclic group. If ζ_f is a primitive f-th root of unity, then we can make these characters correspond to

$$\psi_\nu(\mathfrak{p}) = \zeta_f^\nu.$$

Let $\chi_1, \ldots, \chi_r$ be the characters of $G/\{\mathfrak{p}\}$, i.e. the characters of G which are trivial on $\mathfrak{p}$. Then the products $\chi_\mu \psi_\nu$ constitute all of the character group of G. Hence

$$\prod_{\chi} (1 - \chi(\mathfrak{p})u) = \prod_{\nu=0}^{f-1} (1 - \zeta_f^\nu u)^r = (1 - u^f)^r,$$

thus proving our relation. In terms of L-series, it yields:

Theorem 2. *We have a relation*

$$\zeta_K(s) = \zeta_k(s) \prod_{\chi \neq 1} L_k(s, \chi).$$

Both sides have a simple pole at $s = 1$. The residues must therefore be equal. One knows from elementary analytic number theory that the residues are given by the expressions

$$\rho_K = \frac{(2\pi)^{r_2} h_K R_K}{w_K \sqrt{d_K}} \quad \text{and} \quad \rho_k = \frac{2\pi h_k}{w_k \sqrt{d_k}},$$

where $2r_2 = [K:\mathbf{Q}] = 2h$ because $[K:k] = h$. Thus $r_2 = h$. As usual, w_K is the number of roots of unity in K. From Theorem 2, we therefore obtain the corresponding relation for the residues.

Theorem 3. *Let $\mathfrak{o}$ be the ring of integers in k, and $K = k(j(\mathfrak{o}))$. Then*

$$\rho_K = \rho_k \prod_{\chi \neq 1} \left(\frac{-\pi}{3w\sqrt{d}} \sum_{\mathfrak{a}} \chi(\mathfrak{a}) \log g(\mathfrak{a}^{-1}) \right).$$

We observe that since the residue of the zeta function is not 0, it follows that for any non-trivial character χ of G, the sum

$$\sum_{\mathfrak{a}} \chi(\mathfrak{a}) \log g(\mathfrak{a}^{-1})$$

is also $\neq 0$.

In the next section, we shall do some elementary algebra involving the Frobenius determinant to transform some more the final product in Theorem 3.

The above results are essentially due to Fueter [14], who gets the class number relation implicit in the above, when we substitute the values for the residues of the zeta function. Our exposition follows Siegel [B14], and [B15], §27.3.

§2. THE FROBENIUS DETERMINANT

Let G be a finite abelian group and $\hat{G} = \{\chi\}$ its character group. We have the **Frobenius determinant** relation:

Theorem 5. *Let f be any (complex valued) function on G. Then*

$$\prod_{\chi \in \hat{G}} \sum_{a \in G} \chi(a) f(a^{-1}) = \det_{a,b} f(a^{-1}b).$$

Proof. Let F be the space of functions on G. It is a finite dimensional vector space whose dimension is the order of G. It has two natural bases. First, the characters $\{\chi\}$, and second the functions $\{\delta_b\}$, $b \in G$, where

$$\delta_b(x) = 1 \quad \text{if} \quad x = b$$
$$\delta_b(x) = 0 \quad \text{if} \quad x \neq b.$$

For each $a \in G$ let $T_a f$ be the function such that $T_a f(x) = f(ax)$. Then

$$(T_a\chi)(b) = \chi(ab) = \chi(a)\chi(b),$$

so that

$$T_a\chi = \chi(a)\chi.$$

So χ is an eigenvector of T_a. Let

$$T = \sum_{a \in G} f(a^{-1})T_a.$$

Then T is a linear map on F, and for each character χ, we have

$$T\chi = [\sum_{a \in G} \chi(a)f(a^{-1})]\chi.$$

Therefore χ is an eigenvector of T, and consequently the determinant of T is equal to the product over all χ occurring on the left-hand side of the equality in Theorem 5.

On the other hand, we look at the effect of T on the other basis. We have

$$T_a\delta_b(x) = \delta_b(ax),$$

so that $T_a\delta_b$ is the characteristic function of $a^{-1}b$, and

$$T_a\delta_b = \delta_{a^{-1}b}.$$

Consequently

$$\begin{aligned} T\delta_b &= \sum_{a \in G} f(a^{-1})\delta_{a^{-1}b} \\ &= \sum_{a \in G} f(a^{-1}b)\delta_a. \end{aligned}$$

From this we find an expression for the determinant of T which is precisely the right-hand side in Theorem 5. This proves our theorem.

Theorem 6. *The determinant of Theorem 5 splits into*

$$\det_{a,b} f(ab^{-1}) = [\sum_{a \in G} f(a)] \det_{a,b \neq 1} [f(ab^{-1}) - f(a)].$$

Therefore

$$\prod_{\chi \neq 1} \sum_{a \in G} \chi(a)f(a^{-1}) = \det_{a,b \neq 1} [f(ab^{-1}) - f(a)].$$

Proof. Let $a_1 = 1, \ldots, a_n$ be the elements of G. In the determinant

$$\det f(a_i a_j^{-1}) = \begin{vmatrix} f(a_1 a_1^{-1}) & f(a_1 a_2^{-1}) \cdots & f(a_1 a_n^{-1}) \\ \cdot & \cdot & \cdot \\ \cdot & \cdot & \cdot \\ \cdot & \cdot & \cdot \\ f(a_n a_1^{-1}) & f(a_n a_2^{-1}) \cdots & f(a_n a_n^{-1}) \end{vmatrix}$$

add the last $n - 1$ rows to the first. Then all elements of the new first row are equal to $\sum f(a^{-1}) = \sum f(a)$. Factoring this out yields

$$[\sum_{a \in G} f(a)] \begin{vmatrix} 1 & 1 \quad \cdots & 1 \\ f(a_2 a_1^{-1}) & f(a_2 a_2^{-1}) \cdots & f(a_2 a_n^{-1}) \\ \cdot & \cdot & \cdot \\ \cdot & \cdot & \cdot \\ \cdot & \cdot & \cdot \\ f(a_n a_1^{-1}) & f(a_n a_2^{-1}) \ldots & f(a_n a_n^{-1}) \end{vmatrix}$$

Recall that a_1 is chosen to be 1. Subtract the first column from each one of the other columns. You get the first statement of the theorem.

On the other hand, the function f can be selected so that the elements $\{f(a)\}$, $a \in G$, are algebraically independent over $\mathbf{Q}$, and therefore the factorization given in this first statement for the determinant is applicable in the polynomial ring generated over $\mathbf{Z}$ by the variables $f(a)$. Combining the first statement with Theorem 5 yields the second relation where the product is taken only over $\chi \neq 1$.

§3. APPLICATION TO THE L-SERIES

We apply the determinant of §2 to the case when G is the group of ideal classes in k, and

$$f(\mathfrak{A}) = \log g(\mathfrak{A})$$

where

$$g(\mathfrak{A}) = (2\pi)^{-12} \mathbf{N}\mathfrak{a}^6 |\Delta(\mathfrak{a})|$$

is our previous invariant of the class $\mathfrak{A}$, defined with any ideal $\mathfrak{a}$ in the class. Then

$$f(\mathfrak{A}\mathfrak{B}^{-1}) - f(\mathfrak{A}) = \log \frac{g(\mathfrak{A}\mathfrak{B}^{-1})}{g(\mathfrak{A})}$$

with

$$\frac{g(\mathfrak{A}\mathfrak{B}^{-1})}{g(\mathfrak{A})} = \mathbf{N}\mathfrak{b}^{-6} \frac{|\Delta(\mathfrak{a}\mathfrak{b}^{-1})|}{|\Delta(\mathfrak{a})|} .$$

Recall the Corollary of Theorem 5, Chapter 12, §2 which asserts that the above number is a unit. The product occurring in Theorem 3 can then be interpreted as a regulator of a system of units. [For the computation of the index of the group of inch units in all units, see Kubert–Lang's "Modular Units", Chapter 9, §2.]

22 The Second Limit Formula and L-series

§1. GAUSS SUMS

Let k be a number field and $\mathfrak{o} = \mathfrak{o}_k$ the ring of algebraic integers. Let $\mathfrak{f}$ be an ideal of $\mathfrak{o}$. (Unless otherwise specified, ideal means contained in $\mathfrak{o}$.) We shall consider Gauss sums formed with characters (the generalization to number fields is due to Hecke).

Let χ be a character of the multiplicative group $(\mathfrak{o}/\mathfrak{f})^*$. We extend χ to a function on $\mathfrak{o}/\mathfrak{f}$ by setting $\chi(\alpha) = 0$ if α not prime to $\mathfrak{f}$.

Let $\mathfrak{g}$ be an ideal dividing $\mathfrak{f}$. We have a natural homomorphism

$$\mathfrak{o}/\mathfrak{f} \to \mathfrak{o}/\mathfrak{g}$$

sending $(\mathfrak{o}/\mathfrak{f})^*$ into $(\mathfrak{o}/\mathfrak{g})^*$. If ψ is a character of $(\mathfrak{o}/\mathfrak{g})^*$, then we can define a character χ on $(\mathfrak{o}/\mathfrak{f})^*$ by composing ψ with the natural homomorphism above, and then set $\chi(\alpha) = 0$ if α is not prime to $\mathfrak{f}$. A character χ of $(\mathfrak{o}/\mathfrak{f})^*$ which cannot be obtained by composition with a character ψ as above, for some proper divisor $\mathfrak{g}$ of $\mathfrak{f}$, is called **proper**, and $\mathfrak{f}$ is called its **conductor**. A function on $\mathfrak{o}$ defined as above by a character on $(\mathfrak{o}/\mathfrak{f})^*$ is called a **character modulo** $\mathfrak{f}$.

A character χ modulo $\mathfrak{f}$ is proper if and only if it satisfies the following condition: For each proper divisor $\mathfrak{g}$ of $\mathfrak{f}$ there exists a pair of integers $\lambda, \mu \in \mathfrak{o}$ prime to $\mathfrak{f}$ such that $\lambda \equiv \mu \pmod{\mathfrak{g}}$ and $\chi(\lambda) \neq \chi(\mu)$.

This is immediate from the definition.

Let $\mathfrak{f}$ be an ideal of $\mathfrak{o}$. Let $\mathfrak{d} = \mathfrak{d}_{k/\mathbf{Q}}$ be the different. Recall that if $\mathfrak{o} = \mathfrak{o}_k$ is the ring of algebraic integers, then $\mathfrak{o}^\perp$ is the set of elements $\lambda \in k$ such that

$$\mathrm{Tr}(\lambda\mathfrak{o}) \subset \mathbf{Z},$$

and $\mathfrak{d}^{-1} = \mathfrak{o}^{\perp}$ by definition. The above condition on the trace is equivalent with the condition

$$e^{2\pi i \operatorname{Tr}(\lambda \mathfrak{o})} = 1,$$

which is the reason for the orthogonality sign.

Let γ be a fixed element of k such that $\gamma\mathfrak{f}\mathfrak{d}$ is an ideal prime to $\mathfrak{f}$. Thus $\gamma\mathfrak{d}$ has exact denominator $\mathfrak{f}$.

If $\lambda \in \mathfrak{f}$, then $\operatorname{Tr}(\lambda\gamma) \in \mathbf{Z}$ *and hence*

$$e^{2\pi i \operatorname{Tr}(\lambda\gamma)} = 1.$$

This proves the second assertion in the next identity.

G1. *Let $\lambda \in \mathfrak{o}$. We have*

$$\sum_{z \bmod \mathfrak{f}} e^{2\pi i \operatorname{Tr}(\lambda z \gamma)} = \begin{cases} 0 & \text{if } \lambda \not\equiv 0 \pmod{\mathfrak{f}} \\ \mathbf{N}\mathfrak{f} & \text{if } \lambda \equiv 0 \pmod{\mathfrak{f}}. \end{cases}$$

Proof. Suppose that $\lambda \not\equiv 0 \pmod{\mathfrak{f}}$. The map $z \mapsto z - 1$ permutes the residue classes mod $\mathfrak{f}$, and by the remark before **G1**, we see that the value of the sum is unchanged when we make this permutation. Therefore our sum is equal to

$$e^{-2\pi i \operatorname{Tr}(\lambda\gamma)} \sum_{z \bmod \mathfrak{f}} e^{2\pi i \operatorname{Tr}(\lambda z \gamma)}.$$

But $\operatorname{Tr}(\lambda\gamma)$ is not an integer, otherwise $\lambda\gamma \in \mathfrak{o}^{\perp} = \mathfrak{d}^{-1}$, which contradicts the way we chose γ, and the assumption on λ. Therefore the sum must be 0, thereby proving our property **G1**.

For any character χ modulo $\mathfrak{f}$, we define the **Gauss sum**

$$T_{\gamma}(\chi, \alpha) = \sum_{x \bmod \mathfrak{f}} \chi(x)\, e^{2\pi i \operatorname{Tr}(x\alpha\gamma)}.$$

The γ as subscript to T indicates that the sum depends on γ and also on $\mathfrak{f}$. If $x \equiv y \pmod{\mathfrak{f}}$, then $\operatorname{Tr}(x\alpha\gamma) \equiv \operatorname{Tr}(y\alpha\gamma) \pmod{\mathbf{Z}}$, and hence each term in the sum is well defined. The sum depends on the choice of γ. However, in the applications, it will appear together with a factor which takes away this dependence. Namely, the character χ will arise from a ray class character, and one then verifies that

$$\frac{\bar{\chi}(\gamma\mathfrak{d}\mathfrak{f})}{T_{\gamma}(\chi, 1)}$$

is independent of the choice of γ, as a direct consequence of the next property.

G2. *Let χ be a character modulo $\mathfrak{f}$. If λ is prime to $\mathfrak{f}$, then*

$$T_{\gamma}(\chi, \alpha\lambda) = \bar{\chi}(\lambda) \mathrm{T}_{\gamma}(\chi, \alpha).$$

Proof. The map $x \mapsto x\lambda$ permutes the residue classes $\mathfrak{o}/\mathfrak{f}$, and hence our assertion is obvious. (Note that $\bar{\chi} = \chi^{-1}$.)

G3. *Let χ be a proper character modulo $\mathfrak{f}$. If $\alpha \in \mathfrak{o}$ is not prime to $\mathfrak{f}$, then $T_\gamma(\chi, \alpha) = 0$. If α is prime to $\mathfrak{f}$, then*

$$|T_\gamma(\chi, \alpha)| = \sqrt{\mathbf{N}\mathfrak{f}}.$$

Proof. Suppose that α is not prime to $\mathfrak{f}$, and write

$$(\alpha) = \mathfrak{c}\mathfrak{g}, \qquad \mathfrak{f} = \mathfrak{f}_1\mathfrak{g}$$

where $\mathfrak{g}$ is the greatest common divisor of (α) and $\mathfrak{f}$. Since χ is proper, there exist elements $\lambda, \mu \in \mathfrak{o}$ prime to $\mathfrak{f}$, with $\lambda \equiv \mu \pmod{\mathfrak{f}_1}$ such that $\chi(\lambda) \neq \chi(\mu)$. Then

$$T_\gamma(\chi, \alpha\lambda) = \bar{\chi}(\lambda)T_\gamma(\chi, \alpha) \qquad \text{and} \qquad T_\gamma(\chi, \alpha\mu) = \bar{\chi}(\mu)T_\gamma(\chi, \alpha).$$

But since $\alpha\lambda \equiv \alpha\mu \pmod{\mathfrak{f}}$, we have $\mathrm{Tr}(x\alpha\lambda\gamma) \equiv \mathrm{Tr}(x\alpha\mu\gamma) \pmod{\mathbf{Z}}$, whence $T_\gamma(\chi, \alpha\lambda) = T_\gamma(\chi, \alpha\mu)$. This is a contradiction, which proves the first assertion.

As for the second, for an arbitrary $z \in \mathfrak{o}$ representing a residue class mod $\mathfrak{f}$, we have

$$T_\gamma(\chi, z)\overline{T_\gamma(\chi, z)} = \sum_{\substack{x,y \bmod \mathfrak{f} \\ \text{prime to } \mathfrak{f}}} \chi(x)\bar{\chi}(y)\, e^{2\pi i \mathrm{Tr}((x-y)z\gamma)},$$

and the left side is 0 if z is not prime to $\mathfrak{f}$. We sum over z in $\mathfrak{o}/\mathfrak{f}$. Then from the left-hand side, we get the value

$$\varphi(\mathfrak{f})|T_\gamma(\chi, 1)|^2 = \varphi(\mathfrak{f})|T_\gamma(\chi, \alpha)|^2$$

where $\varphi(\mathfrak{f})$ is the Euler function, i.e. the order of $(\mathfrak{o}/\mathfrak{f})^*$. On the right-hand side, we consider the sum over z as the inner sum. If $x \equiv y \pmod{\mathfrak{f}}$, then each exponential has the value 1, and hence the sum over z, taken for $x \equiv y \pmod{\mathfrak{f}}$, gives a contribution of

$$\varphi(\mathfrak{f})\mathbf{N}\mathfrak{f},$$

since $\mathbf{N}\mathfrak{f}$ is the order of $\mathfrak{o}/\mathfrak{f}$. On the other hand, for the sum taken over $x \not\equiv y \pmod{\mathfrak{f}}$, we apply **G1** to see that we get 0. This proves **G3**.

§2. AN EXPRESSION FOR THE L-SERIES

Again let k be an imaginary quadratic field with $\mathfrak{o}_k = \mathfrak{o}$ and let $\mathfrak{f}$ be an ideal of $\mathfrak{o}$, $\mathfrak{f} \neq \mathfrak{o}$. Let $G_\mathfrak{f} = I(\mathfrak{f})/P_1(\mathfrak{f})$ be the ray class group, where $I(\mathfrak{f})$ denotes the monoid of ideals prime to $\mathfrak{f}$, and $P_1(\mathfrak{f})$ denotes the subset consisting of those principal ideals (α) such that $\alpha \equiv 1 \pmod{\mathfrak{f}}$. Let χ be a character of $G_\mathfrak{f}$.

We define

$$L_\mathfrak{f}(s, \chi) = \sum_{(\mathfrak{a},\mathfrak{f})=1} \frac{\chi(\mathfrak{a})}{\mathbf{N}\mathfrak{a}^s}.$$

Let $\{\mathfrak{A}\}$ be the elements of the ordinary ideal class group $I/P = G$. For each ordinary ideal class $\mathfrak{A}$, let $\mathfrak{b}_{\mathfrak{A}}$ be an ideal in $\mathfrak{A}^{-1}$, prime to $\mathfrak{f}$. Then for each $\mathfrak{a} \in \mathfrak{A}$ prime to $\mathfrak{f}$, the ideal $\mathfrak{a}\mathfrak{b}_{\mathfrak{A}} = (\xi_{\mathfrak{a}})$ is principal, and the association

$$\mathfrak{a} \mapsto (\xi_{\mathfrak{a}})$$

is a bijection between elements of $\mathfrak{A}$ prime to $\mathfrak{f}$, and non-zero principal subideals of $\mathfrak{b}_{\mathfrak{A}}$ prime to $\mathfrak{f}$. We can write

$$\frac{1}{\mathbf{N}\mathfrak{a}} = \frac{\mathbf{N}\mathfrak{b}_{\mathfrak{A}}}{\mathbf{N}\xi_{\mathfrak{a}}}.$$

Let $\mathfrak{b}_{\mathfrak{A}}(\mathfrak{f})$ be the set of non-zero elements of $\mathfrak{b}_{\mathfrak{A}}$ prime to $\mathfrak{f}$. Then

$$L_{\mathfrak{f}}(s, \chi) = \frac{1}{w}\sum_{\mathfrak{A}} \mathbf{N}\mathfrak{b}_{\mathfrak{A}}^{s}\bar{\chi}(\mathfrak{b}_{\mathfrak{A}}) \sum_{\xi \in \mathfrak{b}_{\mathfrak{A}}(\mathfrak{f})} \frac{\chi(\xi)}{\mathbf{N}\xi^{s}} \tag{1}$$

where w is the number of roots of unity in $\mathfrak{o}$. We follow Siegel [B14] in finding an appropriate transformation of this expression. The map which to each element of $\mathfrak{o}$ associates its principal ideal induces an injection

$$(\mathfrak{o}/\mathfrak{f})^{*} \to I(\mathfrak{f})/P_{1}(\mathfrak{f}),$$

and a character of $G_{\mathfrak{f}}$ therefore induces a character of $(\mathfrak{o}/\mathfrak{f})^{*}$. The value $\chi(\xi)$ in the above expression for the L-series can be therefore viewed either as the value of χ on the principal ideal (ξ), or the value of χ on the residue class of ξ in $(\mathfrak{o}/\mathfrak{f})^{*}$.

Lemma 1. *Let χ be a proper character of $G_{\mathfrak{f}}$. Then*

$$L_{\mathfrak{f}}(s, \chi) = \frac{1}{w_{\mathfrak{f}}T_{\gamma}(\bar{\chi}, 1)} \sum_{R} \bar{\chi}(\mathfrak{b}_{R})\mathbf{N}\mathfrak{b}_{R}^{s} \sum_{\xi \in \mathfrak{b}_{R}} e^{2\pi i \operatorname{Tr}(\xi\gamma)} \frac{1}{\mathbf{N}\xi^{s}}$$

where the sum over R is taken over all ray classes $R \in G_{\mathfrak{f}}$; $\mathfrak{b}_{R}$ is a fixed ideal in R prime to $\mathfrak{f}$; $w_{\mathfrak{f}}$ is the number of root of unity in $\mathfrak{o}$ which are $\equiv 1 \pmod{\mathfrak{f}}$; the $\xi \in \mathfrak{b}_{R}$ are of course $\neq 0$; and γ is chosen as in §1, such that $\gamma\mathfrak{f}\mathfrak{d}$ is integral prime to $\mathfrak{f}$.

Proof. Using **G2** and **G3** of the preceding section, we know that

$$T_{\gamma}(\chi, \xi) = \begin{cases} \bar{\chi}(\xi)T_{\gamma}(\chi, 1) & \text{if } (\xi, \mathfrak{f}) = 1 \\ 0 & \text{if } (\xi, \mathfrak{f}) \neq 1. \end{cases}$$

Therefore from (1) we find

$$\begin{aligned} L_{\mathfrak{f}}(s, \chi) &= \frac{1}{w}\sum_{\mathfrak{A}} \bar{\chi}(\mathfrak{b}_{\mathfrak{A}})\mathbf{N}\mathfrak{b}_{\mathfrak{A}}^{s} \sum_{\xi \in \mathfrak{b}_{\mathfrak{A}}} \frac{1}{\mathbf{N}\xi^{s}} \frac{T_{\gamma}(\bar{\chi}, \xi)}{T_{\gamma}(\bar{\chi}, 1)} \\ &= \frac{1}{wT_{\gamma}(\bar{\chi}, 1)} \sum_{\mathfrak{A}} \bar{\chi}(\mathfrak{b}_{\mathfrak{A}})\mathbf{N}\mathfrak{b}_{\mathfrak{A}}^{s} \sum_{\xi \in \mathfrak{b}_{\mathfrak{A}}} \sum_{z \in (\mathfrak{o}/\mathfrak{f})^{*}} \frac{\bar{\chi}(z)e^{2\pi i \operatorname{Tr}(z\xi\gamma)}}{\mathbf{N}\xi^{s}} \\ &= \frac{1}{wT_{\gamma}(\bar{\chi}, 1)} \sum_{\mathfrak{A}} \sum_{z \in (\mathfrak{o}/\mathfrak{f})^{*}} \bar{\chi}(\mathfrak{b}_{\mathfrak{A}})\bar{\chi}(z)\mathbf{N}\mathfrak{b}_{\mathfrak{A}}^{s} \sum_{\xi \in \mathfrak{b}_{\mathfrak{A}}} \frac{e^{2\pi i \operatorname{Tr}(z\xi\gamma)}}{\mathbf{N}\xi^{s}}. \end{aligned}$$

We now observe that the products $\mathfrak{b}_{\mathfrak{a}} z$ represent all the elements R of

$$I(\mathfrak{f})/P_1(\mathfrak{f}) = G_{\mathfrak{f}}$$

exactly $w/w_{\mathfrak{f}}$ times. One sees this by considering the sequence of subgroups

$$I(\mathfrak{f})/P_1(\mathfrak{f}) \supset P(\mathfrak{f})/P_1(\mathfrak{f}) \supset Z/Z_1(\mathfrak{f})$$

where Z is the group of roots of unity in $\mathfrak{o}$, and $Z_1(\mathfrak{f})$ the subgroup of those which are $\equiv 1 \pmod{\mathfrak{f}}$.

We multiply our expression for the L-series by $\mathbf{N}z^s$ and divide by $\mathbf{N}z^s$, to form $\mathbf{N}(z\xi)^s$. The elements $z\xi$ then range over the ideal $z\mathfrak{b}_{\mathfrak{a}}$, as z ranges over $(\mathfrak{o}/\mathfrak{f})^*$ and ξ ranges over $\mathfrak{b}_{\mathfrak{a}}$, always considering non-zero elements, of course. It is now clear that the expression which we obtain for the L-series is equal to that stated in the lemma.

Remark. We made the assumption that k is imaginary quadratic for simplicity. The same arguments prove an analogous expression for the case of an arbitrary number field, and similarly, there is an analogous expression to that of Theorem 1. These can then be used to deal with quadratic real fields, as in the work of Hecke (cf. Siegel [B14]), or to some extent in general number fields, e.g. [33].

Let R be a ray class in $G_{\mathfrak{f}}$. Let $\mathfrak{b}$ be an ideal prime to $\mathfrak{f}$ in the class. We define.

$$E_{\mathfrak{f}}(R, s) = \mathbf{N}(\mathfrak{b}\mathfrak{d}^{-1}\mathfrak{f}^{-1})^s \sum_{\lambda \in \mathfrak{b}\mathfrak{d}^{-1}\mathfrak{f}^{-1}} e^{2\pi i \mathrm{Tr}(\lambda)} \frac{1}{\mathbf{N}\lambda^s}, \tag{2}$$

(where, in such a sum, it is understood that $\lambda \neq 0$). The notation is justified, i.e. the sum on the right does not depend on the choice of ideal $\mathfrak{b}$ prime to $\mathfrak{f}$ in R. Indeed, if $\mathfrak{a}$ is another such ideal, there exist $\mu, \nu \in \mathfrak{o}$ prime to $\mathfrak{f}$ such that $\mu \equiv \nu \pmod{\mathfrak{f}}$ and $\mu\mathfrak{b} = \nu\mathfrak{a}$. The same argument as in the lemma of Chapter 19, §3 shows that the traces in the exponent corresponding to elements in $\mathfrak{b}\mathfrak{d}^{-1}\mathfrak{f}^{-1}$ or $\mathfrak{a}\mathfrak{d}^{-1}\mathfrak{f}^{-1}$ are congruent mod $\mathbf{Z}$. The multiplicativity of the norm shows that the other terms are also independent of the choice of $\mathfrak{b}$.

Theorem 1. *Let χ be a proper character of $G_{\mathfrak{f}}$. Then*

$$L_{\mathfrak{f}}(s, \chi) = \frac{\chi(\gamma\mathfrak{d}\mathfrak{f})}{w_{\mathfrak{f}} T_{\gamma}(\bar{\chi}, 1)} \sum_{R \in G_{\mathfrak{f}}} \bar{\chi}(R) E_{\mathfrak{f}}(R, s).$$

Proof. In Lemma 1 we make the change of variables $\lambda = \xi\gamma$. Then ξ ranges over $\mathfrak{b}$ as λ ranges over $\gamma\mathfrak{b} = \mathfrak{q}\mathfrak{b}\mathfrak{d}^{-1}\mathfrak{f}^{-1}$ if $\mathfrak{q} = \gamma\mathfrak{d}\mathfrak{f}$. Note that $\chi(\mathfrak{q}) = \chi(\gamma\mathfrak{d}\mathfrak{f})$ makes sense since $\mathfrak{q}$ is prime to $\mathfrak{f}$. Substituting in Lemma 1, we note that $R \mapsto R\mathfrak{q}$ permutes the ray classes, and Theorem 1 drops out at once.

Let $\mathfrak{b}$ be an ideal prime to $\mathfrak{f}$ in the ray class R. Let

$$\mathfrak{b}\mathfrak{d}^{-1}\mathfrak{f}^{-1} = [z_1, z_2],$$

and let

$$\tau_R = z_1/z_2 = x + iy, \qquad y > 0.$$

The elements $\lambda \in \mathfrak{b}\mathfrak{d}^{-1}\mathfrak{f}^{-1}$ can be written $mz_1 + nz_2$ with $(m, n) \neq (0, 0)$. As usual, we have the discriminants

$$D(\mathfrak{b}\mathfrak{d}^{-1}\mathfrak{f}^{-1}) = \begin{vmatrix} z_1 & z_1' \\ z_2 & z_2' \end{vmatrix}^2 = -\mathbf{N}z_2^2(2y)^2$$

and also

$$D(\mathfrak{b}\mathfrak{d}^{-1}\mathfrak{f}^{-1}) = \mathbf{N}(\mathfrak{b}\mathfrak{d}^{-1}\mathfrak{f}^{-1})^2 D(\mathfrak{o}).$$

Taking absolute values yields

$$\mathbf{N}z_2 = \frac{\mathbf{N}(\mathfrak{b}\mathfrak{d}^{-1}\mathfrak{f}^{-1})d_k^{\frac{1}{2}}}{2y}.$$

Since $\mathbf{N}\lambda = \mathbf{N}z_2|m\tau_R + n|^2$, we obtain from the definition of Chapter 20, §5:

$$(3) \qquad E_{\mathfrak{f}}(R, s) = \frac{2^s}{d_k^{s/2}} E_{u,v}(\tau_R, s)$$

$$= \frac{2^s}{d_k^{s/2}} \sum_{(m,n)\neq(0,0)} e^{2\pi i(mu+nv)} \frac{y^s}{|m\tau_R + n|^{2s}}$$

where $u = \operatorname{Tr}(z_1)$ and $v = \operatorname{Tr}(z_2)$.

Using the second Kronecker limit formula now gives us the value at $s = 1$ in terms of the Siegel–Ramachandra invariant.

Theorem 2. *Let k be an imaginary quadratic field and $\mathfrak{f}$ an ideal $\neq \mathfrak{o}_k$. Let R be a ray class modulo $\mathfrak{f}$. Let N be the smallest positive integer contained in $\mathfrak{f}$. Then*

$$E_{\mathfrak{f}}(R, 1) = \frac{-2\pi}{d_k^{\frac{1}{2}}\, 6N} \log |g_{\mathfrak{f}}(R)|,$$

where $\mathfrak{b}$ is any ideal in R prime to $\mathfrak{f}$; and $g_{\mathfrak{f}}(R)$ is the Siegel–Ramachandra invariant. In particular, if χ is a proper character of $G_{\mathfrak{f}}$, then

$$L_{\mathfrak{f}}(1, \chi) = \frac{-2\pi\chi(\gamma\mathfrak{d}\mathfrak{f})}{w_{\mathfrak{f}} T_\gamma(\bar{\chi}, 1) d_k^{\frac{1}{2}}\, 6N} \sum_{R \in G_{\mathfrak{f}}} \bar{\chi}(R) \log |g_{\mathfrak{f}}(R)|.$$

It does not seem to be known if the invariant $g_{\mathfrak{f}}(R)$ generates the ray class field modulo $\mathfrak{f}$. The difficulty lies in the fact that the above theorem applies to a *proper* character χ, whereas one needs an analogous statement for a *non-trivial* character. Precisely, one has the following formal result, due to Ramachandra.

Theorem 3. *For each ray class R modulo $\mathfrak{f}$, suppose given an element $\Psi(R)$ in the ray class modulo $\mathfrak{f}$, satisfying the conditions:*

i) $\Psi(R)^{\sigma(S)} = \Psi(RS)$, *for all* $S \in G_{\mathfrak{f}}$, *and where* $\sigma(S)$ *is the Artin automorphism.*

ii) *For any non-trivial character* χ *of* $G_{\mathfrak{f}}$, *we have*

$$\sum_{R \in G_{\mathfrak{f}}} \chi(R) \log |\Psi(R)| \neq 0.$$

Then $\Psi(R)$ *generates the ray class field modulo* $\mathfrak{f}$.

Proof. It suffices to prove that for any R, $\Psi(R)$ is distinct from all its conjugates. By (i), it suffices to prove this for $R = R_0$, the unit class. Suppose that we have some $S \neq R_0$ such that

$$\Psi(R_0 S) = \Psi(R_0).$$

Then $\Psi(RS) = \Psi(R)$ for all R. Let χ be a character of $G_{\mathfrak{f}}$ which is non-trivial on S, and therefore on the subgroup $\langle S \rangle = \{S^i\}$ generated by S. Let $\{R_j\}$ be representatives of the cosets of $G_{\mathfrak{f}}/\langle S \rangle$. Then

$$\begin{aligned} \sum_{R \in G_{\mathfrak{f}}} \chi(R) \log |\Psi(R)| &= \sum_j \sum_i \chi(R_j S^i) \log |\Psi(R_j S^i)| \\ &= \sum_j \sum_i \chi(R_j)\chi(S^i) \log |\Psi(R_j)| \\ &= 0 \end{aligned}$$

because $\sum_i \chi(S^i) = 0$, a contradiction which proves our theorem.

By taking an appropriate product of invariants $g_{\mathfrak{f}/\mathfrak{g}}$ with $\mathfrak{g} \mid \mathfrak{f}$, Ramachandra constructs such invariants $\Psi(R)$, satisfying the hypothesis of the theorem.

Ramachandra also determines the prime factorization of $g_{\mathfrak{f}}(R)$, showing that if $\mathfrak{f}$ is a prime power, say a power of $\mathfrak{p}$, then

$$g_{\mathfrak{f}}(R) \approx \mathfrak{p}^m$$

for some integer m, and if $\mathfrak{f}$ is not a prime power, then $g_{\mathfrak{f}}(R)$ is a unit. For this, he needs arguments similar to those used in the analogous result for the Delta function, together with the finer results of Hasse (reproduced in Deuring [B1]) concerning the prime powers occurring in such values. We shall omit this, merely pointing out the analogies of these cases, and not forgetting the analogy with the simplest case of roots of unity, where we know that if ζ is a primitive N-th root of unity, then $1 - \zeta$ is a unit if N is not a prime power, and otherwise has the obvious order at p.

Hecke also worked out the value of the L-series at $s = 1$ for real quadratic fields (cf. [B8], [B14]). In this case, there is no transcendental term like the log of a transcendental function, but a rational number which it is interesting to determine explicitly. Similar results should hold for number fields. For precise conjectures, cf. Stark's talk at the International Congress in Nice, 1970. The present chapter may be viewed as giving the reader an introduction to these questions, through the first non-trivial case beyond the cyclotomic case.

Appendices

Elliptic Curves in Characteristic p

The two appendices constitute essentially a fifth part of the book, concentrating on results proper to characteristic p. The first appendix gives the basic formulas describing elliptic curves, in general, by algebraic means. The normal forms are due to Deuring [8]. A convenient, complete, systematic tabulation of them and the automorphisms was given by Tate, whose (unpublished) paper is reproduced here. See also [41].

The second appendix relates the trace of the Frobenius endomorphism with the p-th coefficient in the expansion of a differential of first kind. The three basic techniques involved (the arguments on "formal groups" in §1, the Cartier operation, and the Hasse invariant) are logically independent of each other, and the reader can read them in any order he wishes.

We assume that the reader is acquainted with the basic theory of function fields in one variable, e.g. the Riemann–Roch theorem, used on fields of genus 1.

Appendix 1
by J. Tate
Algebraic Formulas in Arbitrary Characteristic

§1. GENERALIZED WEIERSTRASS FORM

Let K be a field. An **elliptic curve** over K is a connected algebraic curve A smooth and proper over K, of genus 1. An **abelian variety of dimension** 1 **over** K is the same thing as an elliptic curve A over K furnished with a K-rational point, O. Given such an A, there exist functions x and y on A defined over K such that x (resp. y) has a double (resp. triple) pole at O and no other poles. Moreover, if $\omega \neq 0$ is a given differential of first kind on A and $\omega = dt + \cdots$ is its expansion in terms of a uniformizing parameter at O, one can arrange (by multiplying x and y by constants) that $x = t^{-2} + \cdots$ and $y = -t^{-3} + \cdots$. Then in the projective imbedding defined by $3(O)$ the equation for A is of the form

$$y^2 + a_1xy + a_3y = x^3 + a_2x^2 + a_4x + a_6 \tag{1.1}$$

with $a_i \in K$. Homogeneity: y is of weight 3, x of weight 2, and the a_i of weight i, meaning that if we replace ω by $u\omega$, then x is replaced by $u^{-2}x$, y by $u^{-3}y$, etc.

If we are given an equation of the form (1.1), we define associated quantities $b_2, b_4, b_6, b_8, c_4, c_6, \Delta$, and j by the following formulas:

$$\begin{aligned} &b_2 = a_1^2 + 4a_2, \qquad b_4 = a_1a_3 + 2a_4, \qquad b_6 = a_3^2 + 4a_6 \\ &b_8 = a_1^2a_6 - a_1a_3a_4 + 4a_2a_6 + a_2a_3^2 - a_4^2 \end{aligned} \tag{1.2}$$

$$c_4 = b_2^2 - 24b_4 \qquad c_6 = -b_2^3 + 36b_2b_4 - 216b_6 \tag{1.3}$$

$$\Delta = -b_2^2b_8 - 8b_4^3 - 27b_6^2 + 9b_2b_4b_6 \tag{1.4}$$

$$j = \frac{c_4^3}{\Delta} \qquad \text{(if } \Delta \text{ is invertible).} \tag{1.5}$$

These quantities are related by the identities

$$4b_8 = b_2b_6 - b_4^2, \qquad \text{and} \qquad 1728\Delta = c_4^3 - c_6^2. \tag{1.6}$$

If the characteristic is $\neq 2$ or 3 and we put

$$\eta = y + \frac{a_1x + a_3}{2}, \qquad \text{and} \qquad \xi = x + \frac{b_2}{12}, \tag{1.7}$$

then equation (1.1) becomes

$$\eta^2 = x^3 + \frac{b_2}{4}x^2 + \frac{b_4}{2}x + \frac{b_6}{4} = \xi^3 - \frac{c_4}{48}\xi - \frac{c_6}{864}. \tag{1.8}$$

The relation to the classical Weierstrass theory is given by

$$\begin{array}{lll} \xi = \wp(u) & c_4 = 12g_2 & \Delta = g_2^3 - 27g_3^2 \\ 2\eta = \wp'(u) & c_6 = 216g_3 & j = 1728J, \end{array} \tag{1.9}$$

and $\omega = \dfrac{d\xi}{2\eta} = du$ (see below).

Some of the first facts to be proved are summarized by the following theorems:

Theorem 1. *The plane cubic curve* (1.1) *is smooth (and hence defines an abelian variety A of dimension one over K with the point O at infinity as origin) if and only if $\Delta \neq 0$, in which case the differential of first kind ω we started with is given by*

$$\omega = \frac{dx}{2y + a_1x + a_3} = \frac{dx}{F_y} = -\frac{dy}{F_x} = \frac{dy}{3x^2 + 2a_2x + a_4 - a_1y}, \tag{1.10}$$

where

$$F(X, Y) = Y^2 + a_1XY + a_3Y - X^3 - a_2X^2 - a_4X - a_6 \tag{1.11}$$

is the equation of the curve.

Theorem 2. *Let A and A' be two abelian varieties of dimension one over K, given by equations of the form* (1.1), *and let j and j' be their "invariants". Then A and A' are isomorphic over some extension field of K if and only if $j = j'$, in which case they are isomorphic over a separable extension of degree dividing* 24, *and indeed of degree* 2, *if $j \neq 0$ or* 1728.

Theorem 3. *For each $j \in K$, there exists an abelian variety A of dimension one over K with invariant j. Indeed if $j \neq 0$ or* 1728, *such as A is given by the equation*

$$y^2 + xy = x^3 - \frac{36}{j - 1728}x - \frac{1}{j - 1728}, \tag{1.12}$$

for which

$$c_4 = c_6 = \frac{j}{j - 1728} \qquad \textit{and} \qquad \Delta = \frac{j^2}{(j - 1728)^3}.$$

Theorem 4. *The group of automorphisms of an abelian variety of dimension one is finite, of order dividing 24, and if $j \neq 0$ or 1728, it is of order 2, generated by $x \mapsto x$ and $y \mapsto -y - a_1x - a_3$* (i.e., *by $P \mapsto -P$*).

These theorems, and indeed more precise versions of them than we have bothered to state, can be proved by straightforward computations, once one analyzes the most general allowable coordinate change in (1.1). This is done as follows. Suppose A and A' are abelian varieties of dimension one over K, given by equations $y^2 + a_1xy + \cdots$ and $y'^2 + a_1'x'y' + \cdots$, and suppose $f\colon A' \xrightarrow{\sim} A$ is an isomorphism defined over K. Then there are elements $u \in K^*$ and $r, s, t \in K$ such that

$$x \circ f = u^2x' + r \qquad y \circ f = u^3y' + su^2x' + t \qquad \omega \circ f = u^{-1}\omega'. \tag{1.13}$$

The coefficients a_i' are related to the a_i as follows:

$$\begin{aligned}
ua_1' &= a_1 + 2s\\
u^2a_2' &= a_2 - sa_1 + 3r - s^2\\
u^3a_3' &= a_3 + ra_1 + 2t = F_y(r,t)\\
u^4a_4' &= a_4 - sa_3 + 2ra_2 - (t + rs)a_1 + 3r^2 - 2st = -F_x(r,t) - sF_y(r,t)\\
u^6a_6' &= a_6 + ra_4 + r^2a_2 + r^3 - ta_3 - t^2 - rta_1 = -F(r,t).
\end{aligned} \tag{1.14}$$

For the b_i' we have

$$\begin{aligned}
u^2b_2' &= b_2 + 12r\\
u^4b_4' &= b_4 + rb_2 + 6r^2\\
u^6b_6' &= b_6 + 2rb_4 + r^2b_2 + 4r^3\\
u^8b_8' &= b_8 + 3rb_6 + 3r^2b_4 + r^3b_2 + 3r^4.
\end{aligned} \tag{1.15}$$

For the c_i' and Δ one then finds

$$u^4c_4' = c_4 \qquad u^6c_6' = c_6 \qquad u^{12}\Delta' = \Delta. \tag{1.16}$$

Hence $j' = j$ is indeed invariant; $j(A)$ depends only on the isomorphism class of A, not on the particular choice of an equation (1.1) defining A'.

§2. CANONICAL FORMS

Let p be the characteristic of our ground field K. The easy case is $\boldsymbol{p \neq 2, 3}$: Then we can always choose coordinates so that A is given by the equation

$$y^2 = x^3 + a_4x + a_6, \qquad \text{with} \qquad \omega = \frac{dx}{2y}, \tag{2.1}$$

and

(2.2) $$c_4 = -48a_4, \qquad c_6 = -864a_6, \qquad \Delta = -16(4a_4^3 + 27a_6^2).$$

Since any curve of the form (1.1) is smooth at the infinite point 0, such a curve is smooth everywhere if and only if the polynomials F, F_x, and F_y have no common zero. In the case of an equation of the form (2.1) with $p \neq 2$, this condition amounts to the non-existence of a common root of the polynomials $G(X) = x^3 + a_4x + a_6$ and $G'(X) = 3x^2 + a_4$, and since $\Delta = 16 \cdot \text{discr. } G(X)$, the condition in this case is just $\Delta \neq 0$, as claimed in Theorem 1.

Let A and A' be given by equations of the form (2.1) with the same invariant $j = j'$. The isomorphisms $f\colon A' \xrightarrow{\sim} A$ are given simply by

(2.3) $$x \circ f = u^2x' \qquad y \circ f = u^3y',$$

where u is such that $u^4a_4' = a_4$ and $u^6a_6' = a_6$.

Suppose $j \neq 0$, 1728 (i.e. $a_4 \neq 0$, $a_6 \neq 0$). Then A and A' are isomorphic if and only if $a_4a_6'/a_4'a_6$ is a square; the smallest field over which A and A' become isomorphic is the field obtained by adjoining the square root of that quantity to K. The automorphisms of A are given by $u = \pm 1$.

Suppose $j = 1728$ (i.e., $a_6 = 0$). Then A and A' are isomorphic over K if and only if $a_4/a_4' \in (K^*)^4$. The automorphisms of A are given by $u^4 = 1$. A typical curve of this type is given by $y^2 = x^3 - x$.

Suppose $j = 0$ (i.e., $a_4 = 0$). Then $A \cong A'$ over K if and only if $a_6/a_6' \in (K^*)^6$, the automorphisms are given by $u^6 = 1$, and a typical curve is $y^2 = x^3 - 1$.

Now suppose $p = 3$. In this case (and more generally if $p \neq 2$) we can always write A in the form

(2.4) $$y^2 = x^3 + a_2x^2 + a_4x + a_6 = G(x), \qquad \text{say,}$$
$$\omega = -\frac{dx}{y}.$$

Using the fact that $p = 3$, we find

(2.5) $$b_2 = a_2, \qquad b_4 = -a_4, \qquad b_6 = a_6, \qquad b_8 = -a_4^2 + a_2a_6$$
$$c_4 = a_2^2, \qquad c_6 = -a_2^3, \qquad \Delta = a_2^2a_4^2 - a_2^3a_6 - a_4^3.$$

Here again Δ is the discriminant of $G(X)$, up to an invertible factor, so $\Delta \neq 0$ is the condition for smoothness.

Suppose A and A' of form (2.4) with $j = j'$.

Suppose $j \neq 0$ (i.e., $a_2 \neq 0$). Then we can make the term in x disappear, getting the reduced form

(2.6) $$y^2 = x^3 + a_2x^2 + a_6, \qquad \Delta = -a_2^3a_6, \qquad j = -a_2^3/a_6.$$

An isomorphism $f: A' \xrightarrow{\sim} A$ is given by

(2.7) $$x \circ f = u^2 x', \qquad y \circ f = u^3 y'$$

where $u^2 a_2' = a_2$. Hence $A' \simeq A$ if and only if $a_2/a_2' \in (K^*)^2$, and the automorphisms of A correspond to $u = \pm 1$.

Suppose $j = 0$ (i.e., $a_2 = 0$). Reduced form:

(2.8) $$y^2 = x^3 + a_4 x + a_6, \qquad \Delta = -a_4^3, \qquad \omega = \frac{dy}{a_4}.$$

Isomorphisms:

(2.9) $$x \circ f = u^2 x' + r, \qquad y \circ f = u^3 y'$$

with

$$u^4 a_4' = a_4, \qquad u^6 a_6' = a_6 + r a_4 + r^3.$$

Hence A and A' are isomorphic if and only if $(a_4/a_4') \in (K^*)^4$ and $(a_4/a_4')^{\frac{3}{2}} a_6' - a_6$ is of the form $r^2 + r a_4$. This is always so over a separable extension of degree dividing 12. The automorphisms of A are given by the pairs (u, r) such that:

(2.10) $$\begin{aligned} &\text{either} \quad r^3 + a_4 r = 0 \quad &&\text{and} \quad u = \pm 1, \\ &\text{or} \quad r^3 + a_4 r + 2a_6 = 0 \quad &&\text{and} \quad u = \pm i, \end{aligned}$$

where $i^2 = -1$. Over the separable closure of K, they form a group of order 12, the twisted product of C_4 (cyclic group of order 4) and C_3 with C_3 the normal subgroup acted on by elements of C_4 in the unique non-trivial way—conjugation of C_3 by a generator of C_4 is the map carrying elements of C_3 into their inverses.

A typical curve of this type is $y^2 = x^3 - x$, the automorphisms being given by $u^4 = 1$, $r^3 - r = 0$ (i.e., $r \in \mathbf{F}_3$) in this case.

Last case, $p = 2$. Here we have $u a_1' = a_1$ (see 1.14) and $c_4 = b_2^2 = a_1^4$ (see (1.2) and (1.3)). Hence we have $j = 0 \Leftrightarrow a_1 = 0$, and separate cases accordingly.

Suppose $a_1 \neq 0$ (i.e., $j \neq 0$). Then choosing suitably r, s, and t, we can achieve $a_1 = 1$, $a_3 = 0$, $a_4 = 0$. Hence A is given by an equation of the form

(2.11) $$y^2 + xy = x^3 + a_2 x^2 + a_6, \qquad \text{with} \quad \omega = \frac{dx}{x},$$

and

$$b_2 = 1, \quad b_4 = b_6 = 0, \quad b_8 = a_6, \quad c_4 = 1, \quad \Delta = a_6, \quad j = \frac{1}{a_6}.$$

$F_x = y + x^2$, and $F_y = x$ have their only common zero at $x = y = 0$, and this is on the curve if and only if $a_6 = \Delta = 0$. Hence $\Delta \neq 0$ is condition for smoothness.

Isomorphisms:

$$x \circ f = x', \qquad y \circ f = y' + sx'$$

with

$$a_2' = a_2 + s^2 - s, \qquad a_6' = a_6. \tag{2.12}$$

Two curves A and A' with the same j are isomorphic if and only if $a_2' - a_2$ is of the form $s^2 - s$, which is true over a separable extension of K of degree $\leqq 2$. The group of automorphisms of A has two elements, corresponding to $s = 0, 1$. A typical curve is $y^2 + xy = x^3 + (1/j)$.

Suppose $a_1 = 0$ (i.e., $j = 0$). Choosing r suitably we can arrange that $a_2 = 0$, so A is given by

$$y^2 + a_3y = x^3 + a_4x + a_6, \qquad \text{with} \quad \omega = \frac{dx}{a_3}, \tag{2.13}$$

and

$$b_2 = b_4 = 0, \qquad b_6 = a_3^2, \qquad b_8 = a_4^2, \qquad \Delta = a_3^4, \qquad j = 0.$$

Since $F_x = x^2 + a_4$ and $F_y = a_3$, the curve is smooth if and only if $a_3 \neq 0$, i.e., $\Delta \neq 0$. Two curves A and A' with the same j are isomorphic if and only if the following equations are soluble in u, s, and t:

$$\begin{aligned} u^3a_3' &= a_3 \\ u^4a_4' &= a_4 + sa_3 + s^4 \\ u^6a_6' &= a_6 + s^2a_4 + ta_3 + s^6 + t^2. \end{aligned} \tag{2.14}$$

This is always so over a separable extension of K of degree $\leqq 24$. A typical curve of this type is

$$y^2 - y = x^3. \tag{2.15}$$

Its group of automorphisms (over the separable closure of K) is of order 24, the elements corresponding to triples (u, s, t) such that

$$u^3 = 1, \quad s^4 + s = 0, \qquad \text{and} \qquad t^2 + t + s^3 + s^2 = 0.$$

It is isomorphic to the twisted direct product of a cyclic group of order 3 with a quaternion group. The quaternion group is the normal subgroup, and is acted on by the group of order 3 in the obvious way.

§3. EXPANSIONS NEAR O; THE FORMAL GROUP.

Let A be defined by a Weierstrass equation (1.1). Let

$$z = -\frac{x}{y}, \quad w = -\frac{1}{y}, \qquad \text{so} \qquad x = \frac{z}{w}, \quad y = -\frac{1}{w}. \tag{3.1}$$

The equation for A in the affine (z, w)-plane is

$$w = z^3 + a_1zw + a_2z^2w + a_3w^2 + a_4zw^2 + a_6w^3. \tag{3.2}$$

The point O is given by $(z, w) = (0, 0)$, and z is a local parameter at O. From (3.2) we get the formal expansion

$$\begin{aligned} w &= z^3 + a_1z^4 + (a_1^2 + a_2)z^5 + (a_1^3 + 2a_1a_2 + a_3)z^6 + \\ &\qquad (a_1^4 + 3a_1^2a_2 + 3a_1a_3 + a_2^2 + a_4)z^7 + \dots \\ &= z^3(1 + A_1z + A_2z^2 + \dots), \end{aligned} \tag{3.3}$$

where A_n is a polynomial of weight n in the a_i with positive integral coefficients. From (3.3) and (3.1) we get

$$\begin{aligned} x &= z^{-2} - a_1z^{-1} - a_2 - a_3z - (a_4 + a_1a_3)z^2 + \cdots, \\ y &= - z^{-1}x = - z^{-3} + a_1z^{-2} + \cdots, \end{aligned} \tag{3.4}$$

as the formal expansion of x and y. Clearly, the coefficients of these expansions have coefficients in $\mathbf{Z}[a_1, a_2, a_3, a_4, a_6]$. The same is true for the expansion of the invariant differential ω:

$$\omega = H(z)dz \tag{3.5}$$

where $H(z)$ is given by

$$\begin{aligned} H(z) = 1 + a_1z + (a_1^2 + a_2)z^2 + (a_1^3 + 2a_1a_2 + 2a_3)z^3 \\ + (a_1^4 + 3a_1^2a_2 + 6a_1a_3 + a_2^2 + 2a_4)z^4 + \dots \end{aligned}$$

because

$$\begin{aligned} \frac{\omega}{dz} = \frac{dx/dz}{2y + a_1x + a_3} &= \frac{- 2z^{-3} + \dots}{- 2z^{-3} + \dots} \\ = \frac{dy/dz}{3x^2 + 2a_2x + a_4 - a_1y} &= \frac{- 3z^{-4} + \dots}{- 3z^{-4} + \dots} \end{aligned}$$

has coefficients in $\mathbf{Z}[\frac{1}{2}, a_1, \dots, a_6]$, but also in $\mathbf{Z}[\frac{1}{3}, a_1, \dots, a_6]$.

Finally, if $P_3 = P_1 + P_2$ and $P_i = (z_i, w_i)$, then we can express $z_3 = F(z_1, z_2)$ as a formal power series in z_1 and z_2, with coefficients in $\mathbf{Z}[a_1, \dots, a_6]$. The expansion begins

$$\begin{aligned} F(z_1, z_2) = z_1 + z_2 - a_1z_1z_2 - a_2(z_1^2z_2 + z_1z_2^2) \\ - 2a_3(z_1^3z_2 + z_1z_2^3) + (a_1a_2 - 3a_3)z_1^2z_2^2 + \dots. \end{aligned} \tag{3.6}$$

This is the "formal group on one parameter" associated with A.

For each integer $n \geqq 1$ we have, formally,

$$z(nP) = \psi_n(z(P)), \tag{3.7}$$

where the series ψ_n are defined inductively by

$$\psi_1(z) = z, \qquad \psi_{n+1}(z) = F(z, \psi_n(z)). \tag{3.8}$$

For example, we have

(3.9) $$\psi_2(z) = 2z - a_1z^2 - 2a_2z^3 + (a_1a_2 - 7a_3)z^4 + \cdots$$

and

(3.10) $$\psi_3(z) = 3z - 3a_1z^2 + (a_1 - 8a_2)z^3 + 3(4a_1a_2 - 13a_3)z^4 + \cdots.$$

In characteristic $p > 0$, the series ψ_p is of the form

$$\psi_p(z) = c_1z^{p^h} + c_2z^{2p^h} + c_3z^{3p^h} + \dots$$

with $c_1 \neq 0$, where h is an integer equal to 1 or 2, because the isogeny

$$p\delta : A \to A$$

is of degree p^2, and is not separable. This means that $z \circ p\delta$ lies in the inseparable subfield of degree p or p^2 of the function field of A, whence our assertion follows.

EXERCISE

Let $p = \text{char}(K)$ be arbitrary, let $j \in K$ with $j \neq 0$ or 1728, and let A_j denote the abelian variety of dimension 1 over K given by the equation (1.12), i.e.,

$$y^2 + xy = x^3 - \frac{36}{j - 1728}x - \frac{1}{j - 1728}.$$

Show that for each separable quadratic extension L of K there exists an abelian variety $A_{j,L}$ of dimension one over K such that $A_{j,L}$ is isomorphic to A_j over L, but not over K, and $A_{j,L}$ is uniquely determined up to isomorphism by j and L. Show also that (denoting by $A(K)$ the group of points on A rational over K) we have

$$A_{j,L}(K) = \{P \in A_j(L) | \sigma P = -P\},$$

where σ is the non-trivial automorphism of L/K, (and where

$$-P = (x, -y - a_1x - a_3) \quad \text{if} \quad P = (x, y)).$$

Appendix 2
The Trace of Frobenius and the Differential of First Kind

§1. THE TRACE OF FROBENIUS

Theorem 1. *Let A be an elliptic curve defined over the prime field $\mathbf{F}_p$ of characteristic p, let t be a local parameter at the origin in the function field $\mathbf{F}_p(A)$. Let ω be a differential of first kind in $\mathbf{F}_p(A)$, with expansion*

$$\omega = \sum_{\nu=1}^{\infty} c_\nu t^\nu \frac{dt}{t}$$

normalized such that $c_1 = 1$. Let $\pi = \pi_p$ be the Frobenius endomorphism of A. Then

$$\omega \circ \pi' = c_p \omega, \quad \text{and} \quad t \circ (p\delta) \equiv c_p t^p \pmod{t^{2p}}.$$

Proof. We lift an equation for the elliptic curve to the integers. Thus it is useful to write $\bar{A}$ for the curve in characteristic p, and A for its lifting. We do this in a naive way, by lifting the coefficients in a Weierstrass equation if $p \neq 2, 3$, or in a normalized equation otherwise. We let $\bar{t}$ be the parameter at the origin $\bar{O}$, and let t be a parameter at the origin O of A, reducing to $\bar{t}$. Then

$$\bar{\omega} = \sum_{\nu=1}^{\infty} \bar{c}_\nu \bar{t}^\nu \frac{d\bar{t}}{\bar{t}},$$

and the differential form ω on A has the expansion

$$\omega = \sum_{\nu=1}^{\infty} c_\nu t^\nu \frac{dt}{t} = h(t)dt$$

with $c_1 \equiv 1 \pmod{p}$.

On the one hand, we have

$$\omega \circ (p\delta) = p\omega = ph(t)dt.$$

Let $\mathfrak{o} = \mathbf{Z}_{(p)}$ be the local ring of $\mathbf{Z}$ at p. There are power series $U(t), V(t) \in \mathfrak{o}[[t]]$ such that

$$t \circ (p\delta) = U(t^p) + pV(t).$$

So on the other hand, we find

$$ph(t) = h(U(t^p) + pV(t))(U'(t^p)pt^{p-1} + pV'(t)). \tag{1}$$

Let $\pi + \pi' = f_p$. Since

$$\bar{t} \circ \pi' = f_p \bar{t} + \cdots \qquad \text{and} \qquad \bar{t} \circ \pi = \bar{t}^p,$$

and since $\pi\pi' = p\bar{\delta}$, we see that

$$U(t^p) \equiv g_p t^p \pmod{t^{2p}}, \qquad \text{with} \qquad g_p \equiv f_p \pmod{p}. \tag{2}$$

We divide (1) by p. We then read it mod p, as well as mod t^p, and look at the coefficient of t^{p-1}. The term $h(U(t^p) + V(t))$ is then congruent to 1. The constant term of $U'(t^p)$ is g_p, and $V'(t)$ has no term of degree $p - 1$. Comparing coefficients of t^{p-1}, we find the desired congruence

$$c_p \equiv f_p \pmod{p}.$$

This proves our theorem.

The above proof is due to Tate, and generalizes to formal groups. The reader will find another proof using the Weierstrass normal form in Manin [30].

All further sections of this appendix take place in characteristic p. Whereas in the first section, we considered a reduced elliptic curve over the prime field, we now work quite generally with any elliptic curve in characteristic p.

§2. DUALITY

Let K be the function field of an elliptic curve in characteristic p, over an algebraically closed constant field k_0. Let $\{P\}$ range over the points of A in k_0 (or in other words, the places of K over k_0). We let K_P denote the completion at P. An adele ξ of K is an element of the cartesian product $\prod K_P$, such that the component ξ_P is P-integral for almost all P. The group of adeles is denoted by $\mathbf{A}$. There is a pairing between differential forms of K and adeles, given by

$$(\omega, \xi) \mapsto \langle \omega, \xi \rangle = \sum_P \operatorname{res}_P(\xi_P \omega).$$

***Theorem* 2.** *Let ω be a differential of the first kind in K. Let Q be an arbitrary point of A, let t be a local parameter at Q, and let ω have the expansion*

$$\omega = \sum_{\nu=1}^{\infty} c_\nu t^\nu \frac{dt}{t},$$

normalized so that $c_1 = 1$. Then for any adele ξ, we have

$$\langle \omega, \xi^p \rangle = c_p^p \langle \omega, \xi \rangle^p.$$

Proof. We assume that the reader is acquainted with Weil's proof of the Riemann–Roch theorem (given in the books on algebraic functions by Artin, Chevalley, or Lang). We let $\mathbf{A}(0)$ be the group of integral adeles (i.e. adeles ξ such that ξ_P is P-integral for all P). Weil's proof of the Riemann–Roch theorem shows among other things that

$$[\mathbf{A} : \mathbf{A}(0) + K] = 1,$$

where the brackets mean dimension of the factor space $\mathbf{A}/(\mathbf{A}(0) + K)$ over the constant field k_0. Therefore the adele

$$\eta = (\ldots, 0, 1/t, 0, \ldots)$$

having 0 at all components except Q, where $\eta_Q = 1/t$, generates this factor space. Since ω is of the first kind, both sides of the formula in the theorem are equal to 0 when ξ lies in $\mathbf{A}(0) + K$. Furthermore, both sides are p-power linear with respect to constants. Hence it suffices to prove the formula when $\xi = \eta$. But in this case, the formula is obvious.

§3. THE TATE TRACE

This section is preliminary to the next section on the Cartier operator, and gives lemmas on purely inseparable extensions of degree p. Let K be a field of characteristic p, and let x be algebraic, purely inseparable over K, so that x^p is an element of K, but $x \notin K$. An element of $K(x)$ can be written uniquely in the form

$$y = y_0 + y_1 x + \cdots + y_{p-1} x^{p-1}, \qquad y_i \in K.$$

We define a substitute for the trace by letting

$$S_x(y) = y_{p-1},$$

and derive properties of S_x as in Tate [42].

We note first that for $0 \leqq i \leqq p - 1$, we have

$$y_i = S_x(yx^{p-1-i}),$$

whence

$$y = \sum_{i=0}^{p-1} S_x(yx^{p-1-i}).$$

Furthermore, S_x is K-linear, and hence linear with respect to p-th powers in $K(x)$.

If $f(X)$ is a polynomial in a variable X over K, we let as usual $f'(X)$ be its formal derivative. Then the map

$$f(x) \mapsto f'(x)$$

is immediately verified to be well defined, because if $f(x) = 0$, then $f(X)$ is divisible by $X^p - a$, with $a = x^p$. Hence $f'(x) = 0$. It follows at once that this map is a derivation of $K(x)$, and is the unique derivation trivial on K, mapping x onto 1. We denote this derivation by D_x. If an element $y \in K(x)$ is expressed as above, then

$$D_x(y) = y_1 + 2y_2x + \cdots + (p-1)y_{p-1}x^{p-2}.$$

A power x^i $(0 \leqq i \leqq p-2)$ can be "integrated", and we see that an element $y \in K(x)$ can be written in the form $y = D_x z$ for some $z \in K(x)$ if and only if $y_{p-1} = 0$. We have the following properties, the first of which is immediate.

S1. $S_x D_x = 0$.

S2. $S_x(y^{p-1}D_x y) = (D_x y)^p$, *or equivalently*, $S_x(D_x y/y) = (D_x y/y)^p$.

Proof. Let R be the set of elements y in $K(x)$ for which **S2** is true. We observe that R is the kernel of the additive map

$$y \mapsto S_x(D_x y/y) - (D_x y/y)^p.$$

The non-zero elements of R form a multiplicative group. Furthermore, if $y \in R$, then $y + 1 \in R$, because $(y+1)^{p-1}D_x y - y^{p-1}D_x y$ consists of terms which can be integrated, so that

$$S_x((y+1)^{p-1}D_x y) = S_x(y^p D_x y),$$

and $D_x y = D_x(y+1)$. Finally, if $y, z \in R$ and $z = 0$, then

$$y + z = z(z^{-1} + 1) \in R.$$

Therefore R is a field containing K and x, thereby proving our assertion.

S3. *Let* $K(x) = K(w)$. *Then* $S_w(z) = S_x(z(D_x w)^{1-p})$ *for all* $z \in K$, *or in other words,*

$$S_x(zD_x w) = S_w(z)(D_x w)^p.$$

Proof. Both sides of the formula are K-linear with respect to the variable z. Hence it suffices to prove the formula when $z = w^i$, and $0 \leqq i \leqq p-1$, or equivalently, it suffices to prove

$$(D_x w)^p S_w(w^i) = S_x(w^i D_x w).$$

If $i < p - 1$, then $w^i D_x w = D(w^{i+1}/(i+1))$ (i.e. $w^i D_x w$ can be integrated), and both sides are equal to 0. If $i = p - 1$, then the left-hand side is equal to $(D_x w)^p$, which is equal to the right-hand side by **S2**. This proves our property.

In the next section, we interpret **S3** more naturally in terms of differential forms.

§4. THE CARTIER OPERATOR

Let k_0 be a perfect field of characteristic $p > 0$ and let $k_0(t)$ be a purely transcendental extension in one variable t. Then $k_0(t)^{1/p} = k_0(t^{1/p})$. Similarly, as we have already seen in Chapter IX, §4, if K is a function field in one variable over k, then K has a unique purely inseparable extension of degree p, namely $K^{1/p}$. Looking at this in another way, we see that K^p is the unique subfield of K over which K is purely inseparable of degree p. If x is an element of K such that $x \notin K^p$, then $K = K^p(x)$.

Let $x \in K$. We denote by dx the functional on derivations D of K, trivial on k_0, given by the pairing

$$(dx, D) \mapsto Dx.$$

A differential form $\omega = ydx$ is therefore the functional whose value at D is yDx, also denoted by $\langle \omega, D \rangle$. If K is a function field of one variable over the perfect constant field k_0, and if $x \notin K^p$, then there exists a unique derivation $D = D_x$ of K, trivial on k_0, such that $Dx = 1$. An arbitrary differential form of K can then be written as ydx for some $y \in K$, or in other words

$$\omega = (y_0^p + y_1^p x + \cdots + y_{p-1}^p x^{p-1})\, dx. \qquad y_i \in K.$$

We define the **Cartier operator** C on differential forms by letting

$$C\omega = y_{p-1}\, dx.$$

In terms of the Tate trace, this is merely

$$C(ydx) = S_x(y)^{1/p}\, dx.$$

Formula **S3** shows that this value $C\omega$ is independent of the representation of the differential form, i.e. we get the same value if we write the form as zdw for some $w \notin K^p$.

The Cartier operator is obviously additive, and it is linear with respect to the prime field. The following properties will be immediate from what we already know, and the definitions. Let $z \in K$ be arbitrary.

C1. $$C(z^p\omega) = zC\omega.$$

C2. $$C(dz) = 0.$$

C3. $$C\left(\frac{dz}{z}\right) = \frac{dz}{z}.$$

C4. $$C(z^{p-1}\,dz) = dz.$$

C5. $$C(z^{n-1}\,dz) = 0, \qquad \text{if} \quad (n, p) = 1,$$

The first property is obvious since S_x is K^p-linear. If $z \in K^p$, then $dz = 0$, and if $z \notin K^p$, then we apply the definition of the Cartier operator to the forms dz or $z^{p-1}\,dz$ directly, substituting z for x in the definition, in order to obtain **C2** and **C4.** Property **C3** then follows from **C4** and **C1**, while **C5** follows from the fact that $z^{n-1}\,dz = d(z^n/n)$, and **C2.**

It is useful to decompose a differential form $\omega = ydx$ as a sum

$$\boxed{\omega = df + g^p \frac{dx}{x}}$$

with some elements $f, g \in K$. The existence of such a decomposition is obvious since terms $y_i^p x^i$ with $0 \leqq i \leqq p - 2$ can be integrated. The uniqueness is equally clear. When ω is so written, then

$$\boxed{C\omega = g\frac{dx}{x}.}$$

C6. *If ω is regular at a place of K over k_0, then $C\omega$ is also regular at this place.*

Proof. We can take for x a local parameter at the given place. In the expression $\omega = ydx$, all the coefficients y_i^p must then also be regular at x, for otherwise, $y_i^p x^i$ has a pole of order $mp - i$ for some integer $m_i \geqq 0$, and there cannot be any cancellation of such poles among $y_0^p + \cdots + y_{p-1}^p x^{p-1}$.

We observe that if we make a constant field extension of our function field, then the definition of the Cartier operator remains the same, and we may assume without loss of generality that the constant field k_0 is algebraically closed.

C7. *Let P be a place of K over k_0. Then*

$$\operatorname{res}_P C\omega = (\operatorname{res}_P \omega)^{1/p}.$$

Proof. We select x to be a local parameter at the place. We then write

$$\omega = df + g^p x^p \frac{dx}{x}$$

so that $C\omega = g dx$. Expanding g in powers of x, say

$$g = c_{-m}x^{-m} + \cdots + c_0 + c_1 x + \cdots,$$

we see that $\operatorname{res}_P \omega = c_{-1}^p$. Taking the p-th root yields precisely $\operatorname{res}_P C\omega$.

In terms of the duality between differential forms and adeles, **C7** can be expressed by the formula

$$\boxed{\langle C\omega, \xi\rangle^p = \langle \omega, \xi^p\rangle,}$$

taking **C1** into account.

Theorem 3. *Let K be the function field of a curve of genus* 1 *(an elliptic curve) over an algebraically closed constant field k_0 of characteristic p. Let ω be a differential of the first kind in K. Let x be a local parameter in K for some place of K, and expand*

$$\omega = \sum_{n=1}^{\infty} c_n x^n \frac{dx}{x},$$

with $c_n \in k_0$. Then $c_1 \neq 0$, and if we normalize ω so that $c_1 = 1$, then $C\omega = c_p \omega$.

Proof. By **C6** we know that $C\omega = c\omega$ for some constant c. On the other hand, the Cartier operator is clearly continuous for the topology induced on K by the discrete valuation arising from the place, and consequently by **C4** and **C5** we find

$$C\omega = \sum c_{np} x^n \frac{dx}{x}.$$

This yields $cc_1 = c_p$. We cannot have $c_1 = 0$, for otherwise the differential of first kind would have a zero at the place, whence would have a zero at every place since it is invariant under translations. This proves our theorem.

The same argument also gives the relations

$$c_{np} = c_p c_n.$$

Atkin and Swinnerton-Dyer had found such congruence relations, and conjectured higher ones. Serre observed that applying the Cartier operator could be used for a proof. For the higher ones, cf. Cartier's talk at the International Congress of Mathematicians, 1970, Tome 2, pp. 291–299.

Theorem 4. *Let K be a function field in one variable over an algebraically closed constant field of characteristic p, and let ω be a non-zero differential form in K. Then:*

i) *We have $C\omega = 0$ if and only if there exists $z \in K$ such that $\omega = dz$.*
ii) *We have $C\omega = \omega$ if and only if there exists $z \in K$ such that $\omega = dz/z$.*

Proof. The above two statements amount to the converses of properties **C2** and **C3**. As to the first, if $C\omega = 0$, then from the decomposition

$$\omega = df + g^p\, dx/x,$$

we conclude that $g = 0$, whence $\omega = df$. The second is somewhat harder to prove, and amounts to showing that there is some $z \in K$ such that for any derivation D of K over the constants, we have $\langle \omega, D \rangle = Dz/z$. If $\omega = ydx$, it suffices to prove this relation for $D = D_x$, and our problem amounts to showing that the element yDx of K is a logarithmic derivative. To show this about an element $w \in K$, it suffices to prove that there exists an element $z \in K$ such that

$$(w + D)z = 0,$$

because in that case, $wz + Dz = 0$ and

$$w = -\frac{Dz}{z} = \frac{Dz^{p-1}}{z^{p-1}}.$$

If $w \in K$, we denote by $L(w)$ the linear map equal to multiplication by w.

Lemma. *Let K be a field of characteristic p, let D be a derivation of K, and $w \in K$. Then*

$$(L(w) + D)^p = L(w)^p + D^p + L(D^{p-1}w).$$

Before proving the lemma, we show how it implies the second part of Theorem 4. We set $w = yDx$ with our previous notation, and $D = D_x$. Then $D_x^p = 0$. From the decomposition $w = df + g^p dx/x$, we see at once from the definitions that

$$\langle C\omega, D \rangle^p = -D^{p-1}\langle \omega, D \rangle.$$

From the hypothesis $C\omega = \omega$, it follows that $a^p = -D^{p-1}a$, whence

$$(L(w) + D)^p = 0.$$

This proves what we wanted.

There remains for us to prove the lemma. Let u, v be elements of a ring (not necessarily commutative), of characteristic p. We let $L = L(u)$ and $R = R(u)$ be left and right multiplication by u respectively. Then L and R commute, and

$$(L - R)^{p-1} = \sum_{i=0}^{p-1} L^i R^{p-i},$$

as one sees by using the geometric series formally on $(L - R)^p/(L - R)$, say. If t is a new variable, we have

$$(tu + v)^p = t^p u^p + v^p + \sum_{i=1}^{p-1} c_i(u, v)t^i,$$

with appropriate coefficients $c_i(u, v)$.

Replacing t by $t + h$, expanding out, and looking at the coefficient of h (i.e. differentiating with respect to t), we obtain

$$\sum_{i=0}^{p-1} (tu + v)^i u(tu + v)^{p-1-i} = \sum_{i=1}^{p-1} ic_i(u, v)t^{i-1}.$$

Writing $Ad\,u$ for the operator such that

$$(Ad\,u)(v) = uv - vu = (L - R)(v),$$

we now see that

$$(Ad(tu + v))^{p-1}(u) = \sum_{i=1}^{p-1} ic_i(u, v)t^{i-1}$$

In the ring of endomorphisms of K (as additive group), substitute $u = L(w)$ and $v = D$. From the formula

$$[L(w) + D, L(z)] = L(Dz),$$

we see that

$$(Ad(tL(w) + D))^{p-1}(L(w)) = L(D^{p-1}w),$$

and in particular that this expression is independent of t. This implies that

$$c_i(L(w), D) = 0, \qquad \text{for} \quad i > 1.$$

Finally, putting $t = 1$, we obtain for $u = L(w)$, $v = D$,

$$(u + v)^p = u^p + v^p + c_1(u, v),$$

and we have just seen that $c_1(u, v) = L(D^{p-1}w)$. This proves the lemma, and thus also concludes the proof of the theorem.

In this section, we essentially followed Cartier's paper "Sur la rationalité des diviseurs en geométrie algébrique," *Bulletin Soc. Math. France* (1958), pp. 177–251.

Let A be an elliptic curve defined over the prime field $\mathbf{F}_p$. If A and its Frobenius endomorphism are obtained by reduction mod p as in §1, then we now see that $c_p = \overline{f_p}$, and hence the residue class of f_p can be determined from the local expansion of ω at any point. If one wishes to avoid reduction mod p, one can use the discussion of Hasse invariants in the next section instead. In any case, we see that for A defined over $\mathbf{F}_p$, we have

$$\boxed{C\omega = \omega \circ \pi'}$$

The Cartier operator is the transpose of Frobenius.

In particular, we have $C\omega = 0$ if and only if π' is purely inseparable, which means that the curve is supersingular, i.e. has no point of order p.

On the other hand, suppose that the curve is not supersingular. Then c_p is the reduction of an ordinary integer ν mod p, with $1 \leqq \nu \leqq p - 1$. Write $\nu = b^{1-p}$ for some constant b. Then the basic formalism of the Cartier operator shows that

$$C(b\omega) = b\omega.$$

In other words, we can normalize the differential of first kind so that it is fixed by the Cartier operator. We now see that the two cases of Theorem 4 correspond to the singular and supersingular cases respectively, i.e. *the differential of first kind (suitably normalized) is logarithmic or exact according as the elliptic curve is singular or supersingular.*

5. THE HASSE INVARIANT

In this section, we follow Hasse [18] *and Hasse-Witt* [20].

Let k_0 be an algebraically closed field of characteristic p, and let K be the function field of an elliptic curve A over k_0. In other words, K is a function field in one variable, of genus 1. *We fix a point Q of A in k_0 (i.e. a place of K over k_0), and we let t be a local parameter of Q in K. If $\mathfrak{a}$ is a divisor of K, we let $\mathscr{L}(\mathfrak{a})$ be the k_0-vector space of functions $z \in K$ such that $(z) \geqq -\mathfrak{a}$. In particular, $\mathscr{L}(pQ)$ is the vector space of functions having at most a pole of order p at Q.*

By the Riemann–Roch theorem, for any positive integer m, the space $\mathscr{L}(mQ)$ has dimension m. Again by the Riemann–Roch theorem, for each $m \geqq 2$, there exists a function in $\mathscr{L}(mQ)$ having a pole of order exactly m at Q, and consequently there exists a function x_m whose expansion at Q (as a power series in t) is like

$$x_m \equiv \frac{1}{t^m} \quad \left(\mathrm{mod}\ \frac{1}{t}\right).$$

In particular, there exists a function $y \in \mathscr{L}(pQ)$ such that

$$y = \frac{1}{t^p} - \frac{a}{t} + \cdots$$

at Q, with some constant a. Since the difference of two such functions has at most a pole of order 1 at Q and no other pole, it must be constant, and we see that y is uniquely determined modulo constants. We call such a function y a **Hasse function** of K (or on A). The constant a is uniquely determined by the choice of parameter t.

***Theorem* 5.** *Let ω be a differential of first kind on K, with expansion at Q given by*

$$\omega = \sum_{\nu=1}^{\infty} c_\nu t^\nu \frac{dt}{t},$$

normalized such that $c_1 = 1$. Let $-a$ be the residue of the Hasse function y as above, with respect to the parameter t. Then $a = c_p$.

Proof. The only possible residue of $y\omega$ is at Q, and is equal to $c_p - a$. It is also equal to 0, whence our theorem follows.

The constant c_p is called the **Hasse invariant** at Q with respect to t. If we change the parameter t by a constant factor b, then c_p changes to $c_p b^{1-p}$.

The Hasse invariant arising in the above fashion is related directly to the existence of points of order p on A. If such a point exists, then the isogeny $p\delta$ breaks up into a separable part of degree p, and a purely inseparable part of degree p. The separable part is unramified, and hence A has an unramified covering of degree p (i.e. K has an unramified extension of degree p). Conversely, if such an unramified extension of K exists, then it has genus 1 (say by the Hurwitz genus formula). Let $\lambda\colon B \to A$ be the corresponding covering of elliptic curves, normalized so that $\lambda(O) = O$. Then λ is a homomorphism with kernel of order p. Indeed, if P_1, P_2 are points of B, then the divisor

$$(P_1) + (P_2) - (P_1 + P_2) - (O)$$

is the divisor of a function on B, whence its image

$$(\lambda P_1) + (\lambda P_2) - (\lambda(P_1 + P_2)) - (O)$$

is the divisor of a function on A (the norm, as is clear from elementary valuation theory). Hence by the Riemann–Roch theorem on A, we get

$$\lambda(P_1 + P_2) = \lambda P_1 + \lambda P_2,$$

and λ is a homomorphism.

[Actually, the fact just proved follows from very general properties of abelian varieties due to Weil, that any rational map of one abelian variety into another is a homomorphism followed by a translation.]

It is clear that the kernel of our homomorphism λ has order p.

The Hasse function will give us a natural way of constructing unramified extensions when the Hasse invariant is $\neq 0$.

For each place P of K we again let K_P be the completion of K (isomorphic to the power series field over k_0, in a local parameter at P). By additive Kummer theory in characteristic p (Artin–Schreier theory) a cyclic extension of K of degree p is obtained by adjoining the roots of a polynomial $X^p - X - z$, with $z \in K$. We let

$$\wp X = X^p - X,$$

and write $\wp^{-1}z$ for any root. An extension of K is unramified at P if and only if it splits completely at P (because we took k_0 algebraically closed), and hence it is unramified at P if and only if $z \in \wp K_P$. Let

$$U = \bigcap_P (\wp K_P \cap K).$$

Then $U \supset \wp K$, and the unramified extensions of K are precisely those obtained by adjoining $\wp$-th roots of elements of U to K. (In fact, $U/\wp K$ is dual to the

Galois group of the maximal unramified extension of K of exponent p, but we won't need this.)

Theorem 6. *The additive group $\mathscr{L}(pQ) \cap \wp K_Q$ is contained in U, and the inclusion induces an isomorphism*

$$(\mathscr{L}(pQ) \cap \wp K_Q)/k_0 \approx U/\wp K.$$

Proof. If an element z of K is integral at P, then a root α of $X^p - X - z = f(X)$ is unramified at P because $f'(\alpha) = -1$ is a unit. This proves the desired inclusion relation. If $z \in \mathscr{L}(pQ) \cap \wp K$, then there exists $x \in K$ such that $z = x^p - x$. Hence x is P-integral for all $P \neq Q$. If x has a pole at Q, then this pole has at most order 1, and hence x is constant, whence z is constant. This proves that the homomorphism of $\mathscr{L}(pQ) \cap K_Q$ into $U/\wp K$ has kernel equal to the constant field k_0. Finally, given an element $z \in U$, we wish to prove that there exists an element $w \in \mathscr{L}(pQ)$ such that $z \equiv w \pmod{\wp K}$. First take $P \neq Q$. Since $z \in \wp K_P$, if z is not integral at P, then z has a pole of order pm at P for some positive integer m, say

$$z = \frac{a}{u^{pm}} + \cdots,$$

where u is a local parameter at P. By the Riemann–Roch theorem, there exists $x \in \mathscr{L}(mP + nQ)$ for some large n such that

$$x = \frac{a^{1/p}}{u^m} + \cdots.$$

Then $z - \wp x$ has a pole of smaller order than z at P. After repeating the above procedure, we may assume without loss of generality that z has a pole only at Q. Since $z \in \wp K_Q$, it follows that z has an expansion of the form

$$z = \frac{b}{t^{pm}} + \cdots$$

at Q, with some positive integer m, and a constant b. If $m = 1$, we are done. If $m > 1$, there exists an element $x \in \mathscr{L}(mQ)$ such that

$$x = \frac{b^{1/p}}{t^m} + \cdots.$$

Hence $z - \wp x$ has a lower order pole at Q than z. Again inductively, we can finally achieve that $z \in \mathscr{L}(pQ)$. This proves our theorem.

Theorem 7. *The following conditions are equivalent:*

i) *The space $\mathscr{L}(pQ) \cap \wp K_Q$ is equal to the constant field.*
ii) *There exists no cyclic unramified extension of K of degree p.*
iii) *The Hasse invariant at Q is equal to* 0.

Proof. By Theorem 6, the cyclic unramified extensions of K are obtained by $\wp$-th roots of elements of $\mathscr{L}(pQ) \cap \wp K_Q$. Hence i) implies ii). If the Hasse invariant a is not 0, let b be a constant such that $b^{1-p} = a$, and let $z = b^p y$, where y is the Hasse function,

$$y = \frac{1}{t^p} - \frac{a}{t} + \cdots.$$

Then $z \in \mathscr{L}(pQ)$, and also $z \in \wp K_Q$, because

$$\wp k_0[[t]] = k_0[[t]],$$

i.e. every equation $X^p - X - v = 0$ with a power series $v \in k_0[[t]]$ has a root in $k_0[[t]]$. A $\wp$-th root of z generates an unramified extension of degree p, by Theorem 6, thus proving that ii) implies iii). Finally, assume iii). If $\mathscr{L}(pQ) \cap \wp K_Q$ contains a non-constant z, then z can be written $x^p - x$ with some $x \in \wp K_Q$. Expanding x as a power series in t, we see that z has an expansion

$$z = \frac{b^p}{t^p} - \frac{b}{t} + \cdots$$

with some constant $b \neq 0$. Dividing by b^p yields the Hasse function, and shows that the Hasse invariant is not 0, thus proving our theorem.

The above arguments also prove:

Theorem 8. *Assume that the Hasse invariant is not* 0. *Then modulo constants, there exists a unique non-zero function*

$$z \in \mathscr{L}(pQ) \cap \wp K_Q.$$

This function has the expansion

$$z = \frac{b^p}{t^p} - \frac{b}{t} + \cdots$$

for some constant b, and the cyclic unramified extension of K of degree p is equal to $K(\wp^{-1}z)$.

Over the prime field, we may now summarize the results obtained, identifying possible definitions of the element c_p.

Theorem 9. *Let A be an elliptic curve defined over the prime field $\mathbf{F}_p$. Let ω be a differential of the first kind in the function field $\mathbf{F}_p(A)$. Let Q be a rational point of A in $\mathbf{F}_p$, and t a local parameter at Q in $\mathbf{F}_p(A)$. Let*

$$\omega = \sum_{\nu=1}^{\infty} c_\nu t^\nu \frac{dt}{t},$$

normalized so that $c_1 = 1$. Let $\pi = \pi_p$ be the Frobenius endomorphism of A over $\mathbf{F}_p$, and let C be the Cartier operator. Then:

$$C\omega = \omega \circ \pi' = c_p \omega.$$

If $y \in \mathscr{L}(pQ)$ has the expansion

$$y = \frac{1}{t^p} - \frac{a}{t} + \cdots,$$

then $a = c_p$. The curve A is supersingular if and only if $c_p = 0$. We also have the expansion

$$t \circ (p\delta) \equiv c_p t^p \pmod{t^{2p}}.$$

The information in Theorem 9 puts together Theorem 1 of §1, Theorem 3 of §2, and Theorem 5 of §5, which relate the various possible definitions of the Hasse invariant.

Bibliography

BOOKS AND MONOGRAPHS

[B1] M. Deuring, "Die Klassenkörper der Komplexen Multiplikation," *Enzyklopädie der Math. Wiss.* Band I, 2. Teil, Heft 10, Teil II.

[B2] R. Fricke, *Die Elliptischen Funktionen und ihre Anwendungen*, Vol. One (1916) and Vol. Two (1922), Teubner, Leipzig, Berlin.

[B3] ——, *Analytisch-Funktionentheoretische Vorlesungen*, Teubner Verlag, Leipzig, 1900.

[B4] R. Fricke and F. Klein, *Vorlesungen über die Theorie der Automorphen Funktionen I, II*, Teubner Verlag, 1897 and 1912.

[B5] R. Fueter, *Vorlesungen über die Singulären Moduln und die Komplexe Multiplikation der Elliptischen Funktionen*, Teubner, Leipzig, Berlin, 1924.

[B6] Y. Ihara, *On Congruence Monodromy Problems*, Vols. I and II, Chapter V, University of Tokyo, 1968.

[B7] S. Lang, *Algebraic Number Theory*, Addison Wesley, Reading, Mass., 1970.

[B8] C. Meyer, *Die Berechnung der Klassenzahl abelscher Körper über quadratischen Zahlkörper*, Akadamie Verlag, Berlin, 1957.

[B9] P. Roquette, "Analytic theory of elliptic functions over local fields," *Hamb. Math. Einzelschriften*, Neue Folgen, Heft 1, 1970.

[B10] J. P. Serre, *Cours d'Arithmetique*, Presses Universitaires de France, 1970.

[B11] ——, *Abelian ℓ-adic Representations and Elliptic Curves*, Benjamin, Reading, Mass., 1968.

[B12] G. Shimura, *Introduction to the Arithmetic Theory of Automorphic Functions*, Iwanami Shoten and Princeton University Press, 1971.

[B13] G. Shimura and Y. Taniyama, *Complex Multiplication of Abelian Varieties and its Applications to Number Theory*, Math. Soc. Japan, 1961.

[B14] C. L. Siegel, *Lectures on advanced analytic number theory*, Tata Institute, 1961.

[B15] ——, Analytische Zahlentheorie II, Course at the University of Göttingen, 1963–1964, notes by K. Kurten and G. Kohler.

[B16] H. Weber, *Lehrbuch der Algebra*, Vol. III, reprinted from the second edition (1908), Chelsea, New York.

[B17] Seminar on Complex Multiplication, Lecture Notes in Mathematics No. 21, Springer-Verlag, Berlin, Heidelberg, New York 1966 (from the Institute Seminar by Borel, Chowla, Herz, Iwasawa, Serre, held in 1957–1958).

ARTICLES

[1] T. Asai, "On a certain function analogous to $\log|\eta(z)|$," *Nagoya Math. J.* **40** (1970), pp. 193–211.

[2] P. Deligne, "Hodge Structures," Publication *IHES*, 1971.

[3] ——, "Variétés abéliennes ordinaires sur un corps fini," *Invent. Math.* **8** (1969), pp. 238–243.

[4] M. Deuring, "Die Typen der Multiplikatorenringe elliptischer Funktionenkörper," *Abh. Math. Sem. Hamb.* (1941), pp. 197–272.

[5] ——, "Teilbarkeitseigenschaften der singularen Moduln der elliptischen Funktionen und die Diskriminante der Klassengleichung," *Commentarii Math. Helv.* **19** (1946), pp. 74–82.

[6] ——, "Die Struktur der elliptischen Funktionenkörper und die Klassenkörper der imaginären quadratischen Zahlkörper," *Math. Ann.* **124** (1952), pp. 393–426.

[7] ——, "Die Anzahl der Typen von Maximalordnungen einer definiten Quaternionenalgebra mit primer Grundzahl," *Jahrsbericht Deutschen Math. Ver.* **54** (1944), pp. 24–41.

[8] ——, "Invarianten und Normalformen elliptischer Funktionenkörper," *Math. Zeitschr.* **47** (1941), pp. 47–56.

[9] ——, "Zur Theorie der Moduln algebraischer Funktionenkörper, *Math. Zeitschr.* **46** (1940), pp. 34–46.

[10] ——, "Zur Theorie der elliptischen Funktionenkörper," *Hamb. Abh.* **15** (1942), pp. 211–261.

[11] ——, "Algebraische Begrundung der komplexen Multiplikation," *Hamb. Abh.* **16** (1946), pp. 32–47.

[12] ——, "Reduktion algebräischer Funktionenkörper nach Primdivisoren des Konstantenkörpers," *Math. Zeitschr.* **47** (1942), pp. 643–654.

[13] ——, "Die Zetafunktion einer algebräischen Kurve vom Geschlechte Eins," four papers in *Nachrichten Akad. Wiss. Göttingen:*
i) 1953, pp. 85–94.
ii) 1955, pp. 13–42.
iii) 1956, pp. 37–76.
iv) 1957, pp. 55–80.

[14] R. FUETER, "Die verallgemeinerte Kroneckersche Grenzformel und ihre Anwendung auf die Berechung der Klassenzahl," *Rend. Palermo* **29** (1910), pp. 380–395.

[15] H. HASSE, "Beweis des Analogons der Riemannschen Vermutung für die Artinschen und F. K. Schmidtschen Kongruenzzetafunktionen in gewissen elliptischen Fällen," *Nachr. Ges. Wiss. Göttingen, Math.-Phys. K.* (1933), pp. 253–262.

[16] ——, "Abstrakte Begrundung der komplexen Multiplikation und Riemannsche Vermutung in Funktionenkörpern," *Abh. Math. Sem. Hamb.* **10** (1934), 325–348.

[17] ——, "Zur Theorie der abstrakten elliptischen Funktionenkörper," *J. Reine angew. Math.* **175** (1936), pp. 55–62, 69–88, 193–208.

[18] ——, "Existenz separabler zyklischer unverzweigter Erweiterungskörper vom Primzahlgrade p über elliptschen Funktionenkörpern der Charaketeristik p," *J. Reine Angew. Math.* **172** (1934), pp. 77–85.

[19] ——, "Neue Begrundung der komplexen Multiplikation, I, II," *J. Reine Angew. Math.* **157** (1927), pp. 115–139 and **165** (1931), pp. 64–88.

[20] H. HASSE and E. WITT, "Zyklische unverzweigte Erweiterungskörper vom Primzahlgrade p über einem algebräischen Funktionenkörper der Charakteristik p," *Mon Math. Physik* **43** (1936), pp. 477–492.

[21] H. HECKE, "Zur Theorie der elliptischen Modulfunktionen," *Math. Ann.* **97** (1926), pp. 210–242.

[22] J. IGUSA, "Kroneckerian model of fields of elliptic modular functions," *Am. J. Math.* **81** (1959), pp. 561–577.

[23] ——, "On the transformation theory of elliptic functions," *Am. J. Math.* **81** (1959), pp. 436–452.

[24] ——, "On the algebraic theory of elliptic modular functions," *J. Math. Soc. Japan*, **20** (1968), pp. 96–106.

[25] ——, "Fibre systems of Jacobian Varieties III (Fibre systems of elliptic curves)," *Am. J. Math.* **81** (1959), pp. 453–476.

[26] Y. IHARA, "Hecke polynomials as congruence zeta functions in elliptic modular case," *Ann. Math.* **85** (1967), pp. 267–295.

[27] K. KOIZUMI and G. SHIMURA, "On specializations of abelian varieties," *Scientific Papers of the College of General Education*, University of Tokyo **9** (1959), pp. 187–211.

[28a] S. LANG, "Isogenous generic elliptic curves," *Am. J. Math.* 1972.

[28b] ——, "Frobenius automorphisms of modular function fields," *Am. J. Math.* 1973.

[29] I. MCDONALD, "Affine root systems and Dedekind's η-function," *Invent. Math.* **15** (1972), pp. 91–143.

[30] J. MANIN, "The Hasse-Witt matrix of an algebraic curve," *AMS Translations* 2, **45**, pp. 245–264.

[31] I. Pjateckii-Shapiro and I. Shafarevič, "Galois Theory of transcendental extensions and uniformization," *Izvestia Akad. Nauk SSSR Ser. Math.* **30** (1966), pp. 671–704, *AMS Translation Series* 2 **69** (1968), pp. 111–145.

[32] I. Pjateckii-Shapiro, "On reduction modulo a prime of fields of modular functions," *Izv. Akad. Nauk SSSR Ser. Math.* **32** (1968), *AMS translation* **2** (No. 6, 1968), pp. 1213–1222.

[33] K. Ramachandra, "Some applications of Kronecker's limit formulas," *Ann. Math.* **80** (1964), pp. 104–148.

[34] ——, "On the class number of relative abelian fields," *J. Reine angew. Math.* **236** (1969), pp. 1–10.

[35] J.-P. Serre, "Groupes de Lie ℓ-adiques attachés aux courbes élliptiques, Colloque, Clermont-Ferrand," *Les tendances géometriques en algèbre et théorie des nombres*, pp. 1964.

[36] ——, "Sur les groupes de Galois attachés aux groupes p-divisibles," *Proceed. Conf. on Local Fields*, Springer-Verlag, 1967, pp. 113–131.

[37] J.-P. Serre and J. Tate, "Good reduction of abelian varieties," *Ann. Math.* **88** (1968), pp. 492–517.

[38] G. Shimura, "Correspondances modulaires et les fonctions zeta de courbes algébriques," *J. Math. Soc. Japan* **10** (1958), pp. 1–28.

[39] ——, "Reduction of algebraic varieties with respect to a discrete valuation of the basic field," *Am. J. Math.* **77** (1955), pp. 134–176.

[40] ——, "A reciprocity law in non-solvable extensions," *J. Reine angew. Math.* **221** (1966), pp. 209–220.

[41] J. Tate, "The arithmetic of elliptic curves, Colloquium lectures," *AMS*, Dartmouth, 1972.

[42] ——, "Genus change in purely inseparable extensions of function fields," *Proc. AMS*, **3** (1952), pp. 400–406.

Index

Graduate Texts in Mathematics

continued from page ii

48 SACHS/WU. General Relativity for Mathematicians.
49 GRUENBERG/WEIR. Linear Geometry. 2nd ed.
50 EDWARDS. Fermat's Last Theorem.
51 KLINGENBERG. A Course in Differential Geometry.
52 HARTSHORNE. Algebraic Geometry.
53 MANIN. A Course in Mathematical Logic.
54 GRAVER/WATKINS. Combinatorics with Emphasis on the Theory of Graphs.
55 BROWN/PEARCY. Introduction to Operator Theory I: Elements of Functional Analysis.
56 MASSEY. Algebraic Topology: An Introduction.
57 CROWELL/FOX. Introduction to Knot Theory.
58 KOBLITZ. p-adic Numbers, p-adic Analysis, and Zeta-Functions. 2nd ed.
59 LANG. Cyclotomic Fields.
60 ARNOLD. Mathematical Methods in Classical Mechanics.
61 WHITEHEAD. Elements of Homotopy Theory.
62 KARGAPOLOV/MERZIJAKOV. Fundamentals of the Theory of Groups.
63 BOLLABÁS. Graph Theory.
64 EDWARDS. Fourier Series. Vol. I. 2nd ed.
65 WELLS. Differential Analysis on Complex Manifolds. 2nd ed.
66 WATERHOUSE. Introduction to Affine Group Schemes.
67 SERRE. Local Fields.
68 WEIDMANN. Linear Operators in Hilbert Spaces.
69 LANG. Cyclotomic Fields II.
70 MASSEY. Singular Homology Theory.
71 FARKAS/KRA. Riemann Surfaces.
72 STILLWELL. Classical Topology and Combinatorial Group Theory.
73 HUNGERFORD. Algebra.
74 DAVENPORT. Multiplicative Number Theory. 2nd ed.
75 HOCHSCHILD. Basic Theory of Algebraic Groups and Lie Algebras.
76 IITAKA. Algebraic Geometry.
77 HECKE. Lectures on the Theory of Algebraic Numbers.
78 BURRIS/SANKAPPANAVAR. A Course in Universal Algebra.
79 WALTERS. An Introduction to Ergodic Theory.
80 ROBINSON. A Course in the Theory of Groups.
81 FORSTER. Lectures on Riemann Surfaces.
82 BOTT/TU. Differential Forms in Algebraic Topology.
83 WASHINGTON. Introduction to Cyclotomic Fields.
84 IRELAND/ROSEN. A Classical Introduction to Modern Number Theory.
85 EDWARDS. Fourier Series: Vol. II. 2nd ed.
86 VAN LINT. Introduction to Coding Theory.
87 BROWN. Cohomology of Groups.
88 PIERCE. Associative Algebras.
89 LANG. Introduction to Algrebraic and Abelian Functions. 2nd ed.
90 BRØNDSTED. An Introduction to Convex Polytopes.
91 BEARDON. On the Geometry of Discrete Groups.
92 DIESTEL. Sequences and Series in Banach Spaces.

93 Dubrovin/Fomenko/Novikov. Modern Geometry — Methods and Applications Vol. I.
94 Warner. Foundations of Differentiable Manifolds and Lie Groups.
95 Shiryayev. Probability, Statistics, and Random Processes.
96 Conway. A Course in Functional Analysis.
97 Koblitz. Introduction in Elliptic Curves and Modular Forms.
98 Bröcker/tom Dieck. Representations of Compact Lie Groups.
99 Grove/Benson. Finite Reflection Groups. 2nd ed.
100 Berg/Christensen/Ressel. Harmonic Analysis on Semigroups: Theory of Positive Definite and Related Functions.
101 Edwards. Galois Theory.
102 Varadarajan. Lie Groups, Lie Algebras and Their Representations.
103 Lang. Complex Analysis. 2nd ed.
104 Dubrovin/Fomenko/Novikov. Modern Geometry — Methods and Applications Vol. II.
105 Lang. $SL_2(\mathbf{R})$.
106 Silverman. The Arithmetic of Elliptic Curves.
107 Olver. Applications of Lie Groups to Differential Equations.
108 Range. Holomorphic Functions and Integral Representations in Several Complex Variables.
109 Lehto. Univalent Functions and Teichmüller Spaces.
110 Lang. Algebraic Number Theory.
111 Husemöller. Elliptic Curves.
112 Lang. Elliptic Functions.